Hazmatology
The Science of Hazardous Materials

Hazmatology: The Science of Hazardous Materials,
Five-Volume Set
9781138316072

Volume One - Chronicles of Incidents and Response
9781138316096

Volume Two - Standard of Care and Hazmat Planning
9781138316768

Volume Three - Applied Chemistry and Physics
9781138316522

Volume Four - Common Sense Emergency Response
9781138316782

Volume Five - Hazmat Team Spotlight
9781138316812

Hazmat Team Spotlight

Robert A. Burke

CRC Press is an imprint of the
Taylor & Francis Group, an informa business

CRC Press
Taylor & Francis Group
6000 Broken Sound Parkway NW, Suite 300
Boca Raton, FL 33487-2742

Printed on acid-free paper

International Standard Book Number-13: 978-1-138-31681-2 (Hardback)

Visit the Taylor & Francis Web site at
http://www.taylorandfrancis.com

and the CRC Press Web site at
http://www.crcpress.com

Dedication

Volume Five

Ron Gore

Jacksonville, Florida Fire Department is the home of the first hazardous materials response team organized in the United States. During the late 1970s, Chief Yarborough of the Jacksonville Fire Department envisioned the need to deal with hazardous materials response in a trained and organized manner. This came directly from the man so many of us credit as the "Godfather of Hazmat". "It wasn't my idea", said retired Captain Ron Gore during a recent visit to Fire Station 7. In town for a reunion of Jacksonville Fire and Rescue Department's (JFRD) original Hazmat Team members, Gore shared how the specialty team's concept originated with Fire Chief Russell Yarbrough in the 1970s. While Ron Gore doesn't want to take credit, even though it was Chief Yarbrough's idea, it was Captain Ron Gore who made it happen. Not just in Jacksonville, but all across the United States. Ron Gore is one

of the most influential people in the world of hazmat and he has touched thousands of people during his training sessions and Ron, you are the undisputed "Godfather of Hazmat" in the American Fire Service. I for one do not think you get enough credit for what you have contributed to all of us in Hazmat Response.

Contents

Preface

Volume 5 takes an in-depth look at hazardous materials teams across the United States and Canada. There are hundreds, if not thousands of hazardous materials teams and thousands of hazardous materials team members. Among the teams I have visited, there are differences from one team to another, but the basic operational procedures are similar. Hazardous materials incidents can occur anywhere in the country. Levels of hazardous materials response vary depending on whether the incident occurs in an urban or rural area. Resources also vary widely between locations and impacts on individual agencies will be presented.

Reasons for establishing hazmat teams in jurisdictions differ from major incidents to dealing with potential hazardous in a community along with many other reasons. It is interesting to see all the different types, configurations, and colors of vehicles. Staffing differs in many jurisdictions along with training requirements for personnel.

Additional information is provided on preferred equipment and resources; target hazards within response jurisdictions; innovations in procedures, equipment, and operations; and finally major incidents that have occurred in jurisdictions of the teams covered in this volume.

Acknowledgements

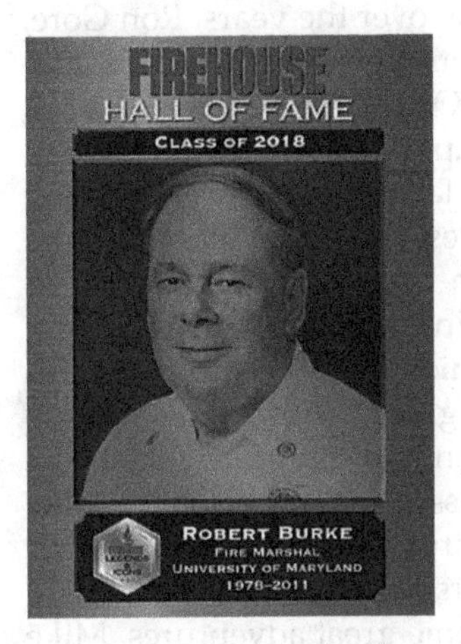

I thank the many fire departments and members across the United States and Canada that I have visited and became friends with during my visits to their departments over the years. I also thank the firefighters from classes I have attended as a student and taught for the National Fire Academy, Maryland Fire and Rescue Institute and Community College of Baltimore County since 1988. Learning is a two-way street, and I have learned much from the students as well. I thank the many friends I have met during the 40 plus years in the fire, EMS, hazardous materials and emergency management fields. There are those who I have not seen for a while; some are no longer with us, but once a friend, always a friend.

I express my thanks to *Firehouse Magazine* for allowing me to write stories about hazardous materials for 33 years and counting. During those years, I have had the pleasure of writing under every editor of the magazine including founder Dennis Smith who gave me the chance to be published for the first time. I also thank Firehouse editors, Janet Kimmerly, Barbara Dunleavy, Jeff Barrington, Harvey Eisner, Tim Sendelbach and Peter Mathews for their support over the years. When I read my first copy of *Firehouse Magazine* in the late 1970s, I was hooked. My dream was to someday go to Baltimore to attend a Firehouse Expo. Never did I dream I would not only attend an expo but teach at numerous expos, write for the magazine and in 2018 be inducted into the Firehouse Hall of Fame. To be placed in a fraternity with sixteen of the people who had an enormous impact on the fire service and who I looked up to my entire career was very humbling.

Several people have been my mentors and have impacted my life and career. When I worked with the State Fire Marshall of Nebraska, Wally Barnett allowed me to accomplish things in the State Fire Marshal's Office

Brent Boydston, Chief Bentonville, AR Fire Department.

that I otherwise would not have. Because of his ability to let his employees reach their potential, I was able to write for *Firehouse Magazine*, become a contributing editor, teach for the National Fire Academy and other things too numerous to mention. He was proud when I gave him a copy of my first book. I owe much of my success in the fire service to the opportunities Wally gave me. Jan Kuczma and Chris Waters at the National Fire Academy have been mentors to me over the years. Ron Gore, retired Captain from the Jacksonville, FL Fire Department and Owner of Safety Systems, has had a large impact on my life and career. The Jacksonville Hazmat Team was the first emergency services Hazmat Team in the United States. Ron Gore is the Godfather of Hazmat response in the United States.

Former student of mine and current Chief of the Bentonville, AR Fire Department Brent Boydston has been a great friend to me and my family over the years. Rudy Rinas, Gene Ryan and John Eversole of the Chicago Fire Department have been fellow classmates and students. Mike Roeshman and Bill Doty of the Philadelphia Fire Department both former students and retired as Hazmat Chief Officers have remained friends. I used to ride with Bill and together we had some great adventures. Mike showed me Philadelphia historical areas, like the spot where Ben Franklin flew his kite and his post office, which is so obscure today in downtown Philadelphia. I also stood on the spot where Rocky stood at the top of the steps in the movie. These adventures enjoyed in Philadelphia would not have happened without Bill and Mike.

Mike Roeshman Retired Hazmat Chief Philadelphia Fire Department.

Just outside of Philadelphia in Delaware County, Tom Micozzie, Hazmat Coordinator for Delaware County, was also a former student and a great friend. We had many adventures together, and I will never forget his introduction to me of the Galati at Rita's Italian Ice! Rita's Italian Ice was started by a retired Philadelphia firefighter and not long ago one opened up in Lincoln, NE.

Thanks to Richmond In Fire Chief Jerry Purcell, who I met during a visit to Richmond to do a Firehouse story on their 1968 explosion in downtown. As a result of

William, "Bill" Doty retired Hazmat Chief Philadelphia Fire Department.

the Richmond story being published I was able to locate and become friends with blast survivor Jack Bales. More recently I visited to do another story on their hazmat team and propane training. Thanks to new friend Ron Huffman who traveled to Richmond to conduct the propane training utilizing water injection to control liquid propane leaks. The article appeared in the September 2019 *Firehouse Magazine.*

Thanks to Tod Allen, Fire Chief in Crete Nebraska who I met when I was researching a train derailment in Crete for another friend Kent Anderson. We have become good friends. Tod is the apparatus operator on Truck one at Station 1 for the Lincoln Nebraska Fire Department. He invited me to come and ride with him, and many adventures later I still go there on a regular basis. I thank all of my friends past and present on "B" Shift at Station 1 for making me feel at home and showing me a good time whenever I am there. Thanks to friend Captain Mark Majors for sharing his experiences with Nebraska Task Force 1 Urban Search and Rescue Team (USAR) and Captain Francisco Martinez Lincoln Hazmat. Finally, I thank Chief Michael Despain and assistant Chief Patrick Borer for their friendship and hospitality while visiting the Lincoln Fire Department on many occasions. This is only the short list—I would have to write a separate book to thank all of you I have met and for the impact you have had on my life over the past 40+ years. You know who you are; I appreciate your friendship and assistance and consider your selves thanked again.

During my year-long book writing adventure that led to *Hazmatology: The Science of Hazardous Materials*, I met and spoke to many people and made new friends. I thank my cousin Dustin Schroeder, Senior Captain at Houston Station 68, and the firefighters and others I met. I also thank Kevin Okonski, Hazmat at Houston Station 22; Ludwig Benner, former NTSB Investigator and developer of several incident management models; Bill Hand, Houston; Richard Arwood; Charles Smith, Memphis; Kevin Saunders, Motivator; Chief Jeff Miller, Butte, MT; and all of the Nebraska Regional Team leaders and members.

Chief Jerry Purcell Richmond, IN Fire Department.

I express my thanks to my cousin Jeanene and her husband Randy for coming all the way from Montana to be with me at the Firehouse Hall of Fame induction. I am also grateful to Brent Boydston, James Rey Milwaukee, Wilbur Hueser and Saskatoon in Canada for the hospitality and tour, and Captain Oscar Robles, Imperial, CA. The list just goes on and on, and there is no room here for everyone, but the rest of you know who you are and I want you to know how much your assistance is appreciated. You are all considered friends, and I hope we will talk and or meet again. Finally, thanks to librarians and historians across the country for your assistance in research, thanks for the memories!

Robert Burke

Special Acknowledgement

When I began thinking about writing a column for *Firehouse Magazine* in September 2017 on *Hazmatology: The Science of Hazardous Materials,* I envisioned that the science would likely produce Hazmatologists. While visiting the Houston Fire Department Hazmat Team in October of 2018 I had the distinct pleasure of meeting a Hazmatologist Tony Janke. Throughout the day I spent time with Tony, Kevin Okonski and the Houston Hazmat Team members that were present. No longer did I have to envision what a Hazmatologist would be like, I had just met a whole team of them. Even though I had only met one shift of the team, there is no doubt the entire team is cut from the same cloth. Tony was the one that took the part of being a Hazmatologist personally and wore his Hazmatologist tag on his shirt. However, I am sure all of the other team members take Hazmatology just as seriously. What I saw the remainder of the day was a well-oiled hazmat machine, everyone knowing their jobs and performing them automatically not needing prompting from anyone else. Thank you Houston Hazmat letting me inside your workings and accomplishments.

Author

Robert A. Burke was born in Beatrice and grew up in Lincoln, Nebraska; graduated from high school in Dundee, Illinois; and earned an AA in Fire Protection Technology from Catonsville Community College, Baltimore County, Maryland (now Community College of Baltimore County) and a BS in Fire Administration from the University of Maryland. He has also pursued his graduate work at the University of Baltimore in Public Administration. Mr Burke has attended numerous classes at the National Fire Academy in Emmitsburg, Maryland, and additional classes on firefighting, hazardous materials and Weapons of Mass Destruction at Oklahoma State University; Maryland Fire and Rescue Institute; Texas A & M University, College Station, Texas; the Center for Domestic Preparedness in Anniston, Alabama; and others.

Mr. Burke has over 40 years' experience in the emergency services as a career and volunteer firefighter, and has served as a Lieutenant for the Anne Arundel County, Maryland Fire Department; an assistant fire chief for the Verdigris Fire Protection District in Claremore, Oklahoma; Deputy State Fire Marshal in the State of Nebraska; a private fire protection and hazardous materials consultant; and an exercise and training officer for the Chemical Stockpile Emergency Preparedness Program (CSEPP) for the Maryland Emergency Management Agency; and retired as the Fire Marshal for the University of Maryland. He has served on several volunteer fire companies, including West Dundee, Illinois; Carpentersville, Illinois; Sierra Volunteer Fire Department, Chaves County, New Mexico; Ord, Nebraska; and Earleigh Heights Volunteer Fire Company in Severna Park, Maryland, which is a part of the Anne Arundel County, Fire Department, Maryland.

Mr. Burke has been a Certified Hazardous Materials Specialist (CFPS) by the National Fire Protection Association (NFPA) and certified by the National Board on Fire Service Professional Qualifications as a Fire Instructor III, Fire Inspector, Hazardous Materials Incident Commander, Fire Inspector III and Plans Examiner II. He served on the NFPA technical committee for NFPA 45 Fire Protection for Laboratories Using Chemicals for 10 years. He has been qualified as an expert witness for arson trials as well.

Mr. Burke retired as an adjunct instructor at the National Fire Academy in Emmitsburg, Maryland in April 2018 after 30 years. He taught Hazardous Materials, Weapons of Mass Destruction and Fire Protection curriculums. He taught at his Alma Mater Community College of Baltimore County, Catonsville Campus and Howard County Community College in Maryland. He has had articles published in various fire service trade magazines for the past 33 years. Mr. Burke is currently a contributing editor for *Firehouse Magazine*, with a bimonthly column titled "Hazmatology," and he has had numerous articles published in *Firehouse, Fire Chief, Fire Engineering* and *Nebraska Smoke Eater* magazines. He was inducted into the Firehouse Hall of Fame in October 2018 in Nashville, TN. Mr Burke has also been recognized as a subject matter specialist for hazardous materials and been interviewed by newspapers, radio and television about incidents that have occurred in local communities including Fox Television in New York City live during a tank farm fire on Staten Island.

Mr. Burke has been a presenter at Firehouse Expo in Baltimore, MD and Nashville, TN numerous times, most recently in 2017. He gave a presentation at the EPA Region III SERC/LEPC Conference in Norfolk, Virginia, in November 1994 and a presentation at the 1996 Environmental and Industrial Fire Safety Seminar, Baltimore, Maryland, on DOT ERG. He was a speaker at the 1996 International Hazardous Materials Spills Conference on June 26, 1996, in New Orleans, Louisiana; a speaker at the Fifth Annual1996 Environmental and Industrial Fire Safety Seminar in Baltimore, Maryland, sponsored by Baltimore City Fire Department; and at LEPC, an instructor for Hazmat Chemistry, August 1999, at Hazmat Expo 2000 in Las Vegas, Nevada. He also delivered a Keynote presentation at the Western Canadian Hazardous Materials Symposium Saskatoon, Saskatchewan, Canada, in 2008.

Mr. Burke has developed several CD-ROM-based training programs, including the Emergency Response Guide Book, Hazardous Materials and Terrorism Awareness for Dispatchers and 911 Operators, Hazardous Materials and Terrorism Awareness for Law Enforcement, Chemistry of Hazardous Materials Course, Chemistry of Hazardous Materials Refresher, Understanding Ethanol, Understanding Liquefied Petroleum Gases, Understanding Cryogenic Liquids, Understanding Chlorine and Understanding Anhydrous Ammonia. He has also developed the "Burke Placard Hazard Chart." He has published seven additional books titled *Hazardous Materials Chemistry for Emergency Responders* (*1st, 2nd and*

3rd Editions, Counterterrorism for Emergency Responders 1st, 2nd and 3rd editions, Fire Protection: Systems and Response and *Hazmat Teams Across America.*

Currently, Mr. Burke serves on the Homestead LEPC in Southeast Nebraska. He also manages a Hazardous Materials section at the Nebraska Firefighters Museum and periodically rides with friends on "B" shift at Station 1, Lincoln Fire Department. He can be reached via email at robert.burke@windstream.net, on Facebook at https://www.facebook.com/RobertAb8731 and through his website: www.hazardousmaterialspage.com.

Volume Five

Hazmat Team Spotlight

> In five words, his guiding principles are: "Prevent harm. Survive. Be nice."
>
> *Chief Alan Brunacini*

Volume 5 provides an in-depth look at selected hazardous materials teams across the United States and Canada. This volume will focus on hazardous materials teams to highlight information that can be shared by other departments and personnel. Incidents that these hazmat teams have experienced may also provide some lessons learned to help other teams in the future. Fire departments have always responded to emergencies involving chemicals. In fact the fire department is our nation's first responder to all types of emergencies including fires, hazardous materials, emergency medical, natural disasters, and acts of terrorism.

Hazardous materials (hazmat) became the buzz word of the 1970s and 1980s. With the passage of the Superfund Amendments and Reauthorization Act (SARA) of 1986 (also known as the Emergency Planning and Community Right-To-Know Act, EPCRA) and continued reauthorization over the years, hazmat took a giant leap forward in the United States Emergency Response Community. Starting with the Jacksonville Fire Department, hazmat teams sprang up across the United States and Canada within a very short period of time.

Hundreds, if not thousands of hazardous materials teams and thousands of hazardous materials team members placed these teams in service from 1978 through the early 2000s.

Reasons for establishing hazmat teams in jurisdictions also vary from major incidents. Hearing about the formation of other teams, to dealing with potential hazardous materials in a community and in response to new Federal Regulations in the 1980s. It is interesting to see all the different types, configurations and colors of vehicles. Staffing also differs in many jurisdictions along with training requirements for personnel. There have been several trends over the years involving regional teams, consolidation, and transition from hazmat teams to Special Operations teams that not only do hazmat but various forms of rescue as well.

The Journey Begins

During the month of May 2004, a new column subject was started on Firehouse.com website titled "Hazmat Team Spotlight". There had been several articles about hazmat teams that I put into the magazine, but this was my first attempt at concentrating on hazmat teams. The idea for the spotlight was born from letters and e-mails from hazmat personnel asking questions about other teams. My intention was to provide a location for hazmat team members to find information about what other hazmat teams are doing around the country; a kind of a place to share information. I published the first spotlight story on the Philadelphia, PA Hazmat Team, which I had covered in *Firehouse Magazine* in January 1999. Information on Philadelphia's team was updated and the first Hazmat Team Spotlight was posted on Firehouse.com.

Following the Philadelphia Hazmat Team Spotlight, I received numerous requests from hazmat teams up and down the East Coast about them wanting their teams in future articles. Over the years I have tried to honor all of the requests as soon as possible, although in some cases it was many years until I had the opportunity to travel to a particular area. Photographic situations are not always available during visits so some photographs are provided by the local hazmat teams and fire departments. But aside from that I do all of my own photography. As I started visiting departments, I found that there were some stories beyond the scope of the website spotlights and some of the team's stories were placed in *Firehouse Magazine*. After a couple of years I would update information on the teams that appeared in the magazine and repost them to the Firehouse.com website, so that others could use them as a resource. During 2008 additions to the website were discontinued and all Hazmat Team articles went directly into *Firehouse Magazine*.

The website had room for many more photographs than the magazine so I had the opportunity to better show (through photography) the assets of the teams that are spotlighted. Since I do not receive travel expenses for the teams I spotlight, I try to plan my visits to teams when I am traveling for another purpose. Additionally, I taught for the National Fire Academy for 30 years (retired in 2018) and tried to visit teams while traveling for teaching jobs. Family vacations also turned into team visits when the opportunity occurred. Some instances I had frequent flyer miles and used those just to visit teams, one in particular that stands out were San Diego, CA, and Yuma, AZ, visits. Team spotlights are listed alphabetically, not in the order of visits for ease of organization and reader searching.

Allegheny County Pennsylvania Green Team Specialized Intervention Team (SIT)

> **Prologue:** *When I received an email from Jim Eaborn Deputy Chief of Allegheny County Hazmat about the Hazmat SIT Team, I have to admit I was skeptical. Only a couple of instances where entry personnel got into trouble and required assistance in the Hot Zone came to mind. With all of the precautions taken when preparing for entry, history has shown it has been a relatively safe operation. However, he invited me to a full drill to show me how the team operates and I am glad I accepted. This team is amazing. Procedures have been well thought out and this team is well prepared and equipped to enter a Hot Zone to make a rescue of downed hazmat personnel. My impression was so far beyond what I expected, I would recommend that anyone who has the chance should journey to Pittsburg, PA and visit the Allegheny County Hazmat Green Team.*

Allegheny County is located in the southwest corner of Pennsylvania and has an approximate population of 1,214,040 in 2020 distributed over 725 miles^2. Pittsburg is the largest city in Allegheny County and has a population of over 294,860 in 2020. Hazardous materials response in Allegheny County including the City of Pittsburg is provided by the Allegheny County Department of Emergency Services Hazmat Team. In addition to hazmat the Department of Emergency Services also provides 911 Communications and Fire Marshal services and operates the county fire training academy. There are over 300 independent volunteer fire companies in the county that provide fire and rescue service to 130 municipalities. Emergency medical services (EMS) response is provided by an independent paid EMS organization which also provides EMS support for the hazardous materials teams. The only career for fire department in the county other than the City of Pittsburg is at the international airport.

Hazmat Team History

The Allegheny County Hazardous Materials Team was formed on June 28, 1988, by the Allegheny County Commissioners. The idea for the formation of a team resulted from training provided by Ron Gore (Jacksonville Florida Fire Department) and owner of Safety Systems, Inc., who presented some of the first hazmat classes in Allegheny County. The Allegheny County Hazardous Materials Team is composed of five teams staffed totally by volunteers in the county and by career personnel in the City of Pittsburg. The Allegheny County Green Specialized Intervention Team services the entire Southern area of the county; the Red Team, East Boroughs Emergency Services Association serves the Eastern areas of the county; the Blue Team, Northeast Allegheny Response Association serves the Northeast and Northwest areas of the county; the Silver Team, North Hills Response Team serves 25 municipalities in Northeastern Central just north of the Ohio and Allegheny Rivers; and the Yellow Team serves the City of Pittsburg.

Each of the teams is somewhat autonomous administrative but all provide assistance throughout the county when needed. Support and funding is provided by the county. When a hazmat incident occurs in the county, the four county teams would be deployed first and the Pittsburg Team would be deployed if they were not on a call in the city. All five Hazardous Materials Teams are certified by Pennsylvania Emergency Management Agency (PEMA) and by the Commonwealth of Pennsylvania. Certified teams must meet certain standard criteria set forth by the state for certification. Certification requirements include having an Emergency Response Plan, personal protective equipment (PPE) Program, and Medical Surveillance Program.

Training requirements for team members: Team Structure and Response, Dispatch and Response, Incident Command System Structure and procurement of approved equipment on the Team Equipment List. There is also a periodic re-certification process. Allegheny County Police provide a bomb squad for the county and technicians train with the hazmat team. The five county hazmat teams receive support from the Department of Emergency Services under the direction of Chief Matt Brown, the County Fire Marshal and Emergency Management Coordinator. All of the teams are dispatched by the County 911 Center.

Hazmat Team

From January 1 to November 30, 2018, the Allegheny County's Hazardous Materials Team collectively responded to 68 calls in 2018. Fuel spills and odor calls are handled by individual fire companies and provide assistance if needed. Engine companies carry four-gas meters. Hazmat will not respond unless a spill is over 55 gallons unless specially called upon to do so. Hazmat Team trucks in the county are equipped with the latest hazardous materials equipment including encapsulated suits, monitoring equipment, and communications capabilities. Fire frequency radios in the hazmat units enable communications with every fire department in Allegheny County. There are approximately 250 technician-level responders on Allegheny County's Hazmat Team. County Hazmat Companies are all one team and each is equipped to handle a hazmat response for

CBRNE. Additionally, each company is equipped to also handle the hazmat risks in their first due areas.

PPE, Equipment, Training

Some of the specific equipment carried by the hazardous materials teams include Chlorine Kits A, B, and C. Portable fire extinguishers include Purple K and Class D agent. PPE used by the Green Team is primarily DuPont Responder for Level A and DuPont CPF for Level B. Self-contained breathing apparatus (SCBA) is manufactured MSA with 1-h bottles and they also have MSA Millieum Cartridge Respirators. In-suit communications system is provided by Conspec VHF and UHF radios that are used in the county with throat mikes and ear pieces. Decontamination is provided by the hazardous materials team. Decontamination is currently being conducted using pressure washing as a primary process. Recent equipment purchased are:

- Thermo Fisher Gemini (County).
- Ground meter with all grounding and bonding cables and clamps. Developed SOG for use.
- Altair SX 4 gas PID's to supplement current cashe of meters.
- LPG water injection equipment.

Green Team

The Green Team initial response area is located in the Southwest part of the county and serves approximately 56 municipalities within the county in an area of roughly 40 miles^2 and a population of 270,000. They respond in conjunction with as many as 70 different volunteer fire departments in their coverage area. Membership on the Green Team is made up of personnel from 16 different organizations including firefighters, emergency medical technicians (EMTs), paramedics, law enforcement, and industry specialists from within the Green Team's response area. The Green Team responds to approximately 8–10 hazardous materials incidents each year. Currently the Green Team has 42 members and is housed at the Broughton Volunteer Fire Company located at 1030 Cochran's Mill Road in South Park Township. All hazmat responders are on call.

Two hazmat vehicles are operated by the Green Team, Specialized Intervention Team 450 Hazmat 1 and 450 Hazmat 2 (Figure 5.1). Hazmat 1 is a 2008 Pierce Velocity with a rescue body. The rear cab is set up as a mini command post. The cab also has "meter" cabinets with charging stations for the meters and extra batteries. The truck has an onboard weather station and light tower. Additionally the outside of the truck has a retractable awning. Hazmat 2 has a roof mounted deck for clearer view of the

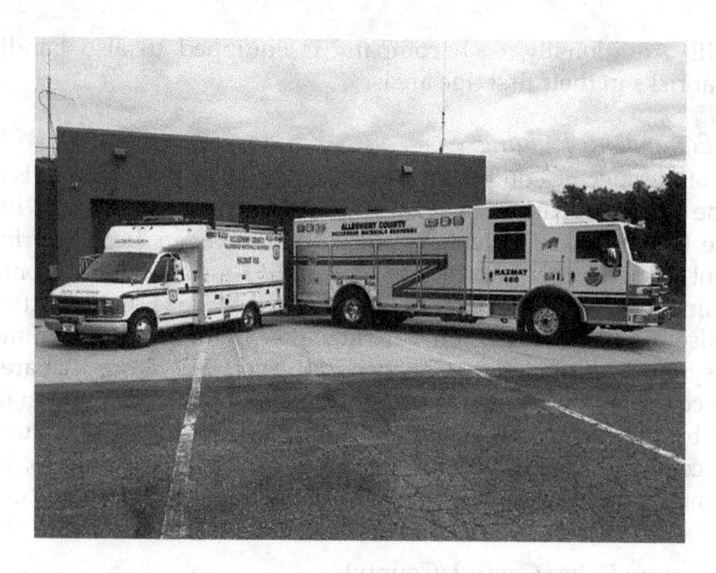

Figure 5.1 Two hazmat vehicles are operated by the Green Team, Specialized Intervention Team 450 Hazmat 1 and 450 Hazmat 2. (Courtesy: Allegheny County Green Team.)

operations for the command staff. This vehicle functions as a support unit with extra equipment, supplies and all of the Midland kits, Chlorine Kits, rail car fittings, and tarps for ammonia leaks.

Training requirements for team members include 168h of training with 24h of annual refresher. Subjects covered are Hazmat Awareness, Operations, Technician, Hazmat Chemistry, Command Post Operating I, II, III, Research, Evacuation, Communications, Command and Control, Radiological Monitoring, LPG, and Pesticide Challenge. Training is provided by the Allegheny County Training Academy and the Pennsylvania State Fire Academy. Most volunteer fire companies in the county require operation-level hazmat training for response personnel. EMS personnel are all trained to the awareness level for hazmat response. Green Team entry personnel now work in teams of three instead of two.

Special Intervention Team (SIT)

One of the unique specialties of the Allegheny County Pennsylvania Green Team is their Rapid Intervention Procedure (Figure 5.2) that was developed by team members for rescue of hazmat team personnel who might become trapped from falling debris or other cause during an entry into the hot zone. The RIT procedure was developed by Deputy Chief Ryan Lattner, Allegheny County Airport Authority, Deputy Chief Kurt Gardner, Allegheny County Specialized Intervention Team, and Commander Jim Eaborn, Allegheny County Specialized Intervention Team.

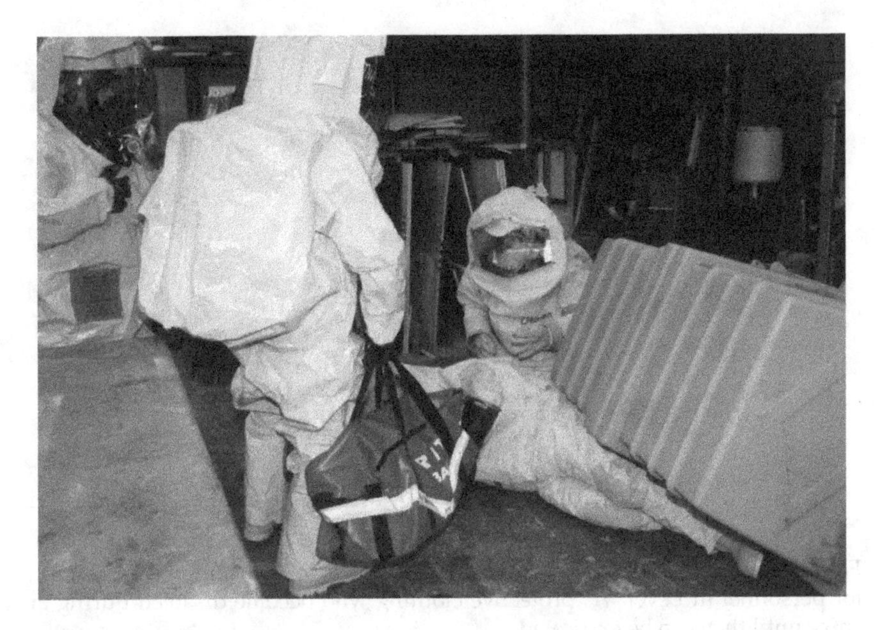

Figure 5.2 One of the unique specialties of the Allegheny County Pennsylvania Green Team is their Special Intervention Procedure.

The primary purpose of the RIT is to provide an emergency air supply for personnel in Level "A" protective clothing who become disabled during an entry until they can be extricated (Figure 5.3), decontaminated and receive medical attention as required by any injuries. The emergency air supply procedure is a last resort and is only used in situations where death or serious injury would occur due to lack of air. The SIT is established, and on stand-by anytime a Level A entry is performed. They are partially suited up and ready for rapid deployment. Special kits are assembled ahead of time and contain a modified SCBA, cutting tools and Chem Tape. Strips of tape are precut and placed on the SCBA tank for quick access. When the team makes an entry, they bring the kit and a skid for removal of the team member when extricated if needed.

When a responder goes down the SIT Team enters the site and locates the team member. They determine if the scene is safe for them to enter and if a rapid extrication is possible. If rapid extrication is not possible, they assess the level of consciousness of the responder that is trapped and their available air supply. When the air supply reaches a critical level the SIT will breach the suit of the responder and supply additional air. A cut approximately 2–3 inches is made in the Level "A" suit. The hose from the rescue SCBA is inserted into the Level "A" suit through the cut. Pre-cut pieces of Chem Tape are used to secure the hose in place and reduce air leakage through the hole.

***Figure* 5.3** The primary purpose of the RIT is to provide an emergency air supply for personnel in Level "A" protective clothing who become disabled during an entry until they can be extricated.

Once the hose is secured, the air supply is turned on slightly so that the suit is not inflated. If the trapped responder is conscious, they are instructed to remove their arms from the sleeves of the Level "A" suit and remove their SCBA mask. If the responder is not conscious, SIT team members remove the mask from the outside of the suit. RIT team members monitor the air supply of the trapped responder and assist with extrication. Once extricated the responder is then removed from the entry site and decontaminated.

> **Author's Note:** *During my visit to the Green Team they staged a drill with a team member trapped by a fallen large shelve inside of a building. I watched the rescue procedures being implemented first hand and it not only went smoothly, but works very effectively.*

Common Sense Decontamination

While gathering updated information on the Green Team for inclusion in this Volume, I learned about their innovative common sense approach to decontamination.

> **Author's Note:** *While preparing for the Hazmatology book project, decontamination was one of the areas that I felt science would back up the idea of taking another look at the way we approach decontamination. Finding a team that was already using "common sense" for decontamination was very refreshing and worth sharing.*

Green Team has created a flowchart to assist them in determining what level of decon is needed based upon circumstances of an incident (Figure 5.4). Decon is another example of a tool in the Hazmat Tool Box that is not always needed. While gathering updated information on Allegheny County Pennsylvania's Green Team for inclusion in Hazmatology, I learned about their innovative common sense approach to decontamination. The complete procedure for Common Sense Decontamination is located in Volume Four.

Using tools based upon need rather than because we can or think we should, only adds additional tasks and stresses that are not necessary. Entering the hot zone in Level A PPE for gases does not result in contamination of the chemical suit. Gases do not stick to a surface. While in the hot zone if personnel do not come in contact with a hazardous material, there is no contamination. Generally we can see the hazardous material if it is a liquid or solid. If a solid is not suspended in the air and does not come in contact with PPE, there is no contamination. Therefore, no decontamination is required.

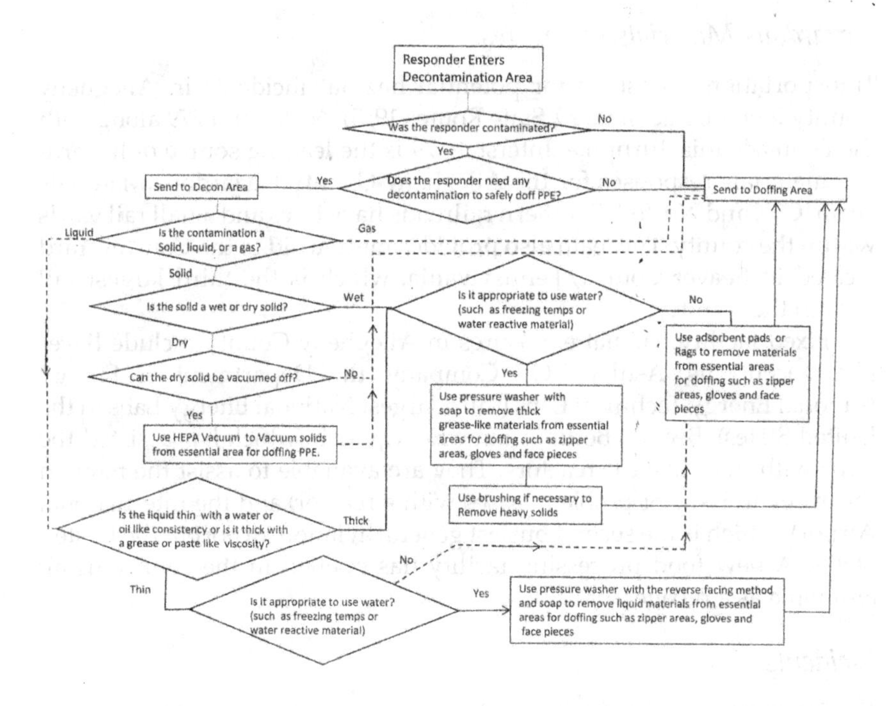

Figure 5.4 Green Team has created a flowchart to assist them in determining what level of decon is needed based upon circumstances of an incident.

Reference Resources

Computer resources are available such as NIOSH Pocket Guide to Chemical Hazards, CHRIS hazardous Chemical Data Manual, Irving Sax Manuals, ACGIH Threshold Limit Values Guidebook, Bretherick's Handbook of Reactive Chemical hazards, Merck Index, Condensed Chemical Dictionary, Crop Protections Handbook, and B.O.E. Emergency Handling of Hazardous Materials in Surface Transportation.

Hard-copy reference materials include the Department of Transportation Emergency Response Guidebook, Jane's Chem-Bio Handbook, Tempest CB-FRG (Chem-Bio) First Responder Guidebook, Matheson Gas Book & First Aid, Emergency Care for Hazmat Exposure, or, Hazmat Injuries, AAR/BOE Field Guide to Tank Car Identification, Bureau of Explosives: American Railroad Emergency Action Guide, GATX Tank Car manual, DOT Radiation emergency Handbook, NFPA Fire Protection Guide of hazardous materials, Firefighters Guide to Hazardous Materials, NFPA Emergency management of Hazardous Materials Incidents and EPA List of Lists, Consolidated List of Chemicals. Each team also has a weather station for monitoring conditions during an incident.

Hazardous Materials Exposures

Transportation exposures for potential hazmat incidents in Allegheny County include Interstate 79, State Routes 19, 51, 60, 88, and 279 along with the Pennsylvania Turnpike. Interstate 79 is the leading source of hazardous materials responses for the Allegheny County Hazardous Materials Team. CSX and Norfolk Southern railroads have lines and small rail yards within the county. The team also provides mutual aid to the Conway Yard located in Beaver County, Pennsylvania, which is the third largest rail yard in the country.

Fixed facility hazmat exposures in Allegheny County include Bayer pharmaceuticals, Ashland Oil Company, the Department of Energy National Energy Technical Labs (third largest National Energy Labs in the United States), Bechtel-Bettis Atomic Energy Lab (which has assisted the navy with their nuclear reactors. They are available to assist the navy in the event an issue or problem arises with a reactor) and the International Airport, which is the second busiest general aviation airport in the United States. A new food processing facility has opened in the county using ammonia as a refrigerant.

Incidents

Ferric Chloride Tanker Leak

Ferric chloride was leaking from an MC 407 parked in a tanker cleaning company parking lot and the truck driver was no where around the vehicle. The hazmat team was called because another driver happened to be picking up another tanker trailer and seen a "substance" on the ground around the trailer then could see the material running out of the tanker. The driver called 9-1-1 and the fire department responded and upon their initial size-up called for the County Hazmat Team to also respond. When the Green Team arrived and reviewed the Bill of Lading it indicated that there was 490 gallons of ferric chloride in the rear compartment, which was leaking. When our Recon team did the initial size-up they determined that the tanker compartment was empty with all 490 gallons on the ground, covering the rear portion of the parking lot and flowing into a stream. Our Entry team initiated damming and diking the product to slow down the flow using soda ash as the damming material.

The trucking company contacted their hazmat contractor who responded to clean up the contaminated areas and other trailers that were sitting in the product. County Hazmat on-scene time was 5.5h. Contractor cleanup cost was in excess of $125,000 over a 2–3-day period removing nine 40 feet^3 roll off boxes of parking lot gravel and soil and also removing the iron sentiment of the ferric chloride from the stream that was directed to be completed by Pennsylvania Department of Environmental Protection.

Liquid Nitrogen Tanker Leak

Chief Eaborn was contacted via telephone by a local fire chief who was dispatched with his department to a leaking tanker trailer. Their size-up revealed that the trailer was loaded with liquid nitrogen. The fire chief was advised that the only current hazard that could be a problem was that the product was an asphyxiate and they should establish a metering perimeter and maintain four gas metering with special attention to the oxygen levels. The trucking company was contacted and dispatched a mechanic whom tightened the flange that was leaking. The County Hazmat Team did not respond to this incident. Pictures were taken by the fire department and was sent to us for initial evaluation of the incident (*Firehouse Magazine*).

Anchorage, Alaska Hazmat Team "The Pride of Alaska"

Prolog: *Both of my visits to Alaska were teaching opportunities through the National Fire Academy. An article for Firehouse Magazine resulted from the first visit in 2005. Another teaching opportunity arose in April of 2018, which would also fulfill a bucket list item for my wife on our 35th Wedding Anniversary. This visit also gave me the opportunity to learn, obtain, and write about the Anchorage Fire Department, especially* *the Great Alaska Earthquake of 1964 (see Volume one). Fortunately it also kept me from the Anchorage earthquake, which occurred on November 30, 2018. During my life time, ironically the only earthquake I experienced was in my basement in Nebraska a couple of years ago. Sitting at my desk in the basement the wind chimes started to ring. Immediately I looked up to close the window, but it was already closed. It took me a few minutes before I realized what had happened and went to the internet to confirm. Sure enough, a quake centered in northern Oklahoma was the cause. Now I refer to all of my indoor wind chimes as my earthquake alarms.*

Anchorage is the largest city in the state of Alaska, comprising more than two-fifths (42%) of the state's population (Figure 5.5). Anchorage is located in *South Central Alaska*. The city is about as far north as Stockholm, Sweden, or St. Petersburg, Russia, and as far West as Hawaii. It lies 290 miles northeast of Kodiak Island, 130 miles south of Mount McKinley, and it is on the upper branches of the Cook Inlet, the northernmost reach of the Pacific Ocean. It is bordered on the east by Chugach Mountain state park. Anchorage stretches from Portage Glacier to Eklutna, encompassing 1,955 miles^2 – about the size of the state of Delaware. Because of the location of mountain ranges, it is only possible to drive north or south out of Anchorage.

Figure 5.5 Anchorage is the largest city in the state of Alaska, comprising more than two-fifths (42%) of the state's population.

Highways only access a small portion of the 515,412 total land area of the state, and many areas are only accessible by air or water. According to the Federal Aviation Administration (FAA) here are 7,933 active pilots, 2,801 airframe and power plant mechanics of which 750 have inspection authorizations, and 9,346 registered aircraft in Alaska. Alaska has 400 public use airports, 282 land-based, 4 heliports (only public use listed this year), 114 seaplane bases, and approximately 747 recorded landing areas (private, public, and military) total. Of course pilots land on many of the thousands of lakes and gravel bars across the state where no constructed facility exists.

Fire Department History

Anchorage, as with most cities in the United States, began with a volunteer fire department organized in 1915. Over the years the department evolved into a combination paid/volunteer department boasting 50 personnel by 1951 when they started providing ambulance service. Finally in 1967, seven volunteer fire departments merged into the fully paid Greater Anchorage Borough Fire Department. The area wide paramedic program began in 1971. Governmental unification occurred in 1975 when the City of Anchorage and the Greater Anchorage Borough unified to become the Municipality of Anchorage.

Today's Modern Department

The Anchorage Fire Department is a career department with 391 uniformed personnel led by Chief Jodie Hettrick. The City of Anchorage has a population of approximately 285,634 in 2020 and covers an area of 100 miles^2 inside the "Anchorage Bowl". The Anchorage Bowl consists of the main populated areas within the municipality including the hillside areas that together are somewhat bowl shaped but not including Eagle River to the north. The bowl is primarily the portion that most visitors would think is the City of Anchorage without knowing that much of the park and uninhabited lands were also part of the city. Volunteer fire departments cover the outlying areas of Anchorage. Emergency medical services (EMS) coverage provided by the Anchorage Fire Department encompasses the entire 1,955 miles^2 of the city. Mutual aid agreements exist between the Anchorage Fire Department, Ted Stevens International Airport Aircraft/Rescue/Fire, Elmendorf Air Force Base Fire Department, Fort Richardson Fire Department, Girdwood Volunteer Fire Department, and Chugiak Volunteer Fire Department.

The Anchorage Fire Department is an ISO Class 1 fire department and operates 14 engine companies, 5 truck companies, 1 rescue, 5 water tankers, 8 medic units, and 12 specialty vehicles from 13 stations located throughout the city. The city is divided into three battalions with a chief over each. EMS paramedic transport service is provided by the Anchorage Fire Department. Seven engine companies provide non-transport Basic Life Support (BLS) and are equipped with a semi-automatic defibrillator to enhance response to cardiac emergencies. Five engine companies provide non-transport advanced life support (ALS) services. The department has 66 trained firefighter/paramedics. Additional services are provided with special teams including dive, foam, hazmat, mountain, and water rescue. Personnel for specialty teams are provided from other service companies when needed. Fire alarms for 2017total were 36,270, broken down: Fire 785, EMS 24,952, Alarms 2,216, Service Calls 7,799, and Hazardous Conditions 518.

Hazmat Team History

Anchorage officially formed their hazardous materials team and placed it in service in 1982 when several members showed an interest in the emerging field of hazmat response.

Hazmat Team

The hazardous materials team responds to an average of 112 hazardous materials calls a year with an additional 478 alarms for gas detection monitoring. Statistics for hazmat responses do include local engine runs for hydrocarbon fuel spills and natural gas leaks. Engine companies carry absorbents for fuel spill and clean-up. The Anchorage Hazardous Materials Team covers the entire 1,955 miles^2 city and is available to respond anywhere needed in the State of Alaska. The team is known as the "Pride of Alaska". The city is also home to two military installations, Elmendorf Air Force base and Fort Richardson Army base, which have their own fire protection, EMS, and hazardous materials response units.

Anchorage does not have the ability to summon mutual aid for hazmat response like departments in the lower 48 states. They therefore need to have the quantity of decontamination equipment and response materials to handle a large-scale incident on their response vehicles. To accomplish this task, they have acquired a 20- and a 24-feet trailers in the past couple of years along with a Ford F350 diesel to tow them. Additional hazmat vehicles include a 2004 KME Custom on a Heavy Rescue Chassis and a 1-ton pickup with a crew cab to tow two three-tent mass decontamination systems in trailers.

Hazmat 1 has a specially designed work area with an exhausted fume hood and extra heaters to accommodate Alaska's cold climate (Figure 5.6). Hazmat 1 is housed at Station One located at 122 E. 4th Avenue (4th and Barrow) which is also the Headquarters Station. Other equipment at

Figure 5.6 Hazmat 1 has a specially designed work area with an exhausted fume hood and extra heaters to accommodate Alaska's cold climate.

Station One includes Engine 1, Medic 1, Truck 1, and Engine 2. The Federal Bureau of Alcohol, Tobacco & Firearms also keeps one of their response vehicles at Station One.

The Anchorage hazmat team is a non-dedicated team staffed with 15 on duty personnel from Station One. All firefighters on the department are certified to the operations level. All Anchorage Firefighters work a 24-h on and 48-h off shift. To request to become a hazmat team member a firefighter must have completed their probationary year with the Anchorage Fire Department. They must obtain technician-level certification prior to assignment to the team.

PPE, Equipment and Training

Personnel Protective Equipment (PPE) for Anchorage Hazmat Team members includes for Level A, Kapplar Responder and Tychem TK Commander EX and for Level B, Kapplar Hooded Coveralls. Respiratory Protection is provided by Draeger SCBA with 60 min bottles.

Air monitoring capabilities include a Hazmat ID infrared spectrometry unit, Ahura first Defender, LTX 312, 412 multi-gas monitors, Manning Systems portable detectors for Chlorine and Ammonia, Mini RAE PPM and PPB PIDs, Multi-Rae, Ludlum and Victoreen radiation survey meters, QuantRad Ranger radionuclide identification, and Travel IR. Terrorist

Agent ID monitoring instruments include Chemical Agent Monitor (CAM), APD 2000, RAID-M, M-8 papers, M-9 badges and 256A Kits.

Training is conducted in house and includes the Environmental Protection Agency (EPA) 40-h technician program. Annual training involves 60–75 h. Ten members of the hazmat team have completed the National Fire Academy Chemistry of Hazardous Materials class and six have completed the Hazmat ALS course. Training and joint exercises are periodically conducted with the U.S. Coast Guard and Pacific Strike Team out of San Francisco. The Anchorage hazmat team also works closely with the National Guard Civil Support Team stationed in Anchorage.

Reference Resources

Research Resources include computer-based programs and hard copy reference books. Computer Software Programs are CAMEO, Chem Knowledge, Micromedex, PEAC-WMD, ERG, NIOSH Pocket Guide, NFPA Hazmat Quick Guide, United States Coast Guard CHIS Manual, and Emergency Handling of Hazardous Materials in Surface Transportation. Hard Copy Reference Books include *ACGIH Guide of Occupational Exposures, CFR 49, Chem-Bio Handbook, CHRIS (Coast Guard Manual), Emergency Action Guides, Emergency Care for Hazardous Materials Exposure, Emergency Handling of Hazardous Materials, Emergency Response Guide Book, Fire Protection Guide to Hazardous Materials, Guide to Selection of Chemical Agent and Toxic Industrial Materials, Detection Equipment for Emergency Responders, Guidelines for Selection of Chemical Protective Clothing, Hawley's Condensed Chemical Dictionary, Hazardous Materials Desk Reference, Hazardous Materials Response Handbook, Hazmat Field Guide, Medical Management of Chemical Casualties, Medical Management of Biological Casualties, Merck Index, NIOSH Pocket Guide, Quick Action Guide to Chemical Clothing, and SAX's Dangerous Properties.*

Hazardous Materials Exposures

Anchorage and most of Alaska are unique in terms of hazardous materials exposures. There are no Interstate Highways or rail systems. There is no land connection with the continental United States except through Canada. There is very little truck transportation of hazardous materials except for local use and primarily hydrocarbon fuels. One major road provides the only access in and out of Anchorage and runs North and South. Most hazardous materials that enter the state come through one of its ports. Anchorage is one of Alaska's busiest ports.

Hazardous materials that come into the Port of Anchorage are loaded onto trucks and railroad cars to be shipped to Anchorage and other parts of the state for end use. Anchorage does not have much industry outside of the port facility and fishing-related businesses. The only chemical industry is oil related and some minimal anhydrous ammonia manufacture. Propane, anhydrous ammonia, chlorine, and petroleum products are the major hazardous materials found in Alaska. Most of the large tank storage of hazardous materials involve jet fuel, as one in every three residents of Alaska is a pilot, and there is a huge volume of private and commercial aircraft activity in the state.

Anhydrous ammonia and chlorine are shipped into the state in 1-ton containers. Propane is shipped in by sea from Seattle and moved by railroad car into Anchorage from the port. Explosives used for construction and mining are brought in by ship and the fire department is required by the Coast Guard to stand by during off-loading. There are limited amounts of radioactive materials, usually related to medical and industrial usage. Sodium hydroxide and sulfuric acid leaks related to the petroleum industry and railroad car leaks of acids and bases make up a large portion of hazardous materials responses. The Anchorage Hazmat Team also responds to clandestine methamphetamine laboratories to assist police in dealing with hazardous chemicals (*Firehouse Magazine*).

Incidents

Great Alaska Earthquake 1964

March 27, 1964, the largest earthquake to hit the United States and the second largest in the world struck southern Alaska (Figure 5.7). The quake registered 9.2 on the Richter scale triggering massive tsunamis that wiped out entire villages and landslides that sent neighborhoods from suburban Anchorage into the ocean. Devastation was catastrophic ultimately killing 131 people with damage costs running upwards of $2.3 billion. Large storage tanks of petroleum fuels were set afire by the quake and had to burn themselves out. Gas mains were ruptured and set afire as well.

On May 20, 1992, a Freon leak occurred at an ice skating rink. Freon is a gas that can displace oxygen in the air and cause asphyxiation. Thirty-three people were exposed to the leaking Freon inside the rink. A worker who entered the compressor room to shut down the refrigeration system where the leak occurred was overcome and died. Two other workers were overcome and unconscious, but were rescued by firefighters and recovered. Following the incident, Alaska State Department of Labor officials issued citations to the mall management that included 48 health and safety violations. Fines related to the health and safety issues amounted to $87,000.

On November 27, 1997, a backhoe operator hit a 500-gallon above ground propane storage tank and knocked off the pressure relief valve. A worker from the propane company suffered frostbite on his hand while

Figure 5.7 The Earthquake caused leaks and fires involving hazardous materials storage facilities. (From U.S. Geological Survey.)

trying to plug the leak in the tank. Propane when stored is at ambient temperature so whatever the outside temperature is, that's what the temperature of the propane is as well. Discharge of a pressurized gas also has the effect of lowering the temperature of the material. The neighborhood was evacuated by firefighters and the air was monitored for propane. Ignition sources were secured and utilities turned off as a precaution. No fire occurred (US Geological Survey).

Anne Arundel County, MD Special Operations

> **Prolog:** *During the late 1980s and the early 1990s, I was a member of the Earleigh Heights Volunteer Fire Company (EHVFC) Station 12, serving as a firefighter, EMT, Engineman, Lieutenant, and Hazmat Team Member. The company is a part of the Anne Arundel County Fire Department. EHVFC was a very busy company with two engines, a heavy squad, brush truck, and BLS ambulance. Five career personnel were assigned to EHVFC, two on a medic unit, an engineman, and two firefighters. As a volunteer, if you wanted to catch the action you needed to be at the fire house. Volunteers carried pagers and would be paged to fill their station when the career first out engine was on an extended call.*

During the time period I was there I attended college at Catonsville Community College and the University of Maryland. I would spend 20–30 h a week at the firehouse. Station 12 is a two story firehouse, and this is where I learned to slide a fire pole, which I did often! EHVFC provided a Hazmat Support Team for the Anne Arundel County Fire Department. The primary duty was decontamination, but the squad also carried some hazardous materials equipment as well. Primarily the hazmat team operations were a career function, but I was the only volunteer allowed to be a part of the hazmat team. There were other support teams in the county and as a hazmat team member I conducted decontamination training for support team members.

Volunteers wore uniforms just like the career personnel and had to meet the same training standards as well. As a result, to the average citizen the volunteer/career response was seamless. As volunteers arrived at the station to spend time responding they would fill out the crew on the first out engine with the career personnel. When numbers reached the proper levels additional apparatus would be placed in service. At certain times Station 12 could supply most of a first alarm assignment in their first due area. Volunteer officers had the same certification requirements as a career officer of the same rank and the same authority on the fire ground and in the station. Between responding to calls and other typical volunteer fire company events, my time at the EHVFC was among the highlights of my career. So, my visit to the AA County Hazmat Team to do the first article was a home coming of sorts.

Anne Arundel (AA) County Maryland is located south of the City of Baltimore and has a combination of career and volunteer fire department led by Chief Trisha Wolford. The fire department covers 588 miles^2 of land and 172 miles^2 of water located on the western shore of the Chesapeake Bay. Anne Arundel County has a population of over 584,909. Fire protection is provided by the Anne Arundel County Fire Department for the entire county except for the Baltimore Washington International Air Port, which is served by its own state fire department, U.S. Army Post Fort George G. Meade, which has its own federal fire department, The United States Naval Academy, which has its own fire department, and the City of Annapolis, which has its own municipal fire department. All fire departments in Anne Arundel County work closely together and have automatic aid in place, so in reality the departments act as one in daily operations. Some automatic aid also occurs with surrounding counties. Anne Arundel County was formed as an original Maryland county in 1650 and has operated under a County Charter government since 1965. Following the formation of the county charter government in 1965 all city governments and services in the county were dissolved except for the

City of Annapolis. All governmental services including fire and police are currently provided by Anne Arundel County.

Fire Department History

Fire Protection in Anne Arundel County has historically been provided by volunteers. Whole families were sometimes involved in the local volunteer fire company. In 1924, the Maryland State Legislature authorized Anne Arundel County Commissioners to appoint a paid "chauffeur and caretaker" for the volunteer stations at Earleigh Heights, Glen Burnie, and Eastport. "Chauffeurs" became county employees in 1932 and were later called "Enginemen".

Additional enginemen were added forming a three platoon shift of 24h on and 48h off. In 1963 the position of fire marshal was established. When the county charter government was created the present day, Anne Arundel County Fire Department was formed. Harry W. Klasmeier (Chief "K") was appointed as the first Fire Administrator and served until 1983. He was instrumental in bringing together the independent volunteer fire companies to form a central county fire department. During 1966 a Central Alarm and Communications Center was established along with a fire prevention bureau and training division.

Today's Modern Department

As Anne Arundel County grew so did demand for emergency services and the department evolved into today's combination county fire department. Many of the fire stations and equipment in AA County are owned by the volunteer companies and staffed by various configurations of career crews.

Co.1 staffed all career with volunteer administrative support – full Career crew.
Co.2 combination station (Woodland Beach VFC) – full Career crew.
Co.3 staffed all career with volunteer administrative support.
Co.4 Career Station (Special Operations station – HazMat and Technical Rescue).
Co.5 Career Station.
Co.6 combination station (Herald Harbor VFC) – full Career crew.
Co.7 combination station (Arundel VFC – PO and FF with Career day work ambulance) – no Career Officer.
Co.8 Career Station.
Co.9 Career Station.
Co.10 Career Station.
Co.11 combination station (Orchard Beach VFC) – full Career crew.

Co.12 combination station (Earleigh Heights VFC) – PO, Paramedic and two FFs with day work officer.
Co.13 combination station (Riviera Beach VFC) – full Career crew.
Co.17 combination station (Arnold VFC) – full Career crew.
Co.18 Career Station.
Co.19 combination station (Cape St. Claire VFC) – full Career crew.
Co.20 combination station (Lake Shore VFC) – full Career crew.
Co.21 Career Station.
Co.22 (NEW) Career Station (day work ambulance and EMS Supervisor).
Co.23 Career Station (Special Operations station – HazMat and Technical Rescue).
Co.26 Career Station.
Co.27 combination station (Maryland City VFC) – full Career crew.
Co.28 combination station (Odenton VFC – PO and day work FF with daytime only Medic Unit).
Co.29 Career Station.
Co.30 Career Station.
Co.31 staffed all Career with Volunteer administrative support.
Co.32 staffed all Career with Volunteer administrative support.
Co.33 combination station (Glen Burnie VFC) – full Career crew.
Co.34 Career day work ambulance – Volunteer staffed.
Co.40 combination station (West Annapolis VFC) – full Career crew.
Co.41 Career Station.
Co.42 combination station (Deale VFC) – full Career crew.

Career and Volunteers work side by side on incidents (see above for specifics). With the exception of Co.34, all Stations have Career staffing (Co.28 has only a PO for 24h).

Under the leadership of Chief Trisha L. Wolford Anne Arundel County Fire Department (AACOFD) boasts over 710 volunteer riding members and 881 career firefighters and EMS personnel. AACOFD, a combination career/volunteer department, provides county wide Fire, EMS, and Rescue services from 32 fire stations. They operate on a daily basis with 31 engines, 9 trucks, 8 squads, and 1 hazardous materials/special operations company. EMS is provided with 40 medic units available, 28 ALS and 12 BLS medic units normally in service. Additionally nine suppression units provide ALS response. Other specialized equipment includes 7 tankers, 1 dive rescue team, 22 brush units, 1 collapse rescue team, 7 fireboats, and POD transporter with 2 pods. Anne Arundel County Fire Department responded to 78,267 during 2019.

Hazmat Team History

Anne Arundel County formed their hazardous materials team in 1987 at Station 23 taking advantage of federal grant money that was available for hazmat team training, development, and equipment. Their first unit was a 1984 Pem Fab/E-One pumper and a 1985 Chevrolet Suburban (Special Unit 23). The first hazmat unit was a 1989 Pem Fab/American Eagle (Figure 5.8), 1,250 gpm pumper with a "squad type" body. The unit was a rear engine design which allowed for a fully functional command cab. This unit responded as an engine company on all types of calls and to hazardous materials incidents as needed. This unit was replaced by Squad 23 (Figure 5.9) which was replaced in 2018 by Squad 4 (Figure 5.10).

Hazmat Team

AA Counties Hazmat Team has experienced growth and reorganization since 2006. Trench rescue was moved from Crofton to the Hazmat fire house in Severna Park, Station 23. During this time period the Hazmat Team made a transition from totally hazmat to a Special Operations format. Six Special Operations (SO) Technicians operated the Special Operations Squad and Quint 23. They respond to Hazmat, Collapse, Trench, Rope, Confined Space, and Swift Water incidents as required. A POD unit was added for Trench Rescue (TR23) which carried the bulk of hazmat equipment. A second POD was added later. During 2016 the

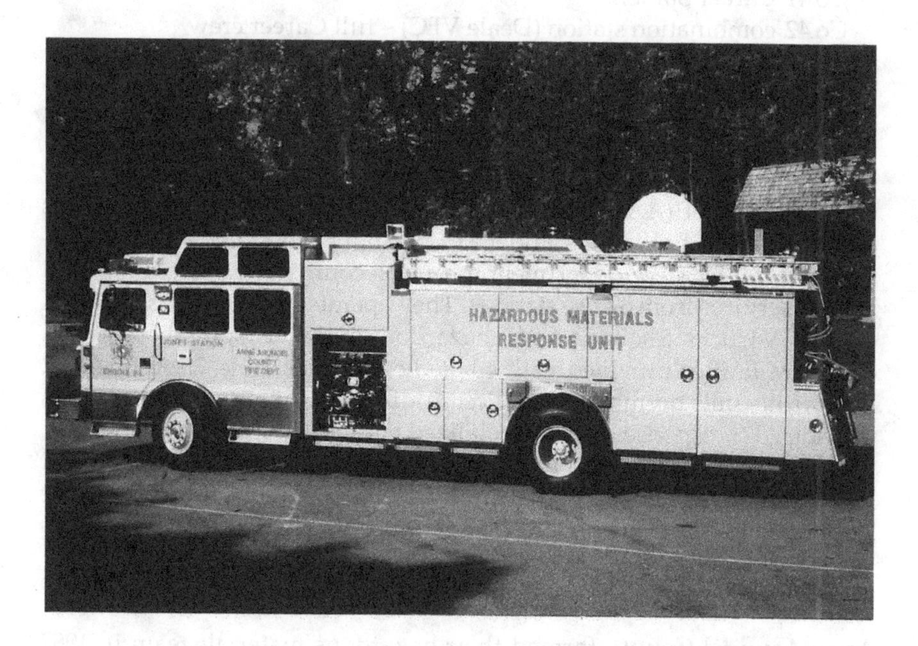

Figure 5.8 The first hazmat unit was a 1989 Pem Fab/American Eagle.

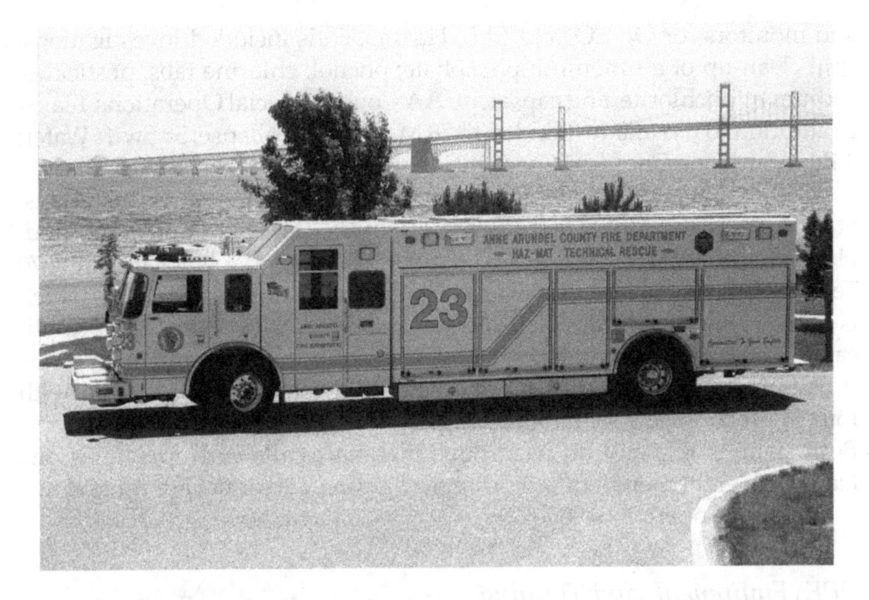

Figure 5.9 Pem Fab Unit was replaced by Squad 23.

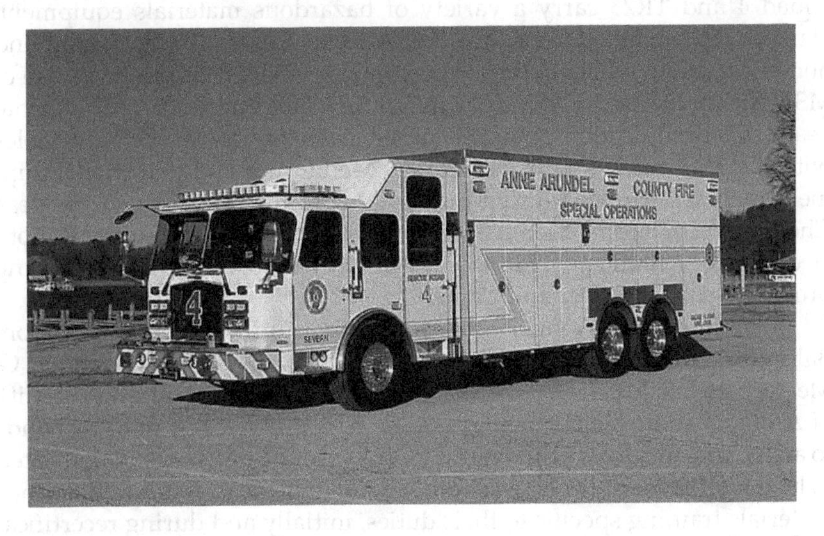

Figure 5.10 Squad 23 was replaced in 2018 by Squad 4. (Courtesy: Battalion Chief Christopher Schaetzly Anne Arundel County Fire Department.)

Special Operations Team was split and assigned to separate stations with the addition of Severn Station 4. The squad was moved to Station 4.

During 2018, AA County Hazmat responded to approximately 125 calls. Hazmat responses do not include gas odors or leaks and only includes fuel spills over 100 gallons. Suppression companies carry absorbent materials and monitors for O_2, CO, and LEL. Hazmat calls included investigations, spill clean-up of ammonium phosphate; phenol; chlorine tabs; pesticides; sodium hypochlorite; and capsaicin. AA County Special Operations Teams in addition to hazmat responded to 78 Marine; 69 Collapse; 55 Swift Water; 3 Rope; and one Confined Space incidents.

AA County today operates out of two Special Operations stations. Station 23, is located at 960 Ritchie Highway, Severna Park, Maryland. Housed at Station 23 are Quint 23 and TR23. Station 4 is located at 7870 Telegraph Road, Severn, Maryland. Housed at Station 4 is Squad 4, which is a 2018 E One walk in rescue body. Squad 4 carries a variety of duplicated hazardous materials gear which enables them to start operations if Station 23 happens to be on another call. Each Station is staffed with four SO Technicians for a total of eight on duty between the two stations. Both stations respond on hazardous materials calls and operate as one team with eight members. All other firefighting personnel are trained to a minimum of operations level.

PPE, Equipment, and Training

Squad 4 and TR23 carry a variety of hazardous materials equipment. Hazmat PPE includes Level A, Level A Flash, Level B encapsulated and non-encapsulated. Respiratory Protection is provided through the use of MSA APR's, 45 Minute MSA for Firefighting, 60 Minute MSA for Hazmat Team. Communications equipment includes Motorola AXP portables with SR65M headsets with throat mic. Monitoring and detection equipment includes Air Monitoring-Industrial Scientific MX4/6, Gas Badges/Chempro/Tiff/Jerome/Gemini and Razor. Research resources is both electronic and hard copy. Electronic is mainly used via the internet using programs such as Wiser and Cameo.

If additional personnel are needed they utilize mutual aid from Baltimore Washington International Airport (BWI), Fort George G. Meade Fire Department, Howard County Fire Department and the City of Annapolis Fire Department. A single ALS unit (closest unit) responds to all hazmat incidents with one additional unit dispatched and dedicated to hazmat if scene entry is required. All ALS personnel receive hazardous materials training specific to their duties, initially and during recertification. All hazmat team members are EMT's and can do their own personnel monitoring for pre and post entry. There are no volunteer members of the hazardous materials team.

The Maryland State Police H.E.A.T Team (Helicopter Rescue) has been replaced by four person MSP crews on their helicopters when they replaced the Dauphin with the AW-139 airframes. In addition to that change, there is the Maryland HART team that utilizes the MD Air National Guard

(MDANG) for aquatic rescue. AA County Special Operations personnel are trained (Technical Rope Rescue) and respond on high angle incidents such as on the Bay Bridge and other calls for service in the county that go beyond slope evacuations. However, they are no longer a part of the State Police Team.

Members of the Hazmat Team are interviewed and selected to be on the team. Team members are required to have a Pro-Board 80h HM Tech Class and one other Special Operations certification. Once these qualifications are completed, they enter the team as a Hazardous Materials Technician (HMT). Daily staffing requirements only allow for one out of the eight team members on a shift to be a HMT. The remaining seven are experienced SO Technicians. An extensive in house training program has been created that takes approximately 1 year to complete. New members achieve benchmarks which increases their capability and status n the team.

The first step is Hazmat Technician where they can function as a member of the hazmat team and work towards completing skills which are basic technical level exercises. Following approximately 6 months of in house training the new member will achieve the SO technician status. This is the point where the members start to work up more sophisticated tasks. This training is often referred to as "below the line". When all task/skills have been completed "below the line" the members must pass an in house practical exam at which point they earn the privilege of wearing the Special Operations patch.

Some team members have also been sent for outside seminars and training including local opportunities, the annual Hazmat Conference in Baltimore, and formal classes offered at the Federal Level through the Department of Homeland Security.

Hazardous Materials Exposures

Hazardous Materials Exposures in Anne Arundel County include, Interstates 95 and 97, 695, the Baltimore Beltway, dissect Anne Arundel County. Other major routes include 295 the Baltimore Washington Parkway, U.S. Highway 50 and Maryland Routes 2, 3, 100 and 648 the Baltimore Annapolis Boulevard. Potential hazmat incidents also involve the Chesapeake Bay, which is a major water shipping route and associated waterways that border Anne Arundel County on the western shore. Some fixed exposures include Baltimore Washington International Airport, Fort George G. Meade, Baltimore Gas & Electric Brandon Shores Power Plant, water treatment plants and others. Anne Arundel County borders Baltimore City on the South side of the city where many of the chemical and petroleum facilities in Baltimore are located.

Incidents

On May 30, 2015, crews responded to an acid leak in the Maryland City/ Laurel area. Upon investigation it was determined that a valve had failed on a 275 gallon poly tank which was leaking Oleum. The surrounding neighborhood was advised to shelter in place while the HazMat crew secured the leak. Once the situation was under control crews remained on location to assist with the transfer and neutralization of the acid (*Firehouse Magazine*).

Anniston, Alabama Hazmat Team

> **Prolog:** *Anniston is home to the Center for Domestic Preparedness, a training center for emergency responders to learn about response to weapons of mass destruction and acts of terrorism. My first visit there was in February 1999. It was during one of five trips I made to the facility for training that I visited the Anniston Fire Department to do a story on their hazmat team.*

Anniston is the county seat of Calhoun County, Alabama. According to population estimates in 2020, the population is 21,223. The City of Anniston has an area of 45.5 miles². Anniston is located in Northeast Alabama just off of Interstate 20, 58 miles East of Birmingham, Alabama, and 80 miles West of Atlanta, Georgia. In 2003, the Anniston Army Depot began the process of destroying nerve agents it had stored over the years. The incinerator was built to destroy the chemical weapons stockpile of Sarin and VX nerve agent and mustard blister agent stored at the depot. The depot, along with associated defense contractors, is now Anniston's largest employer.

Today's Modern Fire Department

Under the leadership of Fire Chief Chris Collins, the Anniston Fire Department has 98 uniformed personnel located at six fire stations, who operate five engine companies, two truck companies, two medic units, one rescue squad/hazardous materials company, and an airport crash truck. In addition to the hazardous materials unit, Anniston Fire Department also maintains an 8 feet × 20 feet trailer with absorbent materials, overpack drums, decontamination tent, water heaters, and other equipment.

Hazmat Team History

Anniston's hazardous materials unit was placed into service for the first time in July of 2004. Prior to that time, they had no hazardous materials response capability. Each engine and truck company carries oil drying material for cleaning up small fuel spills. Larger spills would require the hazmat company response. Anniston worked on establishing hazardous materials capability for 2 years. Money for establishing Anniston's Hazardous Materials Team came from a $1.3 million FEMA grant. They are a top-rated team for a community of their size, and they are the only team covering the 200 miles of Interstate 20 between Birmingham, Alabama, and Atlanta, Georgia.

Hazmat Team

Anniston's response unit is a Pierce 2004 hazardous materials vehicle (Figure 5.10A). It is equipped with a light tower, 40 kW generator, air cascade, command post, and four-man crew cab. Storage space is available on the top along with the light tower. A built-in collapsible stair can be set up to allow access to the top storage and light tower. Compartments have roll-up doors, and some are equipped with pullout trays for easy access to equipment such as self-contained breathing apparatuses (SCBAs) and other equipment. The unit is housed at Fire Station One, located in downtown Anniston at 217 East 17th Street. The hazmat trailer is located at Fire Station Three along with an engine, 100 feet aerial ladder truck, and brush truck.

Figure 5.10A Anniston's response unit is a Pierce 2004 hazardous materials vehicle.

The hazardous materials unit is not staffed full time. When a hazardous materials call is received, the firefighters/hazmat technicians from the other companies in town assemble at the main station where the response force is selected to staff the hazardous materials unit. For an initial hazardous materials call, eight firefighters will respond with the hazardous materials unit. For large calls, off-duty firefighters/technicians are recalled to duty.

PPE and Equipment

Personnel protective equipment (PPE) worn by hazmat team members for Level A is the Kapplar Tychem Responder equipped with flash protection. Protection worn for Level B is the Tychem encapsulated and non-encapsulated suits. Respiratory Protection is provided with MSA SCBA with 1h bottles and 3M 6000 negative pressure cartridge respirators. In-suit communication is provided by Motorola radios. Radios in use are on the 800MHz band with ability to talk countywide to all other response agencies.

Monitoring Instruments & Identification Equipment used by the Anniston Hazmat Team include photo ionization detector (PID), Haz Cat System, AP 2C, RAE Systems five-gas meter, MSA four-gas meters, and TIMS. Terrorist agent monitoring instruments include the APD-2000, which detects chemical warfare agents, pepper spray, and mace, and the RAMP System.

Research Resources

Research materials on board the hazmat unit include *NIOSH Pocket Guide to Chemical Hazards,* DOT ERG, *Emergency Handling of Hazardous Materials in Surface Transportation, NFPA Fire Protection Guide to Hazardous Materials,* CHRIS Manuals, *Jane's Chem-Bio Handbook, DuPont Permeation Guide, Sittig's Handbook of Toxic and Hazardous Chemicals and Carcinogens.* Computer databases include Oreis, CAMEO, Aloha, and Marplot.

Hazardous Materials Exposures

The destruction of the chemical weapon stockpile of Anniston Army Ammunition Depot (ANAD) was completed in 2011. ANAD today is primarily a maintenance facility for the Army. Anniston has the Norfolk Southern Railroad traveling through town, Interstate 20 to the South, an ice plant with anhydrous ammonia, an industrial park, a gasoline tank farm, propane storage tanks, acids, and a high-pressure natural gas transmission pipeline. Anniston has available an equipment trailer with a decontamination tent, water heaters, and other equipment for mass decontamination (*Firehouse Magazine*).

Ashland, KY Regional Hazmat Team Available Resource for Three States

Ashland Kentucky is located on the Southern Bank of the Ohio River in Eastern Kentucky, across from West Virginia, just off of Interstate 64. Known as Kentucky's Country Music Capital, Ashland has a population of approximately 19,792 and covers an area of 10.8 miles^2. Ashland is the largest city in Boyd County and is a part of the fifth largest metropolitan area in Kentucky. Ashland, Kentucky, Huntington, West Virginia, and Ironton, Ohio form a metropolitan tri-state area of more than 300,000 people. Ashland serves as an important economic and medical center for Northeast Kentucky.

Fire Department History

During 1878, Ashland was jarred awake by its first major fire. The fire swept through one of their leading business establishments along Front Street between 15th and 16th streets. The disaster was the catalyst for discussion among the community and its leaders for organizing a fire department and purchasing fire equipment. Little came from these discussions, and the city went on with little or no improvement. Then in 1883, the wooden portions of the Norton Iron Works facility were destroyed by fire. The bucket brigade could not provide sufficient water to hinder the spread and destruction of the flames. The incident spawned discussion for changes in the way firefighting operations were conducted. The city had been hit hard enough, and they were ready for a change. In 1885, the Ashland Fire Department was founded as a volunteer fire department. The new department purchased a hand-drawn hook and ladder wagon and 24 rubber buckets. In later years, the city migrated away from a volunteer department to a full-time, paid organization and purchased steam engines and hoses. Since that time, the department has evolved and expanded into a group of dedicated personnel who not only extinguish fires but perform other specialized operations such as medical assistance to emergency medical technicians (EMTs), code enforcement, fire prevention, fire inspection, public safety education, smoke detector installation, various types of specialized rescue services, and hazardous materials (hazmat) response and mitigation.

Today's Modern Department

Today's modern Ashland Fire Department has an ISO rating of 3 and is led by Chief Greg Ray. The department has 56 uniformed personnel and covers a city area of 12.1 and 160miles^2 for hazardous materials response in the county with an extended population of 50,000. Firefighters work a schedule of 24h on duty and 48h off duty. Ashland Fire Department operates from three fire stations with four engines, one truck, two boats, one hazmat trailer, and a mini-pumper brush unit. In addition to hazmat response, they also have swift-water dive and rope rescue teams. Ashland Fire Department responds to over 1,600 requests for service each year with an average of 60 hazmat calls. Approximately 30% of these are EMS responses with transport provided by Boyd County.

Hazmat Team

Ashland's regional hazardous materials response team was formed in 1992 when the department's leaders saw the need in their jurisdiction for a local and regional response capability. Their first response vehicle was a tractor trailer rig that carried response equipment. Presently, Ashland utilizes a 24 feet Homesteader trailer towed by a step van (Figure 5.11). A hand-drawn decon cart is pre-loaded with decon equipment utilized by operations personnel that have been trained to do decon and is available throughout the county. The hazardous materials team equipment and vehicles respond from Central Station, located at 1021 Carter Avenue in downtown Ashland. Central Station houses fire department administrative offices along with six drive-through bays for apparatus. Hazmat is a part of Special Operations which also includes Rope Rescue, Dive Team, and Swift Water Rescue.

Figure 5.11 Ashland utilizes a 24 foot Homesteader trailer towed by a step van for hazardous materials response.

PPE, Equipment, and Training

Hazmat personnel protective equipment (PPE) is composed of Scott SCBA with one hour bottles. Level A chemical suits are manufactured by Lakeland Industries. Level B chemical suits are manufactured at Kappies Incorporated. They also carry assorted boots and gloves to complete the chemical protective equipment ensemble. Mass decon is accomplished with a Zumro Shelter/Decon Tent. Response equipment includes typical hazmat patching and leak stopping equipment including chlorine kits A & B and other tools and supplies to manage a hazmat incident. Monitoring equipment includes MultiRAE Plus, Mini CO detectors, Haz Mat ID 360 Command System, Draeger CMS and ODS monitors, RAD monitors, and Thermo Eberline ESM detector.

Hazmat team members are required to take a 40 hour technician course in-house developed by the Kentucky Community and Technical College System (KCTCS). All department firefighters are trained to the Technician Level. County firefighters are trained to the Operations Level and to do decon to assist the Ashland Fire Department on hazmat incidents. Some Ashland Fire Department hazmat technicians have attended the Department of Energy's (DOE) Nevada Test Site and attended biological/chemical classes at the Center for Domestic Preparedness (CDP), located in Anniston, AL. Others have attended the explosives classes at New Mexico Tech (EMRTC) and hazmat technician classes put on by the International Association of Firefighters (IAFF). Team members have attended classes put on by some of the local industries as well, including Marathon Refinery, DuPont Chemical, Calgon Chemical, AK Steel, Air Products, and CXS railroad. All these classes resulted in certification for Ashland hazmat team members through their company on particular products or processes they make or perform.

Ashland hazmat personnel also receive quarterly in service training in-house. Quarterly training has included an anhydrous ammonia training tanker with Marathon Oil Refinery. Hazmat team members were tasked with stopping a leak of ammonia simulated with smoke from a smoke machine. Smoke appeared to be ammonia vapors along the ground. Team members went through a full drill using Level A suits. They dressed out, investigated, had members apply the leak kit, sent them through decon, and had a debriefing afterward.

Hazmat responses in Ashland include fuel spills, gas odors, and leaks. Engine companies carry absorbents, CO detectors, and MultiRAEs. If the company officer determines that they need the hazmat team for an initial engine response, they call for the team. The hazmat team is not dedicated, but there are approximately 16 hazmat technicians on duty each shift. There are a total of 52 hazmat technicians in the Ashland Fire Department.

Hazardous Materials Exposures

Ashland has several major U.S. highways in its response area that are transportation routes for hazardous materials, including U.S. Highways 23 and 60 and Interstate Highway 64 along with various state and local roads. CSX Railroad has a line through their response area and barge traffic along with the Ohio River Marine Terminal have hazardous materials as well. Pipelines also transport hazardous materials in and through Boyd County. Major industries that manufacture or use hazardous materials include the Marathon Oil Company's Catlettsburg Refinery. It is the 26th largest refinery in the United States and is a supplier of high-performance aviation fuels to the U.S. Military.

The refinery has its own fire brigade but receives aid from Ashland Fire Department. Other major industries include AK Steel, Special Metals, and Calgon Carbon, Inc., and Mark West Hydrocarbons distributes propane and butane by pipeline, tank truck, and railcar. DuPont Chemical Company produces sulfuric acid, the number one volume chemical in the United States.

Incidents

- Ashland Federal Correctional Institution for a suspicious powder in mail room. Hazmat personnel went in and removed a couple of staff members and deconed them. Samples were taken of the substance for the prison so they could send it off to be tested and identified.
- Another incident involved a response to a barge, where some barge workers were overcome by chemicals used for cleaning a barge on the Ohio River. The injured workers were sent through decon and prepared to be transported by ambulance to the hospital.
- Ashland Fire Department has responded to several residences where a meth lab was present. Most of the time their mission is to enter and test the air quality.
- Multiple responses have occurred several times at the local CSX rail yard for tanker cars leaking hazardous materials. Most incidents involved chlorine. The areas were secured and confirmation of the chemical was made. CSX then brought in a cleanup contractor to stop the leaks and provide any cleanup necessary.
- Clandestine dumping has also been responded to, where several discarded 55 gallon drums were dumped along a road. Efforts were made to identify the chemical(s) and mitigate as necessary. Most of the chemicals identified were … non-hazardous.
- Chlorine leaks have also been reported at the city water treatment and sewage treatment plants. Hazmat personnel were tasked to stop leaks in some of their tanks and contain spills that have occurred.

- On July 30, 2011, a liquid oxygen leak occurred at Air Products, a local industrial facility. Ashland firefighters stood by and monitored the oxygen levels in the area around the plant while the leak was being repaired. Firefighters also sprayed water to keep the cloud from releasing oxygen inside the perimeter of the plant. Oxygen becomes a gas once it hits the atmosphere. The oxygen that was released due to the leak dissipated into the atmosphere, and there was no off-site hazard or environmental impact. Initially, oxygen is literally harvested from the air and placed in tanks through a cryogenic cooling process. The rail lines adjacent to the plant also had been shut down. U.S. 23 had been closed near the facility to minimize the risk of a fire or explosion from the elevated oxygen levels in the atmosphere. The leak that injured one Air Products employee, reported at about 3:15 p.m., was caused by a faulty valve. The employee suffered cryogenic burns on both of his hands and was taken to the Cabell Huntington Hospital burn unit for treatment. He was released the next day. Crews worked through the night to isolate the portion of the plant where the leak occurred so the valve could be replaced.

Aurora, Illinois Hazmat Team

Aurora, Illinois, is a suburb of Chicago on the western edge of the Chicago metropolitan area. It is located along the Fox River in southern Kane County, and portions of the city limits extend into DuPage, Will, and Kendall Counties. Aurora is the second most populous city in the State of Illinois with an estimated 198,870 people in 2020 covering 46 $miles^2$. Before European settlers arrived, a Native American village was located in what is downtown Aurora today, on the banks of the Fox River. The post office was established in 1837 officially creating the town. In 1936, Aurora officially adopted the nickname "City of Lights", because in 1881, it was one of the first sites in the United States to implement an all-electric street lighting system.

Fire Department History

In May of 1856, eight buildings burned on the west side of Broadway between Main Street (now East Galena Blvd.) and Fox Street (now East Downer Place). In spite of this conflagration, it took two more fires, one in a Chicago Burlington & Quincy Railroad storage shed and another destroying three buildings in the business district, to finally bring some action to establish a fire company. An organizational meeting for the new fire company was held on July 1, 1856. The Name "Young America Fire Engine Company No. 1" was chosen. By August of 1856, a fire engine, a hose carriage, and 500 ft of leather hose, were ordered for the sum of $1,100.

A fire company called the "Holly Hose Company" was formed, and the Holly pump was purchased. John Eddy, who eventually was to become Chief Engineer of the Aurora Fire Department, was elected as the first Foreman of the Holly Hose Company.

The new steam fire engine, named "City of Aurora", arrived in Aurora in June of 1869. The steamer required 40 men to pull it on level ground, 80 to pull it uphill. These engines, at the time, were considered as good as any in the market. Horses were frequently rented from a nearby livery stable to pull the steamer until 1874 when the city bought a team of horses to pull the "City of Aurora" Steam fire engine. Eureka Hook and Ladder Company was formed in early 1871. Like the steam fire engine, horses were rented to draw this apparatus to a fire. It was not until 1874 that a new building was built for their equipment. This building was located next door to the No. 1 Fire Station on North Broadway.

Beginnings of a Career Fire Department

On January 28, 1882, the fire department became a partly paid organization under the jurisdiction of the City of Aurora, and the volunteer fire companies were disbanded. The Aurora Fire Company soon became known as Hose Co. No. 1, the Excelsior Fire Company became Hose Co. No. 2, and Holly Hose Company became Hose Co. No. 3. The next great advance in firefighting technology was in the use of "chemicals". Aurora purchased a Muskegon Chemical wagon for Co. No. 1 in September 1892. This engine works on the principal of mixing soda-water and acid to generate carbon dioxide gas which pressurizes the water tanks and forces the water out through the hose.

Because chemical engine did not require any external water supply, it could be ready to attack a fire within seconds of arriving at the scene. (It took the conventional Steamer 3–10 min to get water pumping.) This lead to an increase in the popularity of the chemical engine through the late 1800 and early 1900s. In December 1892, the city bought a 90 feet aerial hook and ladder truck from the Michigan Fire Ladder & Engine Company of Grand Rapids, Michigan, at a cost of $2,300. The truck was known as the "Arrow Aerial Turntable Hook & Ladder" and represented the latest in modern design incorporating improved mechanism for raising a ladder quickly and moving it from one window to another. This apparatus was to have been built expressly for the Chicago World's Fair, but Alderman Messenger got the company to sell it to Aurora and build another one for the fair.

During his years as Fire Chief, Adam Schoeberlein installed a new horse harness system. It was suspended over the horses and released in such a way that the horses were harnessed and out of the barn before the fire bell stopped ringing! When trained, horses were left unhitched in stalls directly behind the apparatus. At the sound of an alarm, the stall doors automatically opened, allowing the horses to run to their assigned places. Before the alarm bell stopped ringing, the well-trained fire horses were out of their stalls and backed up in front of the fire wagons, each horse standing exactly beneath his suspended harness. As the driver pulls a ceiling rope releasing the harnesses onto the backs of the horses, a firemen need only to snap one snap to secure the collar around the horses' necks before the engine is ready to go. The entire process, from the time the bell rang to the time the engine was out of the station, took less than 1 min!

The old No. 1 Hose House was razed in 1894, and a new building was constructed on the North Broadway lot. The police patrol was also housed in Central Fire Station until 1912; the patrol driver and police occupied a small room on the first floor. This building, used as Central Fire Station until 1980, is now home to the Aurora Regional Fire Museum.

The newspaper called the new Central Station, "A building to be proud of - leaving nothing to be desired in cost, appearance, finish or good taste". It was a two-story brick structure surmounted in front with a mosque-like onion dome. A chemical engine and hose cart occupied the south side of the building, and the police patrol wagon and ladder truck were located on the north. Behind the apparatus were eight horse stalls with a hay loft above.

Today's Modern Department

Aurora Fire Department operates out of 9 fire stations strategically located throughout the city. Under the leadership of Fire Chief Gary Krienitz, 195 uniformed personnel provide fire protection for the citizens of Aurora with 12 fire companies. Apparatus includes 9 engines, 3 trucks, and 6 medic units. Additional supporting apparatus for special operations are an airboat, hazmat unit, decon trailer, and an arson investigation unit. Special teams have been created for water rescue and recovery and hazardous materials response. They responded to 19,422 calls in 2019. 81.4 % of those were EMS related.

Hazmat Team

Aurora's hazmat team is located at Central Fire Station, 75 North Broadway Street. Their primary hazmat vehicle is Squad 2 Hazmat/Rescue Squad (Figure 5.12), a 1995 Freightliner Heavy Squad. Rescue 22 is a 2001 Dodge 2500 pick-up hazmat trailer and hazmat decon trailer. Also located at Central Station are Engine1, a 1,500 gpm pumper, Ladder 2, Rescue 1, Medic 1, Water Rescue 23, fireboat, and airboat. Any spill over 5 gallons or less on officers discretion will be handled by hazardous materials response team (HMRT). AFD is a part of Mutual Aid Box Alarm System (MABAS) Division 13 Larger Team. Mutual aid is available from Naperville or other

Figure 5.12 Aurora's hazmat team is located at Central Fire Station, 75 North Broadway Street. Their primary hazmat vehicle is Squad 2 Hazmat/Rescue Squad.

MABAS Team. According to the Chicago Fire Department, Aurora is the only hazmat team they will call for mutual aid. There are usually 4–6 technicians on duty each shift. The HMRT is not dedicated; it is staffed from personnel from Engine 1 and Truck 2 as needed. There are 26 total technicians citywide.

PPE, Equipment, and Training

Hazmat personal protection equipment (PPE) includes (Level A) St. Gobain One suit Pro, DuPont, Tychem and (Level B) Kapplar; TK respiratory protection if provided with Scott 4.5 SCBA, Scott Positive Pressure Air Purifying Respirators (PAPR) and MSA Millennium Air Purifying Respirators (APR); in-suit communications, Motorola radios, Con-space headsets; monitoring and detection equipment, Draeger Xam-2000 4 gas (13); RAE Systems MiniRAE 3000, PID (2), RAE Systems AreaRAE RDK (1); and Smith's Detection LCD 3.3 (2) and Smith's Detection Hazmat ID 360.

The HMRT conducts monthly training drills, including in-suit training, use of monitors, tabletop drills, and department-wide drills. Four Mutual Aid Box Alarm System (MABAS) hazmat drills are held each year for all division personnel. Seven members attended Crude by rail training at the TTCI in Pueblo Colorado. In 2014, HMRT worked with the Naperville Fire Department and MABAS Division 16 on a functional exercise that simulated a crude oil train incident and fire.

Reference Materials

Computer with WiFi, PEAC, Wiser, NIOSH, ERG, and others (Aurora Fire Department).

Baltimore City Fire Department Hazmat Team

Prolog: *It isn't often that I get the opportunity to write about an organization that I have the had the pleasure to work with on a regular basis as Fire Marshal and Hazmat Team Member at the University of Maryland Baltimore. Baltimore City Hazmat Team personnel and firefighters provide our campus with excellent service and we have a very good working relationship between our two organizations. We had our own hazmat team at the University of Maryland, but relied on the Baltimore City Team as a back-up when needed. Often drills were held with the University Team for hazmat exposures at the university. We had many research facilities intermingled with the professional graduate schools at the Baltimore Campus. These included Law, Social Work, Medical, Dental, and Pharmacy Schools.*

Hazardous materials exposures included a multitude of research chemicals scattered throughout hundreds of labs. Specific hazards included a hydrogen generator, high powered MRI machines, radiation sources, military nerve agents, cryogenic liquids, bottled gases, and biological materials including HIV, anthrax, and others. Most times firefighters responding to alarms at our campus were very cautious about chemical, radiological and biological areas and relied on our hazmat personnel to accompany them to the alarm locations.

Baltimore is an independent city in the State of Maryland. (This basically means that Baltimore City is not a part of Baltimore County). The city was established by the Constitution of Maryland in 1729. Baltimore City is the largest independent city in the United States. Baltimore City is situated on the Patapsco River at the southern end of Baltimore County and covers an area of 92.05 miles^2 with a population of approximately 590,479 in 2020, which increases to over 1,000,000 during the daytime. The Inner Harbor connects to the Chesapeake Bay just to the east of the city. The metropolitan area has a population of 2,325,000 in 2020. The city is located 40 miles northeast of Washington, D.C. making it a principle city within the Washington–Baltimore combined statistical area with a population of 9,814,928 in 2019.

Baltimore has the second largest seaport in the Mid-Atlantic. The city's Inner Harbor was once the second leading port of entry for immigrants to the United States, when most arrivals were from Europe. With hundreds of identified districts, Baltimore has been dubbed a "city of neighborhoods". Famous residents have included writers Edgar Allan Poe, Edith Hamilton, Frederick Douglass, Ogden Nash, and H.L. Mencken; jazz musician James "Eubie" Blake; singer Billie Holiday; actor and film makers John Waters and Barry Levinson; and baseball player Babe Ruth. During the War of 1812, Francis Scott Key wrote "The Star-Spangled Banner" in Baltimore after the bombardment of Fort McHenry.

Baltimore has more public statues and monuments per capita than any other city in the country. The city is home to some of the earliest National Register Historic Districts in the nation, including Fell's Point, Federal Hill, and Mount Vernon. These were added to the National Register between 1969 and 1971, soon after historic preservation legislation was passed. Nearly one-third of the city's buildings (over 65,000) are designated as historic in the National Register, which is more than any other U.S. city.

Fire Department History

Organized fire protection in the City of Baltimore began with the formation of the Mechanical Fire Company in 1763. Members of the company were all volunteers. Initially the company was little more than an organized bucket brigade with ladders and other miscellaneous equipment. In 1769, the fire company purchased its first apparatus, built in Holland and made available from a ship's captain who had the "engine" on his ship; it was nicknamed the "Dutchman".

The hand engine was equipped with two pumps and was drawn with ropes by the firefighters to get it to a fire scene. The Mechanical Fire Company was the sole firefighting organization in the city until 1782 when the Union Fire Company was organized. Their first apparatus was also imported from Europe and was similar to the "Dutchman" and was nicknamed "Tick-Tack". Additional companies formed over the next several years included the Friendship, Deptford, Mercantile, and Liberty fire companies.

The volunteer fire companies in the City of Baltimore were organized into the Baltimore United Fire Department in 1834. In 1858, the career Baltimore City Fire Department was formed with 22 fire companies, including three steam engine companies, 17 hand engine companies, and two hook and ladder companies. On February 8, 1904, the Baltimore City Fire Department was faced with its most difficult fire in the history of the city. The fire originated in the basement of a storage facility and was quickly spread by high winds and complicated by the lack of standard hose couplings. Mutual aid fire companies could not hook up to Baltimore City hydrants or hose lines because their hose threads did not match.

Firefighters from 72 fire companies fought the blaze; 38 of the companies responded from Philadelphia, New York, Harrisburg, Chester, Wilmington, Washington, and as far away as Altoona, PA. Over 1,200 firefighters and 200 National Guardsmen fought the fire for 36 h before it was brought under control. The fire destroyed 1,526 buildings spanning 70 city blocks causing over $150,000,000 in damage. One person was killed and 247 were injured. Over 35,000 people lost their places of employment as a result of the fire.

Today's Modern Department

Today's modern Baltimore City Fire Department led by Fire Chief Niles R. Ford is an ISO Class 1 rated fire department. Baltimore City Fire Department operates with over 1,600 uniformed personnel and 69 civilian staff from 38 fire stations located throughout the city organized into six battalions. Each fire station is identified by the engine company that is stationed there except for Station 15 which formerly housed Truck 15 and now houses Engine 33. There are 29 engine companies, 15 truck companies, 21 medic units, 12 critical alert medic units, four engines that are identified as squads, Hazmat 1, Hazmat 2, technical decontamination unit, a dive rescue team, a collapse team, high-angle rescue team, two mobile air cascade units, two fireboats and two fire/rescue boats in active service. Baltimore's decontamination unit is one of five almost identical units in the metropolitan area purchased with Federal Grant Money through a grant applied for by the Baltimore City Fire Department. Additional units are located in Baltimore County, Howard County, Harford County, and Carroll County.

Fireboat #1, commissioned the "John R. Frazier" (named in honor of Bureau Commander John R. Frazier who was instrumental in obtaining funding for the Marine Division during his tenure with the fire department) in August 2007 is the first new steel hull boat purchased by the department in 47 years. Fire/Rescue Boat #1, the department's other first line Marine Unit, is a 30 feet, 1,500 gpm boat commissioned December 2003. The department also has two reserve boats in its fleet. The special teams including the hazmat team are under the Special Operations Command. Baltimore City responds to an average of over 270,000 alarms per year, many of which are EMS related.

Hazmat Team History

Baltimore City Fire Department's hazmat team was formed in 1985 under the direction of now retired Battalion Chief C.B. "Buzz" Melton. The team was formed as a result of newly passed Federal Legislation requiring specialized operations and training for hazardous materials response. Brooklyn Station 35 and Truck 21 was the original home of the hazmat unit (Figure 5.13). It was relocated to the downtown "Super House" Steadman Station at 15 South Eutaw Street during 2001 along with Battalion Chief 6.

Hazmat Team

Also located at Steadman Station are Engine 23, Truck 2, Rescue Boat #1, the technical decontamination unit, dive rescue unit, the collapse unit, Airflex 1, EMS 2, and Medic 1. There are approximately 350 trained hazmat technicians working four shifts (about 87 per shift) with the Baltimore City Fire Department. Firefighters in Baltimore City work two 10h days followed by two 14h nights and then have 4days off. The hazmat team is not dedicated and has one driver assigned to the unit per shift. Acting Lieutenant James E. ("Slim") Stanley is currently the hazardous materials team coordinator. His office is located on the second floor of Steadman Station. All personnel at Steadman Station are hazardous materials technicians, and Engine 23, Truck 2, and Rescue Boat #1 respond with Hazmat 1 as needed depending on the type of alarm received. Personnel at Engine 35, Truck 21, and the four squads in the city are also hazmat technicians and available to assist with hazmat calls. Hazmat 1 responds to an average of 400 hazmat incidents each year (Figure 5.14). Squads, truck companies,

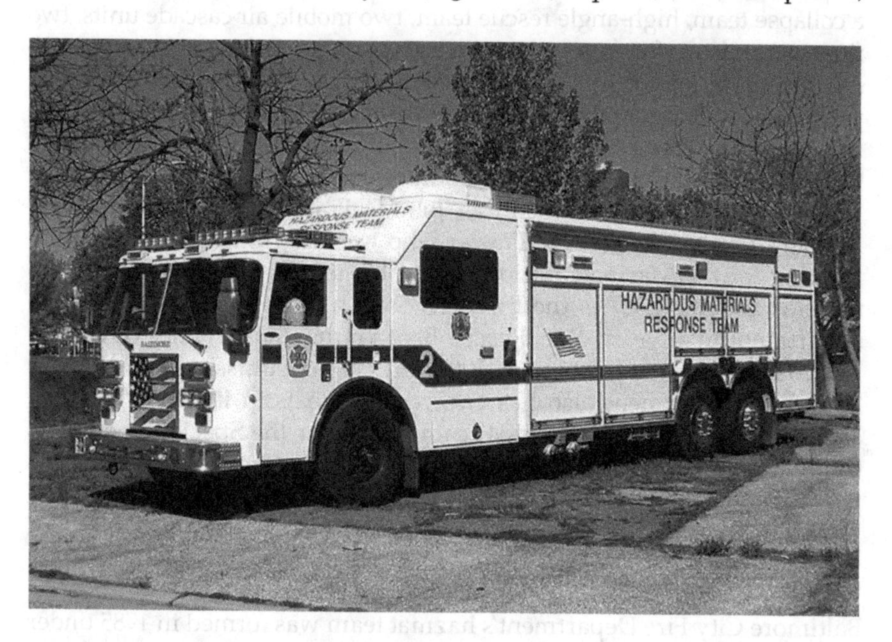

Figure 5.13 Brooklyn Station 35 and Truck 21 was the original home of the hazmat unit. (Courtesy: Baltimore City Fire Department.)

Figure 5.14 Hazmat 1 responds to an average of 400 hazmat incidents each year.

and other first-due engines respond to an average of 2,500 incidents each year involving fuel spills, gas leaks, odor investigations, and carbon monoxide alarms.

Absorbent materials are carried on all apparatus, and first-due companies can handle fuel spills up to 50 gallons but have the option to call in Hazmat 1 anytime they need additional resources. If a spill enters the sewer system, Hazmat 1 is called automatically. Truck companies and squads carry five-gas meters and PIDs (photo ionization detectors). All companies in the city carry radiation meters. All firefighters in the city are trained to a minimum of the Operations Level.

Mutual aid is available from both public and private sources. Anne Arundel County and Baltimore County Fire Departments work together with Baltimore City on a regular basis. Private industry is also available to provide resources to Hazmat 1. The South Baltimore Industrial Mutual Aid Plan (SBIMAP) with 80 member organizations, 60% private and 40% public, are dedicated to assisting each other during emergencies. The organization received a Partnership award from the United States Environmental Protection Agency (EPA) in 1998 for its innovative emergency program. They have been providing emergency equipment, know-how, and other resources in the Baltimore area for over 25 years. SBIMAP has been a model for the formation of other mutual aid organizations across the country. As I spoke with Hazmat Coordinator James Stanley in his office, he pointed to a radio in the room and indicated all he had to do to get a wealth of hazmat resources to an incident scene was to use that radio.

Hazmat 1 is a 2006 Seagraves Rescue Body with an interior walk-through compartment area and command center. Their technical decontamination unit is a 2005 Pierce.

PPE, Equipment, and Training

PPE for body protection is provided with Kapplar and DuPont suits for Level A entry and Kapplar encapsulated and unencapsulated suits for Level B entry. Motorola radios with ear pieces and mikes built into the SCBA face pieces are used for in-suit communications. Hazmat team members are equipped with Draeger SCBA with 1 h bottles, along with Draeger PAPRs and cartridge respirators for respiratory protection at hazmat and terrorist incidents. Firefighters in the city also use Draeger SCBA but with 45 min bottles. Originally they used 1 h bottles like the hazmat team, but they found them too bulky for interior firefighting operations.

Hazmat team personnel receive the Maryland Fire and Rescue Institute (MFRI) 56 h Hazmat Technician course and a 40 h Weapons of Mass Destruction (WMD) course before they are assigned to be active in the unit. Some personnel are sent for additional training at the TTCI Emergency Response Training Center in Pueblo, Colorado; Radiation School at the Nevada Test Site near Las Vegas, Nevada; WMD training at the Center for Domestic Preparedness in Anniston, Alabama; and Incident Command courses at the National Fire Academy in Emmitsburg, Maryland.

Hazardous Materials Exposures

Highway transportation routes in the Baltimore area where hazardous materials are shipped on a regular basis include Interstates 70, 83, 95, 97, 395, 695, and 795. Baltimore is situated on the Patapsco River which is a tributary of the Chesapeake Bay. Hazardous materials in intermodal containers are shipped to and from the city through the Port of Baltimore. Intermodal containers are loaded on highway vehicles and railcars and shipped throughout the Eastern United States. Rail lines located in the city include CSX and Norfolk Southern. The CSX line which runs under the city below Howard Street was the site of a major hazardous materials spill and fire in 2001. Colonial Pipeline Company has a pipeline system which terminates in the Curtis Bay area in the SE side of the city where there is a major concentration of petroleum and chemical facilities. Some of the chemicals found in the industrial areas include uranium-hexaflouride, anhydrous ammonia, chlorine, petroleum products, and ethanol produced at a facility in the city.

Incidents

Howard Street Tunnel Fire

On July 18, 2001, the Baltimore City Fire Department and hazmat team faced one of their largest and most challenging hazmat/fire incidents on record. At approximately 3:08 p.m., 11 cars of a 60-car CSX train derailed while passing through the 1.7 mile Howard Street Tunnel in downtown Baltimore. Eight tank cars on the train contained hazardous materials, and four of those cars derailed. The derailed tank cars contained tripropylene, a flammable liquid, hydrochloric acid, and di (2-ethylhexyl) phthalate, a plasticizer and environmentally hazardous substance. The leaking tripropylene caught fire and burned along with other combustible cargo in other railcars. Civil defense sirens sounded through the city; major highways into Baltimore were closed along with boat traffic in the inner harbor. Public Works Department representatives indicated that this was the first time the sirens had been used that was not a drill or test. (The sirens are tested every Monday at 1:00.)

Shelter in place orders were issued for downtown buildings, although ignored by the University of Maryland Baltimore Campus. Traffic was gridlocked on city streets, people waited for buses that could not get to them, light rail service was restricted because of the proximity to the incident scene, and the Metro subway was closed for a period of time to make sure smoke had not entered the tubes. The second game of an Orioles day–night double header was canceled, downtown stores were closed, and night classes at the University of Baltimore were canceled. Five alarms of fire equipment and approximately 125 firefighters responded to the incident which lasted until July 23rd. Firefighters and hazmat personnel had a difficult time getting to the site of the derailment because of the leaking chemicals, the thick black smoke from the fire, and the fact the incident occurred inside the tunnel. The derailment location was approximately 3/4 of a mile into the tunnel. The situation was further complicated by a 40 inch water main brake in Howard Street above the fire and a resulting power outage that affected approximately 1,200 residents. Twenty-two people were injured including two firefighters who experienced chest pains. Ironically, it was scheduled to be Firefighter Appreciation Day 2001 at Oriole Park at Camden Yards located just a few hundred feet from the opening to the tunnel (*Firehouse Magazine*).

Baltimore County Fire Department Hazmat Team

Prolog: *During my 22 years living in the Baltimore Metropolitan Area I had the opportunity to work with members of the Baltimore County Fire Department on several fronts. Their unique system of career and volunteer resources and interface, and traditions one of the best in Maryland.*

Baltimore County (BC) surrounds the City of Baltimore on the North, East, and West. BC is Maryland's third most populace county with an estimated 828,067 in 2020. Daytime population swells to over 1 million. BC covers a land area of 598 miles^2 and a water area of 83 miles^2. BC borders the State of Pennsylvania on the north, Carroll County on the west, Harford County on the east, Anne Arundel County on the south, Kent County on the southeast, Howard County on the southwest, and the Chesapeake Bay and Patapsco River. BC is part of the Baltimore metropolitan area and Baltimore–Washington metropolitan area. Much of BC is suburban except for the north which is predominately rural with a landscape of rolling hills and deciduous forests.

Fire Department History

The BC Fire Department was formed in June 1881 with seven stations housing various-sized chemical engines. Prior to that time, BC relied on Baltimore City for fire protection. Firefighters were paid, and most of the companies were eventually annexed into Baltimore City. Several paid companies remained in BC, and others were created in the late 1800s. In the early 1900s, volunteer companies sprang up through necessity in unprotected areas of the county. Those companies experienced continued growth and were eventually financially supported by BC. BC firefighters responded to assist Baltimore City during the Great Baltimore Fire of 1904.

Today's Modern Department

The modern-day BC Fire Department has a true "joint" fire service under the command of Chief Joanne R. Rund. Over 1,000 career personnel operate out of 25 stations, and 2,000 volunteers provide service from an additional 29 stations. Firefighters in BC respond to 141,026 incidents in 2019, of which EMS calls made up 71.4%. BC firefighters operate 79 engine companies (28 career and 51 volunteer), 8 truck companies (6 career and 2 volunteer); 2 tower ladders (1 career and 1 volunteer); and 9 heavy rescue squads (all volunteer).

EMS includes 55 advanced life support (ALS) medic units (35 career and 20 volunteer) and 10 ALS engine companies strategically located throughout the county. Other apparatus includes an urban search and rescue (USAR) vehicle, a hazmat unit with two satellite vehicles Company 13 and Company 15, a decon unit, six large-capacity tankers for rural operations, and various brush and squad units. Specialized units include advanced tactical rescue, swift-water rescue, dive rescue, and marine emergency teams. BC also has a year-round fully staffed training academy under the direction of Fire Director Timothy Rostkowski.

Rehab Units

Two volunteer rehab units are available in the county for fire, EMS, and police personnel at the scenes of incidents or for departmental functions. Box 234 Association Inc. and White Marsh VFC provide the rehab units that supply food, drink, and other services as needed throughout BC.

Hazmat Team

BC created its hazardous materials unit under the direction of the then Battalion Chief Edward Crooks. Its first apparatus was a 1965 Seagraves, the department's old Engine 41. Presently, the hazmat team is housed at Brooklandville Station 14, at 10017 Falls Road.

The hazmat unit, Hazmat 114, is a 2010 Rosenbauer-Spartan command cab. Command cab has three laptops, video monitor, color printer/copier/scanner, and work station with extra radios and headsets. The 24 foot rescue body is equipped with a 20 kW generator and onboard air compressor. There are two air reels and 2 electric reels. A 15 feet, 6,000 W light tower provides scene lighting. The generator is also used to charge and condition the onboard meters. A rear slide-out tray contains pump off and grounding/bonding equipment.

HM 114 is supported by a 2002 Freightliner foam unit (Figure 5.15). Foam 14 carries 250 gallons of water, 750 gallons of alcohol-resistant (AR) foam concentrate and protein foam. Also housed at Station 14 are

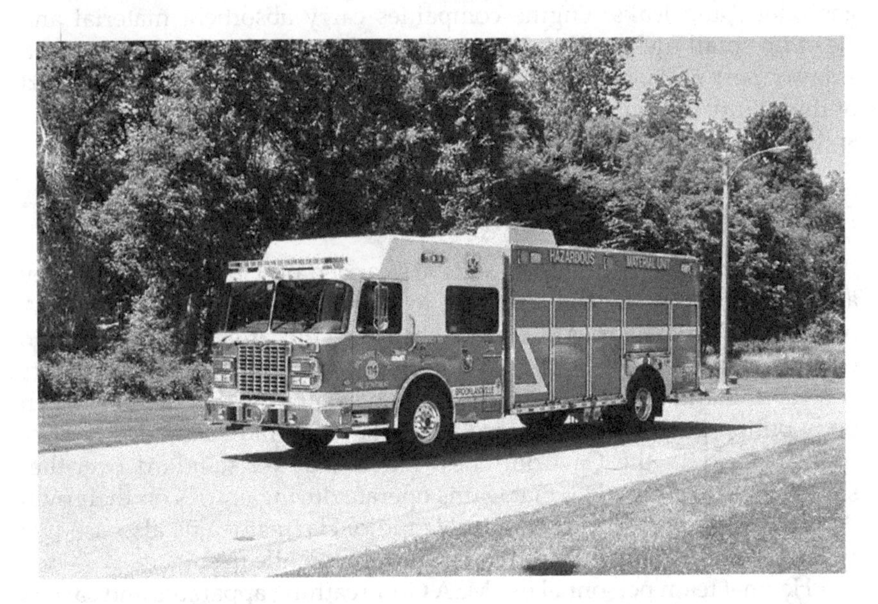

Figure 5.15 Hazmat 114 is a 2010 Rosenbauer-Spartan command cab.

Engine 14, Brush 14, Medic 14, Foam 14, and Air Unit 14. Because of the common practice of blending alcohol with gasoline in various concentrations, AR foam is used for all flammable-liquid fires. The proportion of foam is adjusted for the type of fuel burning. Crash trucks from Baltimore Washington Airport and Martin State Airport are available for mutual aid, but they do not carry AR foam. Decon 54 is a 2006 Pierce Enforcer technical decon unit, one of 6 units in the Baltimore Metropolitan area. Decon 54 has the ability to perform gross and technical decon to large groups of people as well as stretchered patients. The decon unit is housed at Station 54, at 12426 Eastern Blvd.

Two satellite (support) units are at Stations 13 and 15 and use Chevrolets (plumber's body-style trucks). Station 13 is at 6300 Johnnycake Road and Station 15 is at 1056 Old North Point Road. The satellite units carry monitoring and metering equipment, absorbent materials, Level B chemical suits, extra supplies, and manpower to assist the hazmat team when needed. Neither the hazmat team nor the support units are dedicated. A driver is assigned to Station 14 to place the hazmat unit in service. Personnel from Engine 14 provide manpower for the hazmat unit and respond with the hazmat vehicle and engine when necessary. Medic 14 personnel are also cross-trained as hazmat technicians and respond to hazmat calls as needed. Personnel from Engine 13 and Truck 13 place one of the units out of service to provide personnel for Hazmat 13. Personnel from Station 15 place Truck 15 or Engine 15 out of service to place Satellite 15 in service.

BC hazmat responded to 4,328 calls in 2019, which include fuel spills, gas odors, and leaks. Engine companies carry absorbent material and clean up small fuel spills. Larger spills are handled by the east- or west-side support units, and Hazmat 114 responds to spills in the center part of the county. Hazmat 114 also responds throughout the county to larger spills.

PPE, Equipment, and Training

A number of different Level A suits are carried on Hazmat 114, but they are in the process of standardizing to DuPont Tychem TK for Level A operations. Various types of Level B chemical suits are used. In-suit radio communications are provided through CavCom Talk Through Your Ears control unit and QUAD ear set/microphone, paired with a remote switch body button push to talk. Motorola APX 8000 XE radios and the control unit are held in the CavCom chest harness or on standard over-the-shoulder radio straps. Hazmat teams operate during entries on their own dedicated talk group on the Motorola radio. Hazmat teams also use five types of hand signals for basic communication while in suit.

Hazmat team personnel use MSA G-1 breathing apparatus and carries both 45 min and 1-hour bottles. Firefighters throughout the county use

Scotts with 45-min bottles. Hazmat 114 carries air carts to provide air for supplied-air operations. Personnel also have powered air purifying respirators (PAPRs) for incidents where there is plenty of oxygen and the type of hazardous material is known. Chief's vehicles are equipped with portable tablets containing the A2 accountability software which allows for monitoring and communication with the SCBAs being used down range. This allows real-time monitoring of air usage, remaining air, and alarm status of all SCBAs. The system also allows evacuation notification to be sent directly to an individual SCBA or all SCBAs that are in use.

Engine companies carry Industrial Scientific MX4 4-gas meter (CO, LEL, O_2, and H_2S). Truck companies carry MultiRAE Lite 5-gas meters (CO, LEL, O_2, H_2S, and HCN). Engine and truck companies are primary responders for CO and odor complaints. Truck companies also monitor post-fire atmosphere during overhaul. Safety officers have MX4 and Draeger PAC 7000 HCN meter for fire ground and confined-space monitoring. Hazmat units carry MultiRAE Pro 6-gas meters (CO, LEL, O_2, H_2S, HCN) and PID. These meters can pair up with the AreaRAE computer via Bluetooth for remote monitoring. All hazmat units also carry RAE PPb.

Hazmat 114 carries various monitoring equipment. Recently AreaRAEs replaced the aging Safesites. When paired up with the MultiRAE Pros, a total of ten remote monitors can be tracked. These are two ChemPros 100 on the unit with 12 libraries. Engine 14 currently carries a ChemPro in the cab detect off gassing of gear and SCBAs after fires. Also in the cache is the Micro FID, Jerome mercury meter, and colorimetric tubes.

All BC Fire Department units carry some type of meter for atmospheric monitoring. Medic units carry the Draeger Pac 5000 carbon monoxide meter in front of medical bag. Several "sick subject" type of medical calls have actually been because of high levels of CO that were discovered by the CO PAC 5000. Smith ACE-ID Ramen identifier and Hazmat Elite FT-IR identifier round out the detection meters. Both units have the capability to reach back to a chemist if the substance tested doesn't trigger a hit from the library. Hazmat is often requested by police for unknown substances and possible Fentanyl responses.

BC trains its hazardous materials technicians in-house at the county fire training academy using Maryland Fire and Rescue Institute (MFRI) technician training. Approximately 102 personnel across the county are trained as hazardous materials technicians with up to 27 on duty at hazmat stations 13, 14, and 15 on any given shift. Technicians are also a part of the Advanced Technical Rescue Team (ATRT). Personnel at Station 54, where the decon unit is housed, are trained to the hazardous materials operations level. All other firefighting and EMS personnel in the county are trained to the hazardous materials operations level as well. Mutual aid is available from surrounding county teams, including Anne

Arundel, Carroll, Harford, and Howard, along with the cities of Baltimore and Annapolis.

Reference Resources

Three laptop computers are carried on Hazmat 114. The computers have access to WiFi which allows crews to access Wiser, Cameo, HazMaster, and Aloha programs. Sewer, storm management, and water main maps are stored on the computer's hard drives. Tier II facility information is accessible both online and via the stored hard drives. Plume, weather, and research information can be printed out via the nits color printer. Incident Action Plans and research forms are stored on the computer.

Hardcopy research items include NIOSH books and flip charts from Hazmat IQ, Propane IQ, Drug IQ. Several flow sheets are available to assist each hazmat branch member. Several books are in the library for chemical research and transportation vessel identification and leak control.

Hazardous Materials Exposures

The primary hazmat response in the county is for highway transportation incidents. Major highway hazmat exposures in BC include those on Interstates 83, 95, and 795 and Baltimore Beltway 695. CSX is the major railroad that goes through the county. Parts of BC are also situated on Chesapeake Bay, with several rivers that empty into the bay. Most incidents that occur involving the waterways originate on land and enter the waterways; they do include fuel leaks from boats, but they are not frequent. When a waterway is involved with a hazmat spill, the U.S. Coast Guard also responds along with the Maryland Department of Environment spill response team.

Fixed facilities in the county are primarily light industry and a power plant. Fixed facility response is rare there because of aggressive pre-planning and pre-incident safety measures. A major natural gas pipeline runs through the county, but responses involving pipelines are also rare. All four shifts of the BC hazmat team have responded to major hazmat incidents over the past several years.

Incidents

Playground Acid Incident

On April 14, 2007, a 2-year-old boy was critically burned on a slide at a playground at an elementary school. Unknown persons had broken into the Victory Villa Elementary School and removed six bottles of industrial-strength drain cleaner and poured it onto playground equipment.

The major ingredient of the drain cleaner was sulfuric acid. The child suffered second- and third-degree chemical burns on his legs as a result of the incident. Hazmat team members cleaned up the playground equipment to make sure no one else was injured.

Overturned Tanker Ammonium Nitrate Slurry

An incident on February 8, 2008, involved an overturned tanker truck hauling 3,000 gallons of ammonium nitrate slurry and 50 gallons of acetic acid. The vehicle was placarded with an oxidizer placard. Ammonium nitrate is an oxidizer used for making blasting agents. The acetic acid was neutralized by the hazmat team, and hydraulic fluid was offloaded prior to up-righting the tanker.

Train Derailment and Explosion

A train derailment and explosion in 2013 in Rosedale (Figure 5.16). A CSX freight train struck a trash truck causing the train to derail. Several chemicals mixed and reacted causing a large explosion and fire that could be seen from miles. This was a large-scale multiagency unified response. Some of the chemicals involved were hydrofluorosilicic acid, terephthalic acid, and sodium chlorate.

Figure 5.16 Train derailment and explosion in 2013 in Rosedale, SE BC.

Charles County Maryland Special Operations

Charles County Maryland is located in South Central Maryland and borders the Potomac River and the State of Virginia on the South, Maryland counties of St. Mary's on the Southeast, Prince Georges on the North, and Calvert on the East. Charles County is considered part of the Washington D.C. Metropolitan Area. Charles County was established in 1658 and covers an approximate area of 215 miles^2 with an estimated population in 2020 of over 165,000. La Plata is the county seat.

Hazmat Team History

Charles County's Tactical Response Team was organized in December of 2003. The name was changed, and hazmat moved to Special Operations as it is today. Previously, hazmat response was handled by the volunteer fire companies in the county with mutual aid from hazmat teams in Prince George's County; Naval Surface Warfare Center (NSWC) Dahlgren, Virginia; and Indian Head Volunteer Fire Department & Rescue Squad. Many of the volunteer firefighters in Charles County are career firefighters in other departments.

Hazmat Team

On September 11, 2001, a large number of the counties' volunteers were recalled by their career departments. Military agencies were locked down and could not provide mutual aid. This situation left Charles County stripped of its fire protection and hazmat response capabilities. As a result, the county decided they needed to create in-house hazmat response capability utilizing county employees rather than relying on others that might not be available all the time. Applications were initially received from 42 county employees to become a part of the team. The idea was that they would remain county employees with their present jobs and respond when a call for the hazmat team was needed.

A full-time chief was hired to organize and provide leadership for the team. All members of the team are paid for their duties on the team. The hazmat team currently has 28 sworn personnel and 10 volunteers. Personnel from other county departments who are team members include 12 from the Emergency Medical Services, 5 from the County Sheriff's Office, 5 from County 911 Communications, 1 from County Public Facilities, 11 from Special Operations Division. Since all hazmat responses and training are in addition to their regular 40 hour jobs, full-time county personnel are paid overtime for hazmat activities.

Part-time personnel are also utilized to maintain the operational readiness of the team. These include 5 from the Pentagon Force Protection

Agency. Typically there are 4 hazmat technicians on duty during normal business hours, the chief and three from the Emergency Medical Services. Other team members carry pagers and are on call 24/7 and alerted when needed for a response. After hours, a duty officer is on call, and all team members are paged when a response is required. There are approximately 50–60 people trained to the technician level in the county. Charles County hazmat is a FEMA Type 1 CBRN Team capable of responding to a myriad of WMD calls if requested.

Special Operations/hazmat team is a part of Charles County Office of Emergency Services under the direction of Michelle Lilly. John Flier is the Chief of Emergency Services EMS and Special Operations Division. Fire service in Charles County is provided by all volunteer fire companies throughout the county. Emergency Medical Services (EMS) is provided by both career and volunteer personnel and is also a part of the Charles County Office of Emergency Services.

Charles County's hazmat team responded to 265 calls for hazardous materials assistance in 2018. Engine companies in the county carry absorbent materials and four-gas meters detecting CO, O_2, LEL, and SO_2 and handle fuel spills up to approximately 55 gallons. For spills above 55 gallons, the hazmat team is called in. All fire companies in the county carry 4-gas meters which monitor explosive limits (LEL), oxygen (O_2), carbon monoxide (CO), and sulfur dioxide (SO_2).

Charles County's hazardous materials response vehicles include (Figure 5.17) 2004 Pierce hazmat response unit, 2017 Freightliner M2 spill unit, 2015 Ford F-350 Utility, 2003 Navistar technical decontamination trailer, 2008 Ford F-150 mass casualty trailer (Figure 5.18), 2007 Bauer air support trailer, 2013 Ford Interceptor SUV duty officer vehicle, light tower, 2017 Chevrolet Tahoe PPV command vehicle, 2009 Polaris Ranger UTV, and 2012 Polaris Ranger UTV. Bomb squad services are provided by the Maryland State Fire Marshal's Office.

Figure 5.17 Charles Counties' hazardous materials fleet. (Courtesy: Charles County hazardous materials team.)

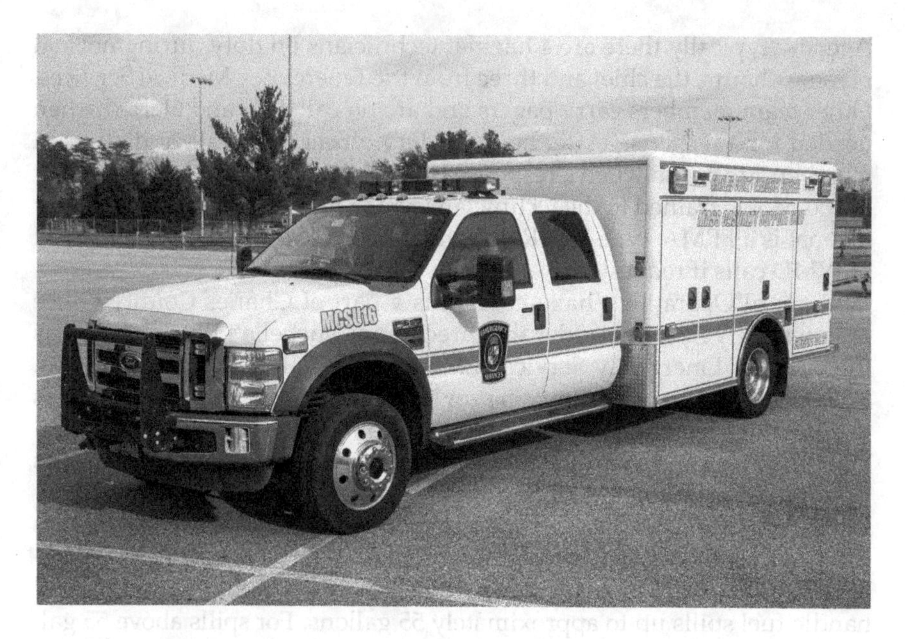

Figure 5.18 2008 Ford F-150 mass casualty trailer.

Mutual aid if needed is available from Prince George's County, Naval Surface Warfare Center (NSWC), and Indianhead Volunteer Fire Department and Rescue Squad.

PPE, Equipment, Training

Personnel protective equipment (PPE) used by Charles County includes Level A TrellChem VPS Flash (Lvl A) and Kappler Z200 (LvlB). Respiratory protection is provided by Scott AV3000HT Face pieces w/Scott 4.5 CBRN. Communications equipment includes Motorola APX 6000XE portables, CavCom communications with chest rig PTT.

Equipment carried on the units is typical plug and patch, spill containment and clean-up, and entry and decontamination equipment. Monitoring and detection instruments include Thermo Scientific TruDefender FTXi, handheld FTIR for chemical identification, MultiRAE Pro 5-gas meter w/gamma, AreaRAE Pro w/gamma, Sensit HXG-3 combustible gas detector, TIF XP-LA refrigeration leak detector, Draeger chemical detection system, chemical classifier, Draeger Clandestine Lab Detection Systems, Draeger colorimetric tubes, Jerome J4005 mercury vapor analyzer; radiation monitors include Tracerco T404 PRD's, WB Johnson Eagle Eye, and BNC SAM 940 radiation isotope identifier.

Maryland Fire and Rescue Institute (MFRI), which also has a training center in La Plata, provides technician-level training for the team – hazmat technician level for all team members. Firefighters in the county are trained to the hazmat operations level. In addition to hazardous materials training, some team members are also trained for rope rescue, confined-space rescue, trench rescue, and dive rescue. All team personnel attend 12-3 hour drills each year.

Hazmat team training requirements include Firefighter I; Hazmat Technician; Emergency Vehicle Operator; Recreational Off highway Vehicle Course; Confined Space Technician; DEA Clandestine Laboratory Certification; or Equivalent, IS-100, IS-200, IS-700, and IS-800. Additionally, Defensive Driving, and Public Safety Sampling Course. Advanced training for Special Operations includes Hazmat Incident Commander, Firefighter 2, Vehicle and Machinery Rescue, Swift Water Rescue, Rope Rescue, Trench Rescue, Advanced Hazmat Life Support, and Highway Emergency Response Specialist. All team personnel attend 12–4 hour drills each year. Firefighters in Charles County are trained to the operations level. MFRI, which also has a training center in La Plata, provides technician level training for the team – hazmat technician level for all team members.

Reference Resources

Reference resources carried on Charles County vehicles for hazmat response are computer software, Cameo, *CHRIS Manual,* Hazmaster G3, Aloha, Marplot, Emergency Response Guide and Clandestine Drug Lab Wizard. Hard copy reference books include *Hazardous Materials Field Guide, Hazardous Material Chemistry, Merck Index, NIOSH Guide for Industrial Chemicals, NIOSH Guide to Chemical Hazards, Emergency Handling of Hazardous Materials in Surface Transportation, Farm Chemicals Handbook, GATX Tank Car Manual, Hawley's Condensed Chemical Dictionary, Sittig's Handbook of Toxic and Hazardous Chemicals and Carcinogens, TLVs and DEIs Guidebook, Matheson Gas Data Handbook, Gardner's Chemical Synonyms and Trade Names, Tomes Plus Advanced Hazmat Life Support, Tempest Chem-Bio First Responder Guide, Medical Management of Biological, Chemical, Radiological Casualties Handbooks, and Jane's Chem-Bio Handbook.*

Hazardous Materials Exposures

Hazardous materials exposures in Charles County Maryland include one railroad line in the county along with several pipelines. Major transportation routes are Maryland 5, 210, and 301 (which also continues on to and from Virginia). The Potomac and Patuxent Rivers also provide transportation exposures for hazardous materials. Major chemical exposures in the county include liquefied petroleum gas, sulfuric acid, chlorine, hydrocarbon fuels, cryogenic liquid oxygen, and nitrogen. A new LNG power plant was added as well as Nammo Energetics, a defense company that manufacturers munitions and rocket motors.

Charles County's Tactical Response Team is a FEMA Type 1 CBRN Team. Much of Charles County's WMD equipment is part of the National Capital Region resources. They also respond in a three county area to Maryland's St. Mary's, Calvert, and Charles counties. Past incidents have included a propane truck, oil spill in the Patuxent River, man with a pipe bomb, clandestine drug lab, and white powder incidents. Thanks to the La Plata Volunteer Fire Department Truck 1 and crew for their support in taking elevated photographs of the Charles County Tactical Response Team Apparatus (*Firehouse Magazine*).

Cheyenne, Wyoming Hazmat Team

Cheyenne is the Capital of Wyoming and the most populous city in the state. Cheyenne is located in Laramie County which is in the Southeast corner of Wyoming just north of the Colorado border in the Mountain Time zone. The population of Cheyenne, which includes all of Laramie County, as of 2020 was 64,165. Cheyenne covers an area of 21.2 $miles^2$. The city of Cheyenne is named after the Cheyenne Indian Tribe which was located in the area when the location of the city was established on July 4, 1867. Previously the area was known as Crow Creek Crossing in the Dakota Territory later the Wyoming Territory before becoming the 44th State in 1890. Cheyenne became a city because of the Union Pacific Railroad presence in the area and is still today a major rail center for the Union Pacific Railroad. The F.E. Warren Air Force Base is located on the West side of Cheyenne and is the oldest continuously operating military installation within the United States Air Force. Starting as U.S. Calvary Fort D.A. Russell in 1867 to protect railroad workers from the Indians the military installation has evolved over the years into today's modern Air Force facility. Warren is home to the 90th Space Wing and Headquarters, 20th Air Force, of Air Force Space Command. They were the first ICBM wing in the United States and today are responsible for Minuteman III missiles located over 12,600 miles on sites in Colorado, Nebraska, and Wyoming.

Fire Department History

The Cheyenne Fire Department was formed just after the turn of the century by the consolidation of existing volunteer fire companies which until that time had provided an unorganized protection service. The consolidation was looked upon as a method to muster, organize, and administer all of the local available fire protection assets as opposed to the fractured effort of the past.

Today's Modern Department

Cheyenne Fire Department operates with 78 firefighters under the direction of Chief Greg Hoggatt. Cheyenne firefighters operate five paramedic engine companies (one squirt), one aerial platform, a hazmat unit, and a command light rescue unit from six fire stations strategically located throughout the city. They do not provide EMS transport; that service is provided by a private company. Cheyenne Fire Department responds to an average of 5,500 alarms per year. Fire protection for Cheyenne airport is provided by the Wyoming Air National Guard Fire Department.

Hazmat Team History

Cheyenne Fire Department officials formed their hazardous materials team and placed it in service during 1994. The team was formed due to an increase in requests for hazmat response and the availability of a grant from the State of Wyoming. Originally 15 members were trained to the technician level to staff the team when needed using the IAFF "Hazardous Materials Training for Team Members" curriculum. When a regional hazmat team response program was developed by the State of Wyoming, the Cheyenne team became RERT Region #7. In addition to the City of Cheyenne, the regional team covers Laramie, Platte, and Goshen counties. Additionally they provide hazmat response to all state offices and facilities in Cheyenne including the governor's office and residence.

Hazmat Team

Cheyenne's hazardous materials unit is a Pierce 2004 and is housed at Station 1 located at 2101 O'Neil Avenue (Figure 5.19). Located on the hazmat unit are an air cascade system and a freezer for cooling vests. Compartments go all the way through from one side to the other, which allows more room for equipment and supplies. The unit is also equipped with pullout trays for easy access to equipment. They have a 15 passenger van and a 16 feet trailer with equipment to respond to an incident with the main hazmat unit. Other apparatus housed at Station 1 includes Engine 1, a 1995 Seagraves 1,500 gpm pumper, and Truck 1, a 2001 Pierce 100 foot aerial platform Quint with a 2,000 gpm pump.

Cheyenne's hazmat team is not a dedicated team. When a request for hazmat response is received, the team's 20 members are activated using a pager system. They currently respond to an average of 350 hazardous

Figure 5.19 Cheyenne's hazardous materials unit is a Pierce 2004 and is housed at Station 1 located at 2101 O'Neil Avenue.

materials calls each year including hydrocarbon fuel spills and gas leaks. Engine companies carry absorbent materials for small spills. All firefighters in the city are trained to the operations level with annual refresher training provided. If mutual aid is required, Warren Air Force Base has a hazmat response unit along with two engines and a truck. Also stationed in Cheyenne is the 84th Civil Support Team of the Wyoming National Guard. Both provide mutual aid to the Cheyenne team upon request.

PPE, Equipment, and Training

Personnel protective equipment provided for the Cheyenne hazmat team for Level A is Kappler Responder, Kappler Tychem TK, and Chemron Chemrel Max. Level B protection includes Kappler CPF System 2's and 3's. Respiratory protection is provided by MSA Stealth SCBA with 60 minute bottles and MSA respirators. Communications equipment includes Motorola Walk-A-Bouts plus Motorola XTS5000R's.

Monitoring instruments and identification equipment carried on Cheyenne's hazmat unit include Hazmat ID chemical identifiers, Draeger Miniwarns, Draeger x-am 3000, Hazmat CAD chemical agent detectors, Draeger CAS kits, Draeger 4-gas detectors, Chemical Badges, Guardian Bio Threat Sampling kits, MultiRAE 2000 and Bio-Threat Kit. Radiation monitors include Inspector radiation monitor and other identification kits carried are HazCat.

Hazmat personnel take an operations course, an 80 hour technician course at Pueblo, Colorado, have monthly in-house training sessions, and send personnel to the National Fire Academy in Emmitsburg, Maryland, for Chemistry and Operating Site Practices. They have also sent personnel to the TTC's Tank Car, Highway, and Incident Commander courses.

Reference Resources

Hard copy reference book research resources carried on the Cheyenne hazmat unit in the command cab are the DOT ERG, NIOSH pocket guides, *Medical Management of Chemical Casualties Handbook, Farm Chemicals Handbook, Hawley's Condensed Chemical Dictionary, Firefighter's Hazmat Reference Book,* Lewis's *Hazardous Chemicals Desk Reference,* Genium's *Handbook of Safety, Health and Environmental Data,* Brethericks' *Handbook of Reactive Chemical Hazards, Chemical Protective Clothing Performance Index, Hazmat Response Handbook, Emergency Handling of Hazardous Materials in Surface Transportation,* the *Merck Index,* HazCat MSDSs, Crop Protection Reference, *GATX Tank and Freight Car Manual,* emergency action guides, CHRIS manuals, and *Dangerous Properties of Industrial Materials.*

Hazardous Materials Exposures

Transportation exposures in Cheyenne include interstates 25 and 80, pipelines, and major Union Pacific and Burlington Northern Santa Fe rail yards. Materials transported cover the range of DOT hazard classes. Radioactive materials are shipped on the interstates as well. Fixed facilities include oil refineries, tank farms, power plants, propane, anhydrous ammonia storage tanks, hydrogen fluoride, and chlorine (*Firehouse Magazine*).

Chicago Hazmat Team "Chicago's Twins"

Don't get excited baseball fans, the Minnesota Twins have not moved to Chicago! However, the Chicago Fire Department has a set of twins of their own in the form of twin hazardous materials response units (Figure 5.20).

Chicago, IL, founded in 1837, is the largest city in the State of Illinois and is located in the northeast part of the state on Lake Michigan near the border with the State of Wisconsin. The City of Chicago is the third most populous city in the United States. As of the 2017 census-estimate, it has population of 2, 716, 450. Chicago is the county seat of Cook county, the second most populous county in the United States. Chicago's metropolitan area is often referred to as "Chicagoland", with a population of nearly 10 million. Chicagoland is the third largest metro area in the United States, fourth largest in North America, and third largest in the world by land area.

Figure 5.20 Chicago Fire Department has a set of twins of their own in the form of twin hazardous materials response units. (Courtesy: Chicago Fire Department.)

Chicago is home to the Cubs, Bears, White Sox, Bulls, and Black Hawks. Important landmarks include Millennium Park, Navy Pier, the Magnificent Mile, the Art Institute of Chicago, Willis (Sears) Tower, Soldier Field, Wrigley Field, Museum of Science and Industry, Lincoln Park Zoo, Buckingham Fountain, and the Water Tower that survived the Great Chicago Fire of 1871.

Fire Department History

Chicago was nothing more than a little frontier village during its beginning days with roughly six-frame buildings and a population estimated at 150. The first building erected in 1831 was a tavern. Mark Beaubien was the proprietor, and he named his establishment the "Sauganash Tavern". Most dwellers lived along the south bank of the Chicago River on South Water Street. The city boundaries were Ohio Street on the north, Jefferson on the west, Jackson on the south, and on the east was Stat Street south of the river: Lake Michigan north of the river. The Sauganash Tavern remained the social center of Chicago until March of 1851, at which time a fire caused its closure.

Chicago's first fire company was formed in 1832 and called the "Washington Volunteers". During 1835, the first fire bucket brigade was formed and called the "Fire Guards Bucket Company". Also formed in 1835 was the "Pioneer Hook and Ladder Company No. 1". Hirman Hugunin was the first Chief of the Chicago Fire Department appointed in 1835. On August 2, 1858, Chicago formed a paid fire department placing Engine Company 3 in service at 225 South Michigan Street. A new second engine was purchased in the fall of 1837 during the formation of the "Tradesman's" Fire Company in December.

The name soon changed to the "Metamora Engine Company No. 2" located on Lake Street east of the river. In 1939, Alexander Lloyd was chosen Chief and held the office for 1 year. Alvin Calhoun was the fifth Chief of the Fire Department, serving from 1839 to 1840. The first fire of any significance occurred on October 27, 1839, on Lake Street near Dearborn. The fire destroyed the Tremont House and 17 other structures as well, with damages amounting to $65,000. September of 1841, the "Chicago Bag and Fire Guard Company", better known as "The Forty Thieves", was formed. The company fought fires with canvas bags, cords, and wrenches protecting life and property for 5 years. The "Citizens Fire Brigade of Chicago" was patterned after "The Forty Thieves" company following the disastrous fire of October 19, 1857, in which 23 people perished and the property damage was near $500,000.

During the next decade, the Chicago Fire Department added to its volunteer companies a total of two bucket companies, three engine companies, and two hose companies. These companies held names such as "The Rough and Ready Bucket Company No. 1", "Red Jackets Engine Company No. 4", "Neptune", and "Hope Hose Company No. 2", which was the last of the volunteer companies. The "Hope Hose Company No. 2" was formed in March of 1848. That company was notorious because of its 1 min and 7 s run of 500 yards with 300 feet of hose and its connection to the water source. Hope Company had a remarkable career of service until it was disbanded in 1859. Dennis J. Swenie was the first paid Chief Engineer of the Chicago Fire Department and served in this capacity until March of 1859 and again from November 10, 1879 to June 1, 1901.

In February of 1855, a deep-toned bell was installed in the new courthouse. In July of 1855 an ordinance was passed, dividing the City into six fire districts. An alarm sounding code was initiated – eight strokes of the bell signaled the alarm; the additional strokes indicated the district. A watchman was continually on duty in the tower. Apart from ringing the bell, the watchman was responsible for handing out flags by day and lanterns by night, used to direct firefighters to the scene of the fire. In those days, there was no gas, so to light the way, torch boys ran ahead of the engine. When an alarm was sounded at night, citizens were responsible for placing lighted candles in their windows, lighting the way for the firefighters.

The "Long John" steam fire engine tested in February 1858 was met with hostility from the volunteer firemen. The volunteers could sense this was the beginning of their extinction. The steam engine "Long John" was put into service on May 1, 1858, at the corner of Adams and Franklin Street. Firefighters of the volunteer hose companies and two paid members, the engineer and assistant engineer, manned the "Long John". The death bell of the volunteer fire department was rung on August 2, 1858, when city council passed the ordinance organizing the paid Chicago Fire Department. The first completely paid company was Engine Company No. 3, located at 225 South Michigan Street. This company was one of fifteen engine and three hook and ladder companies acquired from the volunteer department.

The entire inventory for Chicago's Fire Department in the year 1866 was 11 steamers, 2 hand engines, 13 hose carts, 1 hook and ladder truck, 120 paid members, 125 volunteers, and 53 horses. February 5, 1923 at 12:40 p.m. Chicago becomes the first city in the country to have a completely motorized fire department as the last horse drawn fire engine at Engine 11 quarters (10 E. Hubbard) is placed out of service.

Major Fires in Chicago

The Great Chicago Fire began around 9:00 p.m. on October 8, 1871. Eighteen thousand buildings were destroyed at a cost of $200,000,000. Over one-third of the city was destroyed (Figure 5.21). Over 100,000 people

Figure 5.21 The Great Chicago Fire began around 9:00 p.m. on October 8, 1871. Eighteen thousand buildings were destroyed at a cost of $200,000,000. Over one-third of the city destroyed.

lost their homes. Between 200 and 300 people perished in the fire. In 1997, the Chicago City Council approved a resolution absolving Mrs. O'Leary's cow of all blame for the Great Chicago Fire.

Chicago has experienced some significant loss of life fires over the years that often led to changes in fire codes. The Great Chicago Theatre Fire at the Iroquois Theatre occurred on December 30, 1903. Over 600 people, many of them children, lost their lives in the largest loss of life fire in Chicago's history. On June 5, 1946, a fire in the LaSalle Hotel at LaSalle and Madison resulted in the death of 61 people and First Battalion Chief Eugene Freemon. On December 1, 1958, a fire broke out at the Our Lady of Angels School. Ninety-two children and three nuns lost their lives (Figure 5.22). The most famous non-loss of life fire occurred in 1967 when the newly built "Fire Proof" McCormick place caught fire and basically burned to the ground. Chicago is one of the oldest major organized fire departments in the U.S. established before 1833. As of January 22, 2019, 12 volunteer and 557 career firefighters have perished in the line of duty. At least 20 of the firefighter fatalities involved hazardous materials.

Today's Modern Department

Chicago's fire department is led by Fire Commissioner Richard C. Ford II who was appointed September 2018, by Mayor Rahm Emanuel. The Chicago Fire Department has over 4,500 uniformed personnel and is the

***Figure* 5.22** On December 1, 1958, a fire broke out at the Our Lady of Angels School. Ninety-two children and three nuns lost their lives.

second largest fire department in the United States. Fire department personnel in Chicago respond to over 500,000 calls per year for emergency assistance in a coverage area of 228 miles^2. Chicago located on the western shore of Lake Michigan and has 29 miles of shoreline.

Today firefighters operate from 100 fire stations with 96 engine companies; 61 truck companies; 4 squads; 65 ALS and 15 BLS medic units; 11 crash rescue units between O'Hare and Midway airports; a collapse unit; 2 chemical units, with 100 gallons of foam and 500 lb of dry chemical; 1 fire boat; 2 helicopters; 2 hazardous materials companies; 3 light wagons; a mass decontamination trailer at O'Hare International Airport; and Big Mo, a large volume water supply unit. Helicopters are used for water rescue and dispatched on all still and box alarms for high-rise buildings. Helicopters are used as "eyes" of the incident commander outside of building fires watching for fire spread.

Air Sea Rescue Unit (ASRU)

Chicago's Air Sea Rescue Unit was established in 1965, where they provide search and rescue services for 37 miles of lakefront, an extensive river system, numerous lakefront venues, and the largest harbor system in the U.S. Busy waterfront activity provides the ASRU unique emergency challenges including assistance to boats in distress, water rescues, and air search missions. Divers assigned to the ASRU are trained under public safety rescue diver guidelines specific to cities' needs and particular environments. Pilots are trained in helicopter search and rescue and hoist rescue techniques patterned after nationally recognized standards. ASRU uses (two) Bell 412 EP helicopters to aid in their efforts. The unit also has dive rescue vehicles equipped with the latest available communications devices, full facemask and dry-suit dive equipment, and lighting for night operations. They also have a dedicated swimming pool for training the members.

Hazmat Team History

Chicago formed its hazardous materials team in 1985 under the direction of John Eversole, its first coordinator and "Godfather of Hazmat in Chicago". Eversole felt hazmat was the thing to come and wanted Chicago to be ready to deal with hazmat emergencies in the future.

> ***Author's Note:*** *I had the privilege of attending the National Fire Academy Hazmat II Class (now called Operating Site Practices) with John in the early 1980s when John was still a firefighter about to enter the "world of hazmat" that he would leave a lasting impression upon. John was a member of the Chicago Fire Department from 1969 until*

> *2001 when he retired as Deputy Chief. Sadly, John passed away in May of 2007.*

The original hazmat unit was an old panel truck with tools. It was not a dedicated unit and was placed in service at Engine 42's quarters. Since the team's inception in 1985, there have been just three coordinators at the time this article was originally written in 2005: Chief Eversole, followed by Eugene Ryan (who became Deputy Commissioner of the Operations Bureau), and present coordinator Chief Daniel R. O'Connell. I had the honor of knowing all three coordinators.

> ***Author's Note:*** *Gene Ryan I met through Chicago Firefighter friend Rudy Rinas while he was a hazmat team member. Rudy, Gene, and I attended several classes at the NFA and Illinois Fire Service Institute; I also attended one class at the Center for Domestic Preparedness (CDP) with Gene as well. Daniel R. O'Connell was the coordinator when I first put together this as an article on the Chicago hazmat team in 2005.*

Built by American LaFrance, units 511 and 512 were placed into service in January of 2004. Deputy Commissioner for the Bureau of Operations Eugene Ryan, a former hazardous materials team Coordinator for the fire department having over 18 years of experience in hazmat response, felt for a long time the city needed a second hazmat unit and fought hard along with Fire Commissioner Cortez and Chief of Special Operations Fox to make the second unit a reality.

Hazmat Team

Currently "Chicago's Twins" are located at Engine 22 on the near North side and Engine 60 on the near South side. These units respond to an average of between 700 and 800 hazardous materials calls per year, not counting fuel spills. Units 511 and 512 are equipped the same. Compartments are numbered on the right R 1–6 and on the left L 1–6. There is a walkway inside each unit with compartments inside. A combination crew and command cab is located behind the driver and officer seats in the front. Space above the wheel wells that is usually wasted has been utilized for spare SCBA bottles.

Hazmat Response in Chicago Consists of Three Levels

- **Level I** incident is the minimum initial response to any suspected or potential hazardous materials incident. This level is primarily for investigative activities and/or to mitigate incidents involving small quantities or low potential materials. Level I hazmat response consists of one battalion chief, one engine, one truck, one ambulance, and a HIT task force of one squad and hazardous materials Squad 5-1-1 or 5-1-2.

- **Level II** is a confirmed incident involving a moderate or high potential and/or a large quantity of material or an incident requiring protective clothing above Level D (Structural firefighting clothes) or an incident requiring the need to initiate an evacuation. Level II hazmat response consists of the following units: one Deputy District Chief, one HIT task force team engine, one HIT task force team truck, one Safety Officer, one EMS Field Officer, one Air Mask Support Vehicle, and one command van.
- **Level III** hazmat incidents are rarely called. This would be an extensive incident which may require additional manpower, more supplies than are normally carried by the HIT task force, and an expanded evacuation area. Engine and truck companies carry absorbent materials. Each company will carry approximately 5 gallons of Biosolve, which is a water-based dispersant agent for cleaning up fuel spills.

Units 511 and 512 are dedicated, and normal staffing is seven on each shift with four personnel and one officer set as minimum manning.

Hazmat team members are often referred to by firefighters on the department as "Glow Worms". When responding to a Level 1 or higher hazmat call, a medic unit is assigned to the hazmat team to provide medical monitoring and ALS for hazmat personnel. In addition to the hazmat units, every squad in the city has personnel trained to the technician level. When a Level 1 or higher hazmat response is dispatched, a squad is also dispatched. Within the city, there are also 24 hazmat "Team" engine and truck companies.

Personnel in these engine and truck companies are trained to the technician level just like personnel on 511 and 512. Four squads also have five technicians each and respond as needed to hazardous materials incidents. The Team engines and trucks are dispatched as needed on hazmat alarms and, along with the squad and hazmat unit, become part of the on-scene hazmat team. Engines and trucks that are Team units have a Hazardous Incident Team logo decal on the officer's side of the apparatus.

PPE, Equipment, and Training

Level A suits are DuPont TK and Kappler Responders. Level B suits are Kappler Responder. They use Motorola bone mikes for in-suit communication and are currently trying other products. Hazmat personnel utilize MSA SCBA with 1 hour bottles; firefighters use 30 min bottles.

Monitoring instruments carried on Chicago's hazmat units include Industrial Scientific pH meter, RAE PID "Million", RAE PID "Billion", Saw Mini-Cad, Draeger CDS Kit, radiation monitors, Dragger PAC III Chlorine and Ammonia, APD 2000, Industrial Scientific HCN meter, and others. Video equipment is also carried for videotaping any aspect of the incident deemed necessary. Mark IV antidote kits are carried for response personnel exposed to nerve agents at the scene of a WMD incident. General hazmat equipment carried includes technical decontamination setup; mass decontamination tent; Chlorine Kits A, B, and C; Weather Pack; and various leak and patching equipment.

A tool that is unique to Chicago is called the "JEB" tool. It is a large mallet similar to the type used for carnival strength games. It was at one time used to pound wooden pegs into leaking gas lines to stop leaks. It would take a pretty big guy to swing the mallet as it is 14 inches in diameter and weighs 40 lb! Chicago has tested and hopes to soon place in service a trailer with a mounted 60 inch positive pressure fan for use in mass decontamination. The fan is mounted on a platform with a scissor lift allowing it to be elevated up to 14 feet as needed. A 1¾ inch fire hose is attached creating a large "shower head". According to Chief O'Connell, "it works great". The unit can also be used for misting during firefighter rehab and ventilation for large buildings. Chicago relies on suburban departments Bedford Park and Aurora if they need mutual aid response in the city.

Chicago uses their own 40 h operations class that results in all operations personnel having Level A capability. Over 70% of the firefighters in the city are currently trained to the operations level. In order for a firefighter to get transferred to the hazmat unit, they must have had awareness and operations training and be a firefighter II. Every firefighter in the department has also received Emergency Response to Terrorism: Basic Concepts training.

Manufacturers often approach the Chicago Fire Department hazardous materials team to evaluate new products for hazmat response. One of the first tests these products are exposed to is called the "605 Test". (Hazmat 511 is located at Engine 22, and 605 is the street address of the station.) Firefighters at Station 22 perform the 605 Test. The first test generally leaves the sales personnel with their mouths wide open. A firefighter drops the product on the floor. If the product does not pass the drop test, the sales person is sent on their way. After all, if a product is not firefighter proof, what good is it?

Hazardous Materials Exposures

Chicago, which is centrally located in the United States, is a major rail center with rail yards from several railroads on the city's South side. Almost any chemical that is shipped by rail can be found in Chicago's rail yards at any given time. Interstates 55, 57, 65, 80, 90, 94 go through the city also making it a major transportation center for the trucking industry as well. There are many "mom & pop" chemical companies in the city. Major companies include Clorox, major oil companies, and other companies that manufacture almost any type of chemical recognized. Petroleum barges also navigate the city's rivers and canals as well as Lake Michigan (*Firehouse Magazine*).

Incidents

1897 August 5, Four firefighters killed at an explosion at the Northwestern Grain Elevator, Cook and Water Streets. (See Volume One)
1927 March 11, Two firefighters killed in an explosion at the Draeger Chemical Company, just outside the downtown district.
1940 August 17, Five firefighters perish at an explosion at the Van Schaack Chemical Company, Henderson and Kimball.
1968 February 7, Four firefighters perish at a gasoline tanker explosion & fire at Mickelberry's Food Products company, 301 West Forty-Ninth Place. (See Volume One)
1997 August 4, Level 3 Hazmat, EMS Plan 1 at 735 E 115th Street. Overturned tanker with 200–300 gallons of Sulfur Trioxide.
1998 August 31, 4–11, EMS Plan 1 & Level 2 Hazmat at 2,450 W. Grand, Magnesium Fire.
2000 September 14, 3–11, EMS Plan 1, Level 2 Hazmat at Dearborn & Wacker underground electrical vault. Buildings evacuated, O'Hare Foam Task Force used to extinguish fire.
2001 August 8, 2–11, EMS Plan 1, Level 3 Hazmat at 47th & Dan Ryan Expressway. Tractor trailer truck fire containing azodicarbonamide. CFD awarded US EPA Superfund Team of the Year Award for mitigating this incident.
2002 April 1, 3–11, EMS Plan 1, propane tank explosion, 1 dead, 9 injured at 5 S. Wabash Jewelers Building.

2001 Azodicarbonamide Spill & Fire

Chicago's hazardous materials team has been faced with several major incidents over the past few years. On an extremely hot August 8, 2001, a flatbed truck carrying 3,000 lb of azodicarbonamide rolled over on the Dan Ryan Expressway (Interstate 94) on the city's South side. Rush hour traffic was backed up for miles in both directions as well as on arterial streets. The Red Line of the Chicago Transit Authority rail service also had to be shut down. Azodicarbonamide is a highly flammable solid used to manufacture foam rubber products. Responding firefighters called for a Level III hazardous materials response, and over 130 firefighters and

22 paramedics responded to the incident. Approximately 160 lb of the material spilled and burst into flames creating a yellow vapor cloud and smoke. Containment and cleanup of the highly reactive material proved difficult, and as a result, the Dan Ryan Expressway was shut down for 12h. Approximately 1,500 residents of the heavily populated area were evacuated. Heat exhaustion sent 8 police officers and 13 firefighters to the hospital for treatment.

Clarified Flurry Oil Barge Fire

Another major incident occurred on January 19, 2005, when a 295 foot barge carrying 588,000 gallons of Clarified Flurry Oil caught fire and exploded while refueling on the Chicago Sanitary and Ship channel. Flames from the fire shot over 100 feet in the air at the height of the fire. The Chicago Sanitary and Ship channel is 105 years old and is used as a link between the Great Lakes and the Mississippi River. Clarified Flurry Oil is a by-product of oil refining and is used to make fuel oils. One worker was killed and another injured in the explosion and fire. It is believed the explosion involved a boiler on board the barge that exploded and ignited the oil. The barge sank in the channel approximately 1 hour after the fire began extinguishing the remainder of the fire. Cleanup efforts were underway under the supervision of the Environmental Protection Agency and the United States Coast Guard.

MABAS Chicago's Unique Box Alarm System for Dispatching

Chicago uses a box alarm system for dispatching emergency calls, and the system has been adopted throughout the State of Illinois and some adjoining states for mutual aid response (Figure 5.23). The system outside the city is known as the Mutual Aid Box Alarm System or MABAS. MABAS is an organization of hundreds of fire departments in Illinois and Southern Wisconsin and neighboring states including Missouri, Indiana, Iowa, and Kentucky. MABAS provides an orderly movement of equipment to the scene of fires, accidents, or other incidents or to cover vacant fire stations. Equipment is moved around according to predetermined lists, called "box cards". Each card covers specific types of incidents in specific areas. Small towns may have a single Fire box card; larger towns may have dozens. Chicago's response system is divided into several response phases.

The primary response to an incident is the "Still Alarm" which results in the dispatch of 2 engines, 1 truck, and 1 battalion chief. Next is a "Working Still Alarm" (confirmed working structure fire) that results in the dispatch of an additional squad and command van. A first alarm assignment is referred to as a "Still and Box Alarm", and 4 engines, 2 trucks, 1 tower ladder, 3 battalion chiefs, 1 deputy district chief, 1 squad, 1 command van, and 1 ambulance are dispatched. Extra alarms are added

Figure 5.23 Chicago uses a box alarm system for dispatching emergency calls, and the system has been adopted throughout the State of Illinois and some adjoining states for mutual aid response. (Courtesy: Aurora IL Fire Department.)

in the following order until alarms are reached: 2–11, 3–11, 4–11, and 5–11. All alarms after 5–11 are listed as special alarms, so the highest level of a fire in Chicago is the 5–11 plus various special alarms. There is no limit to the number of special alarms that can be requested. When a 2–11 alarm is called, an additional 4 engines, 2 trucks, 1 tower ladder, 2 battalion chiefs, and 1 district chief are dispatched.

The 3–11 alarm brings an additional 4 engines and deputy fire commissioners to the scene. On request of a 4–11 alarm, an additional 4 engines are dispatched. Dispatching a 5–11 alarm the highest, brings to the scene an additional 4 engines. By the time a 5–11 alarm is called for, there are 20 engine companies, 6 truck companies, and 7 plus chief officers on scene. If additional equipment is needed, a "special alarm" is called for, usually 5 engines at a time and any other equipment needed.

Corpus Christi, Texas Hazmat Response

Prolog: *Corpus Christi Chief Robert Rocha was formerly Senior Deputy Fire Chief for the Kansas City, Kansas Fire Department. During 2010, I met Chief Rocha during a visit to the Kansas City to gather information for an article on the hazmat team. When Chief Rocha took the job in Corpus Christi, he contacted me and suggested the area there had some attractive subjects for an article as well. My trip to Corpus Christi turned out to be a special one. Not only is it a great city to visit right on the Gulf of Mexico, but the fire department is very innovative and impressive. I brought my wife and granddaughter Abby and we had several days to enjoy Corpus Christi. Jim DeVisser was the hazmat team coordinator at the time of our visit and Jim and Chief Rocha along with Jim's wife made sure we had a great time while we were there. We also went home having found three new life time friends and a wonderful place to visit again sometime.*

Corpus Christi is located in South Texas on the Gulf of Mexico. Corpus Christi in Latin means "Body of Christ". The name was given to the settlement and surrounding bay by Spanish explorer Alonso Alvarez de Pineda in 1519, as he discovered the lush semitropical bay on the Catholic feast day of Corpus Christi. The city has been nicknamed "Texas Riviera" and "Sparkling City by the Sea". The estimated population of Corpus Christi today is in excess of 325,000, the 8th largest city in Texas. Corpus Christi also boasts the 6th largest port in the United States. The Port of Corpus Christi was opened in 1926, and the Corpus Christi Naval Air Station was commissioned in 1941. Corpus Christi has a total area of 460.2 miles^2 of which 154.6 miles^2 is on land and 305.6 miles^2 is covered by water.

Fire Department History

Citizens of Corpus Christi organized their first fire company on November 28, 1871, with a meeting to elect officers as their new engine had arrived. Felix Noessel was elected Foreman of the Pioneer Fire Company No. 1. Uniforms of the company would consist of black pants and belt, white flannel or crino shirt, and a glazed cap. On March 10, 1873, the second fire company in Corpus Christi was formed, the Lone Star Hook and Ladder Co. On July 11, 1874, the Pioneer and Lone Star fire companies joined together to form the Corpus Christi Fire Department.

Today's Modern Department

Today's modern Corpus Christi Fire Department is led by Chief Robert Rocha. Corpus Christi Fire Department operates from 18 fire stations covering an area of 170 miles^2. Personnel include 414 firefighters who work shifts of 24h on and 48h off, 30 administrative, and 7 civilians with an operating budget of over $47 million. They have an ISO rating of 2. Responses in 2019 were 54,242 and of those 40,969 were emergency medical service (EMS) calls and 1,058 were fire related. Front line apparatus includes 16 engines, 2 towers and 3 trucks, 1 heavy rescue and 6 vehicle extrication units, and 11 medic units. Specialty apparatus includes 5 brush, a command vehicle, "Big Bertha", a 3,000 gpm monitor, hazmat vehicle, light and air unit, and fire boat. Every front-line apparatus is staffed with a captain, driver, and, ideally, two firefighters.

Hazmat Team History

Hazardous materials response in Corpus Christi began in July 1987. Originally Hazmat was placed under Safety and later moved to Operations. A battalion chief was placed in the position of team coordinator (Figure 5.24). Hazmat was created as a result of the release of methyl isocyanate (MIC) in Bhopal, India, and the passage of Sara Title III. Originally the team was located at Station 9 and used an old ambulance as a response vehicle. Twenty personnel were sent to Texas A&M University for training. In 1990, they converted a beer truck (Lone Star) to

Figure 5.24 Originally Hazmat was placed under Safety and later moved to Operations. A battalion chief was placed in the position of team coordinator. (Courtesy: Corpus Christi Fire Department.)

their response vehicle and equipped it with a Mac Computer and CAMEO software. In 1993 and 1998 Stations 3 & 5 were added to Hazmat.

Hazmat Team

Currently, Station 12, located at 2120 Rand Morgan Road, houses the Hazmat Unit 22 (Figure 5.25). Station 5, located at 3312 Leopard Road, is the Hazmat/Decontamination Station. They respond to Hazmat calls with Engine 5 and a pickup pulling the Decontamination Trailer. Station 3 at 1401 Morgan is also designated as a Hazmat Station. The Hazmat Response Vehicle is a 2001 E-1 with a midship engine purchased with funds from a refinery. Corpus Christi Hazmat responds to around 100 calls each year. This total does not include fuel spills, gas odors, and leaks. Engine companies carry absorbent, Plug & Dike, and a "Football" plugging device. They also use a PRO/pak foam injection and application system to apply any Class A, aqueous film-forming foam (AFFF) or alcohol-resistant foam concentrate fast and easy. It works well on small or contained hydrocarbon fires. They recently used it to extinguish a tanker fire on a bridge as well as other small hydrocarbon fires.

Engine companies respond to fuel leaks and the officer-in-charge makes a judgment call as to when the spill is beyond their control. Two engines and a chief are dispatched on the initial alarm and one of the

Figure 5.25 Currently, Station 12 located at 2120 Rand Morgan Road houses Hazmat Unit 22.

engines dispatched must have a 4 gas meter. Specialized pickup trucks with additional absorbent are dispatched for assistance. The Hazmat Unit is only dispatched if the leak is beyond the company level response, again as determined by the officer-in-charge. Hazmat 12 made 1,100 responses in 2019.

Hazmat 12 is not a dedicated unit. Personnel from Station 12 man the unit when a Hazmat call comes in. Sixteen personnel are desired to be on duty each shift, and the total averages 12–16. There are 48 firefighters and 2 chief officers trained to the Technician Level on the Corpus Christi Fire Department. Mutual aid for Hazmat is available locally from the Naval Air Station Corpus Christi from their Hazmat Team. Support help is also available from the Annaville Volunteer Fire Department. The next level of mutual aid comes from farther away, either San Antonio or Houston about a 3h response time. When mutual aid is required for a refinery fire, Corpus Christi contacts Refinery Terminal Fire Company (RTFC). It is a private company specializing in fire protection for the refineries. We would use them if we had a large refinery fire.

PPE, Equipment and Training

Chemical suit protection is provided by Level A Tychem Reflectors and Level B Tychem SL, Chemrel, BR's, Tyvec Saranex and Lakeland LV94s. BR's are medical suits provided for infection control for things like MRSA. Scott Radio Com IIs are used for in-suit communications. Breathing protection is provided by Scott (3000) SCBA with 1h bottles. They also use cartridge respirators and PAPRs. The light and air unit is utilized for refilling bottles.

Monitoring instruments include AHURA, Gas-ID, Haz-ID, RAD-60, Ludlum, Draeger Tubes, Multirae Plus, RAELINK 3, Multirae and First Defender, and RAELINK 3 MESH. All are carried on Hazmat 12. Eight other stations are equipped with 4 gas and PID monitoring instruments as well. Terrorist agent monitoring instruments include radiological and biological. Corpus Christi Health Department has a Gas Chromatograph. Station 12 has a monitoring instrument shop where maintenance and repairs are made on monitoring instruments by hazmat Team personnel.

Reference Resources

Reference materials carried on HazMat 12 are a combination of online, 3 computers with Cameo, Aloha, Marplot and Wiser, and hard copy reference books. These include ERG 2012 (2016 due out this spring); *James CBRN Response Handbook* 4th edition; *ACGIH TLVs and BEIs; Quick Selection Guide to Chemical Protective Clothing,* 6th edition; *NIOSH Guide to Chemical Hazards and Field Guide to Tank Cars-Association of American Railroads.*

Hazardous materials team members are trained with a 40hour Hazwoper course from various sources, CAMEO, and air monitoring.

They also attend National Fire Academy (NFA) courses. Captains are given command training and chiefs HazMat ICS. Firefighters are all trained to the Operations Level. Drills are held periodically such as simulated chlorine leaks from tank rail cars and 1 ton cylinders at the Corpus Christi water treatment plant.

Hazardous Materials Exposures

Hazardous materials exposures in Corpus Christi include all modes of transportation: highway, rail, water, pipeline, and air. Interstates 37 and 69E, US highways 77 and 181, and Texas Highway 44; the Union Pacific and Kansas City Southern Railroads; Port of Corpus Christi, where barges and ocean going ships are loaded and unloaded; Corpus Christi International Airport (CCIA) and Naval Air Station Corpus Christi; and various pipelines. Fixed facilities consist of a large water treatment plant, with multiple tank cars of chlorine and multiple ton cylinders as a backup.

There are also refineries and support facilities; cryogenic facility, and disposal injection well. The disposal injection well is a hazardous waste disposal site where waste is injected miles into the ground. When full, it is capped and cased. Generally, all facilities accept a wide range of aqueous saline wastes. These wastes include process wastewaters, complex organics, soluble heavy metals, contaminated cooling waters, and non-biodegradable wastes. Acceptance of waste is dependent on several factors, including the amount of suspended solids and the pH. Wastes should be filterable through a 5-micron filter (100 ppm suspended solids maximum), and pH should be between 6.0 and 12.0. In addition to Corpus Christi, there are 4 other such sites located in Vickery, OH; Plaquemine, LA; Deer Park, TX; and Tulsa, OK.

Corpus Christi Naval Base is where the largest helicopter repair facility in the world is located. This will be covered in more detail in a future article. Many other businesses in Corpus Christi utilize typical hazardous materials, such as petroleum products, acids, and others. Markwest Energy Partners gather, process, treat, fractionate, and transport off-gas as well as natural gas and natural gas liquids. If you have ever driven by a refinery you often see flames coming out of stacks on the site. These are burning off the "off-gas", which is a by-product of the refining process that are not economically viable. Markwest processes these by-products into useful products rather than venting them into the environment.

Incidents

In March 2006, two workers were killed during a leak from a tank at the Texas Molecular Ltd. deep well waste facility in Corpus Christi. Fire, EMS, and HazMat responded. HazMat went in with Level A suits, found the victims and recovered the bodies. Air monitoring was conducted to try to establish what was released. It is suspected that two chemicals mixed and were spilled on the ground. The workers were not wearing their respirators. Another incident Corpus Christi HazMat responded to was a propane tanker, MC331, that hit the supports of the Harbor Bridge. The driver was killed instantly and the vehicle badly damaged. The pressure tank was crushed on the front end, causing a dent about 18 inches deep and a tear approximately 3 inches long and 1/8 inch deep. Propane escaping from the container eventually iced over slowing down the leak, which went on for more than 24 h. Investigation showed the cause of the accident was a combination of poor bridge design and the tanker traveling too fast for a curve. The tanker rolled over and skidded along the highway before hitting the concrete bridge supports.

AERO Team (Drones)

Corpus Christi Fire Department purchased two drones and received them in August 2015. I was there visiting Jim DeVisser for an article on their hazmat team for *Firehouse Magazine* and was able to help Jim open the boxes and that was the extent of my knowledge of their Drone program. They were in phase 3 of a 5 step process of obtaining their Certificate of Authorization (COA) or waiver from the Federal Aeronautics Administration (FAA). Standard Operating Procedures (SOP) have been developed for the use of the drones. Plans are to use them for reconnaissance in and around Hazmat scenes. Utilizing the High-Definition (HD) and infrared camera, they can identify locations of leaks, downed civilians, track hazmat personnel on entry, and scene situational awareness. They plan in the future to add air-monitoring capability.

While gathering information for this book project, I once again visited Jim in Corpus Christi to write about the progress they had made in their Drone Program. Jim is now the AERO Team Coordinator for the Corpus Christi Fire Department. Battalion Chief J.D. Johnson is now the Coordinator for the Hazmat Team. AERO stands for Aerial and Emerging Robotics. Corpus Christi's Drone program is off and running. Jim had been sending me Drone photos and video from time to time from hazmat incidents (Figure 5.26), so I knew they were up and running. Any incident commander or outside agency can request the AERO team through Fire Chief Robert Rocha or AERO Team Coordinator Jim DeVisser. Team Drone operators and Chief DeVisser respond to the incident with the drones (Figure 5.27). Jim acts as the mission coordinator and interacts with the requesting IC or outside agency.

So far the AERO Team has responded to requests from the Water, Wastewater, Solid waste, Police, and Marketing departments within the City of Corpus Christi. Outside agencies that have requested assistance

Figure 5.26 Corpus Christi's Drone program is off and running. Jim had been sending me Drone photos and video from time to time from hazmat incidents.

Figure 5.27 Any incident commander or outside agency can request the AERO team through Fire Chief Robert Rocha or AERO Team Coordinator Jim DeVisser. Team Drone operators and Chief DeVisser respond to the incident with the drones.

from the AERO Team include National Weather Service, Nueces and San Patricio County, City of Port Aransas, Emergency Management, and Texas Parks and Wildlife. During my visit, I had the opportunity to respond with the AERO Team at the request of Emergency Management to survey flooding going on to the West of Corpus Christi. This was my first time to witness a drone in operation first hand. They utilized their Leptron RDASS for the flood surveillance. Jim told me they had just received their new DJI Matrice 200 and 210. Maybe I lead a sheltered life on the dusty plains of Nebraska, but I was blown away and realized what an asset drones can be to the emergency services. My next mission is to get my drone license and learn to fly one.

I asked Jim if he envisioned the expansion of the Drone program beyond the fire department to other agencies. His response was: "There are two lines of thought in regards to that. One is an all inclusive program that covers the entire city and the other is everyone does their own thing. It is initially easier to have everyone do their own thing but more expensive and not as efficient in the long run. I foresee police and fire joining together for a public safety team but that is in the future sometime."

Texas A&M University–Corpus Christi, The State of Texas pre-assessment activity (PAA) Designated UAS Test Site. Island University opened on Ward Island in 1947 as the University of Corpus Christi. The first students met in facilities used by the military during World War II to train technicians on radar, what was known as the high-tech revolution of the time. With a new era in technology upon us, the Island University has taken on a distinctive role as the "Home of Unmanned Flight". In collaboration with the Federal Aviation Administration (FAA), test-site designation puts them on the forefront of the business community with hundreds of calls from industry since the announcement was made by the FAA, December 30, 2013. The FAA selected Texas A&M–Corpus Christi to facilitate testing and research of UAS technologies and the safety issues surrounding unmanned flight (Figure 5.28). How ironic this is for me, as I was growing up the talk was all about manned flight, particularly in space. Now the focus seems to be on unmanned flight. While in Corpus Christi, I had the opportunity to visit the facility with Chief DeVisser. It was very interesting to see the work they were doing with evaluating and improving drones already on the market and building innovative changes to drones of their own. In addition to military, public safety and other government agency use, there are many other innovative uses for drones. To name a few: agriculture, oil production monitoring, bridge inspections, power line inspection, wind turbine inspection, and railroad track inspection. Texas A&M's focus is on bringing UAS to America's skies safely and securely (*Firehouse Magazine*).

Figure 5.28 The FAA selected Texas A&M–Corpus Christi to facilitate testing and research of UAS technologies and the safety issues surrounding unmanned flight.

Dayton, Ohio Hazmat Team

Dayton is the sixth largest city in the state of Ohio and the county seat of Montgomery County with a population of 139,756 in 2020 (Dayton's population peaked at 250,000 during the 1950s). The city is located within Ohio's Miami Valley region, just north of Greater Cincinnati and covers an area of 808 $miles^2$. Dayton is in western Ohio approximately 30 miles from the State of Indiana border. The Great Miami and Stillwater rivers flow through the city. Dayton is known as the Birthplace of Aviation and home to Wright-Patterson Air Force Base. The National Museum of the United States Air Force is located in Dayton and it is the birthplace of Orville Wright. Powered flight was invented by the Wright Brothers in Dayton.

Fire Department History

The first fire of consequence was June 30, 1820, destroying Cooper's mills at the corner of Mill and Water (now Monument Ave.) with 4,000 bushels of wheat and 1 ton of wool. There was plenty of water in the mill race under and around the mills but all the tools they had were brought from their homes and possibly not many buckets were to be had at that time, so about all that could be done was to keep the fire from spreading. This fire caused quite an agitation for fire protection, and council bought a lot of leather buckets, and all active men were expected to keep two buckets with his name upon them, at his home and to run to every alarm with the buckets and fight the fire.

Dayton's first volunteer fire company was the Independence Fire Engine and Hose Company No. 1, founded February 20, 1834. Formation of other companies followed in 1846. Three "crack" fire companies in Dayton at the time in addition to the Independence, the Vigilance and Safety companies were formed. Other companies followed including the Oregon's, Pacific's, Miami's, Neptune, and Deluge. While I have made many inquiries about why fire companies were named and what the names meant, I have never been given an answer. However, some of the Dayton fire companies were given names like Wooden Shoes, because the membership was of German decent. Silver Tails was so named because the membership was largely made up of merchants. Plug Uglies was made up largely of some of the wealthier classes because of their habit, then in style, of wearing Plug Hats. During the early 1860s the first steam engine was used on the Huston Hall Fire. Up to that time all of the fire pumps were operated by man power.

Dayton's Paid Fire Department was organized in 1864. It was largely managed by The Committee on Fire of the Dayton City Council. The first fire chief appointed by the council was William Patton. For 15 years the council managed the fire department and it was highly political. The bi-partisan Metropolitan Department was established in 1880 and D.C. Larkin was its first Fire Chief.

Dayton was fortunate in that very few great conflagrations occurred in the city prior to 1900 in terms of financial loss. The greatest fire ever to strike the city was the Turner Opera House Fire on the corner of Main and First Streets. The Victoria Theatre occupied that space afterwards. In terms of loss of life, the burning of the Winthrop House on Ludlow Street near Fourth resulted in two people losing their lives. When the Eighth Ward House at South Main Street near Washington Street burned, a mother and her five children perished. February 1, 1900, marked the second largest fire in terms of financial loss on the corner of First and Foundry Streets. It was bitter cold with a high wind blowing. Flames spread rapidly, and several buildings were on fire. Help was summoned from Columbus, Cincinnati, and Springfield but before Cincinnati and Columbus could arrive the fire was brought under control. Around 1907 the entire fire department fleet became motorized.

There is no record that I could find of volunteers who may have died in the line of duty in Dayton. It appears the first career firefighter to die was James Bear, who died on May 16, 1900, after falling from an apparatus responding to a call the night before (Figure 5.29).

Figure 5.29 It appears the first career firefighter to die was James Bear, who died on May 16, 1900, after falling from an apparatus responding to a call the night before. (Courtesy: Dayton, OH Fire Department.)

Today's Modern Department

Dayton Fire Department under the leadership of Chief Jeff Lykins, 350 personnel operate a modern fleet of fire apparatus from 12 fire stations strategically located throughout the city. Firefighters operate 10 engines, 4 ladders, 2 rescue units, air truck, trench rescue, 2 boats, and 2 hazmat units. Emergency medical service (EMS) calls are handled by 7 medic units staffed by a combination of firefighters and civilian emergency medical technician (EMT) and paramedics. All fire apparatus are equipped and staffed as advanced life support (ALS). Dayton Fire Department responded to 39,013 calls for service in 2017. Of those, 30,340 were EMS related, and 5,053 were fires.

Hazmat Team History

Dayton's Hazmat Team was created in 1982. Fire chiefs in Montgomery and Greene Counties planned, trained, and started the team. Dayton Fire Department housed and staffed the team. Jurisdictions had liaison members available to help on-duty people. Prior to 1982, County Health was responsible for hazmat response. The team is funded by per-capita assessment from each county. Their first hazmat vehicle was a Civil Defense REO. Current Dayton's team is a part of the Ohio Regional Hazmat Team System. There are 10 teams in the system.

Hazmat Team

Dayton's Hazmat Team is housed in Station 11 at 145 Warren Street along with Engine 11, Truck 11, Medic 11, District Chief, and Incident Support Unit (Figure 5.30). Hazmat 1, Hazmat 2, Decon Trailer pulled by Freight Liner van. Also a Gator carried in the trailer. Two foam trailers with 600 gallons of aqueous film forming foam (AFFF)/AR concentrate and application appliances in the near future. At one time Dayton had an old Soda Pop truck they used for a hazmat unit. They liked it so much that they bought a new one, a Freight Liner in service today. Dayton's Hazmat Team responded to 162 hazardous materials Incidents in 2017. That figure includes other companies responses to fuel spills, gas leaks, and odor complaints.

Engine companies carry absorbents and handle hydrocarbon fuel spills of 25 gallons or less. Anything over 25 gallons becomes a hazmat team response. If mutual aid is needed, resources are available from

Figure 5.30 Dayton's Hazmat Team is housed in Station 11 at 145 Warren Street along with Engine 11, Truck 11, Medic 11, District Chief, and Incident Support Unit.

Wright-Patterson Air Force Base (Wright-Patterson Air Force Base is located less than 1 mile from the city of Fairborn, Ohio. Dayton is within 5 miles of the installation. The base is one of the largest and most important bases in the U.S. Air Force). Rickenbacker Air National Guard Base (Rickenbacker Base, located in Franklin and Pickaway counties in a rural/ residential area, 12 miles southeast of downtown Columbus, OH). The 52nd Civil Support Team, Columbus, OH. Each shift there are 15–16 technicians scheduled, Station 11 has 4–6, and Station 2 has 4. The hazmat team is not dedicated. The unit is cross staffed with Engine 11 personnel. Medic 11 is staffed with ALS Technicians. Dayton's Hazmat Team has 45 total Hazmat Technicians on the department. Training requirements for hazmat team members are Hazmat/WMD Technician.

Incidents

Miamisburg Train Derailment & Phosphorus Fire

Dayton's hazardous materials team responded to the Miamisburg, OH, train derailment on July 6, 1986. In the beginning stages, this was a phosphorus tank care on fire that released a toxic cloud over the city of Miamisburg, 30,000 people were evacuated. Air monitoring and protecting the public was the primary mission in the early stages of the incident. Air was pumped into the burning phosphorus tank care to accelerate combustion and the fire was allowed to burn itself out. Once the fires were extinguished, continued air monitoring occurred around the site and entries were made with Level B personal protective equipment (PPE). There were no injuries among hazmat team members (Dayton Fire Department).

Denver Colorado Hazmat Team

Denver is the capital and most populous city in the State of Colorado. Denver is located in the *South Platte River Valley* on the *High Plains* approximately 15 miles east of the *Front Range* of the *Southern Rocky Mountains.* Denver is *nicknamed The Mile-High City* because its official *elevation* is exactly 1 mile (5,280 feet or approximately 1,609 m) above sea level. In 2020 the population of Denver was estimated to be 734,134 making it the 19th most populated city in the United States.

Fire Department History

Denver Fire Department began as an all-volunteer department on March 25, 1866, with the formation of Hook and Ladder Company No. 1. Initial firefighting was accomplished with the use of bucket brigades. In 1867 the first hand-operated engine was purchased and used until 1872 (the apparatus is currently housed in the Denver Fire Museum located in historic Denver Fire Station No. 1, at 1326 Tremont Place in downtown Denver). Fleet management originated in 1880 as part of the original 11 "professional" members hired when numerous volunteer horse companies combined to form the Denver Fire Department. That first rank of "machinist" indicated the firefighter was a highly skilled machinist, mechanic, and blacksmith. That tradition continues to this day. Mechanics hired today are required to possess a journeyman's full range of skills and knowledge. In 1881 the Denver Fire Department became a paid department with the appointment of a fire marshal, 2 engineers, 2 stokers, 6 captains, 6 horsemen, 4 ladder-men and 7 janitors. The first motorized apparatus was purchased in 1909, and the last 3 horse-drawn apparatus were retired in 1925. By 1946, there were 22 fire stations in the City of Denver.

Today's Modern Department

Today's fully career Denver Fire Department is led by Chief Todd Bower with 900 uniformed personnel serving a geographical area of 154 miles^2. Each day there are approximately 190 suppression personnel on duty divided into 6 response districts with Denver International Airport (DIA) a separate district and a total of 900 on the department. Each district operates under a District Chief assigned to oversee the operations of up to 8 pieces of apparatus. District Chiefs are the incident commander when two or more pieces of apparatus are at an incident scene. Denver Fire Department is organized into six functional divisions: Operations, Fire Prevention & Investigation, Technical Services, Administration, Safety & Training, and D.I.A (Denver International Airport) (Figure 5.31). All Denver fire stations participate in fire prevention by inspection and preplanning all commercial occupancies in their first due areas. They have an Insurance Services Office (ISO) rating of Class 2 and an average response time anywhere in the city of 4 minutes or less.

Denver firefighters respond to over 100,000 calls per year. They average 3.4 structure fires per day throughout the city. Of their annual responses, approximately 66% are medical related and 14% are false alarms. Denver Fire Department operates 26 engine companies, 14 truck companies (1-tower ladder at the airport), 1-heavy rescue, dive team, collapse team, hazmat team, and decontamination trailer from 38 stations. Thirty-three of the stations are suppression stations in the city and five are located at the DIA. There are five stations staffed with a total of 30 firefighters on duty 24/7 who are certified in Aircraft Rescue Fire Fighting (AFR). Apparatus includes 7 Crash Rescue Vehicles, each carrying 4,500 gallons of water, 600 gallons of aqueous film-forming foam (AFFF), and 500 lb of dry chemical extinguishing agent, 1 structural engine a structural truck,

Figure 5.31 Denver Fire Department is organized into six functional divisions: Operations, Fire Prevention & Investigation, Technical Services, Administration, Safety & Training, and Denver International Airport (DIA). (Courtesy: Denver Fire Department.)

2 low profile vehicles, a chief's vehicle, a special operations apparatus, two air stairs vehicles, and a snow cat. Airport units are not allowed to leave the airport so the resources are not available to the rest of the city. Each engine and truck company responds with four personnel that work a 24 hour on and 48 hour off shift schedule.

Denver Fire Department also provides fire response to the city of Glendale, city of Sheridan, and Skyline Fire Protection District. Ambulance service is provided by the City of Denver Health Department. The fire department responds to medical calls as first responders trained to the emergency medical technician (EMT) level, but they do not transport patients.

Hazmat Team History

Denver Fire Department formed its Hazardous Materials Team in 1985 over concerns that there was an increase in hazardous materials transportation through the city. Initially the team was staffed by other companies, and in 1998 the team became dedicated with 4 personnel assigned full time to the hazardous materials unit.

Hazmat Team

Engine 6 is also housed with HAMER 1, and all personnel are trained to the technician level. Engine 6 is the primary back-up for HAMER 1 when assistance is needed. Four additional hazmat technicians are assigned to Rescue 1 and respond with the hazmat unit as needed as their secondary back-up. Rescue 1 is assigned to Station 11 located at 40 W. Second Avenue. Decontamination is provided by Engine and Truck 9 assigned to Station 10 located at 4400 Brighton with their 8 personnel. Additionally, there are on average 28 other hazmat personnel on duty at other stations throughout the city.

Hazmat personnel receive 80 hours of training for technician level. All other Denver Firefighters are trained to the Hazardous Materials Operations Level. The Hazmat Coordinator and Hazmat Chief are on call 24/7. A full hazmat response in the city includes HAMER 1, Engine & Truck 9 for decontamination, Rescue 1, HAMER 40, two additional engines & 1 truck, a District Chief, Public Information Officer (PIO), Air-Light Unit, and other specialty units as deemed necessary by the type of incident. Tiered responses may also be dispatched depending on the information available from the 911 call or reports from the incident commander on scene.

The hazmat unit HAMER 1 (Figure 5.32) is housed at Station 6, located at 1300 Blake Street on the Southwest side of downtown Denver. Along with HAMER 1, Station 6 houses Engine 6 and HAMER 3, a dump truck

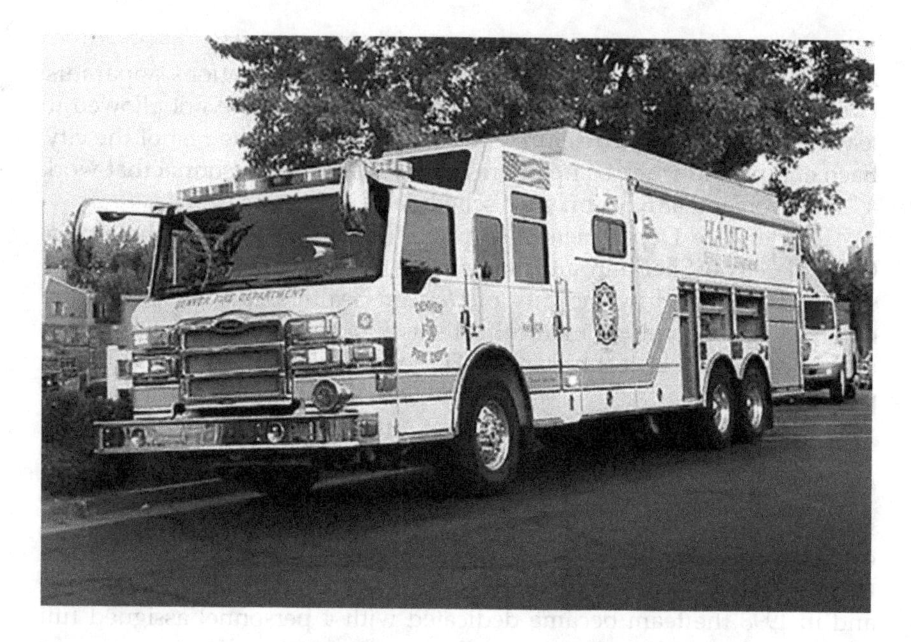

Figure 5.32 Hazmat unit HAMER 1. (Courtesy: Denver Fire Department.)

full of dirt used as needed on hazmat incident scenes for diking and other purposes. HAMER 3 was acquired from the City and County of Denver Public Works Division through an inter-city transfer (Figure 5.33). Having the dirt readily available provides a valuable resource for the team in dealing with hazmat spills. Bomb squad services are provided by the Denver Police Department, but the hazmat team works closely with the police on clandestine drug labs. Police and fire departments work well together utilizing an integrated Incident Command System (ICS). They work with the FBI and Bureau of Alcohol, Tobacco & Firearms (ATF) as well. Hazmat team members assist the fire department fire prevention bureau on inspections of hazardous materials in fixed facilities located in the city.

Each year between 35% and 40% of the hazmat responses involves flammable or combustible materials spills and leaks; many from motor vehicle accidents. This total includes Engine Company responses to fuel spills with absorbents carried on the engines. Another 13%–20% of the calls involve accidents with hazardous materials potential. The balance of the calls are divided among chemical releases, biological hazards, radioactive conditions, bomb removal, and other hazardous conditions. Denver's Hazmat Team will respond outside the city on mutual aid if called. When mutual aid is needed for the hazmat team in Denver, several area teams are available for assistance including Buckley Air Force Base. Help is also

Figure 5.33 HAMER 3 was acquired from the City and County of Denver Public Works Division through an inter-city transfer.

available from the Colorado National Guard Civil Support Team located in near-by Aurora, Colorado. Denver has established a second hazmat team located at the DIA. This team is not dedicated but rather staffed by other airport firefighters.

PPE, Equipment and Training

Chemical protective clothing includes Kappler–Lakeland for Level A and Tyvec, Fire Chem and Lakeland for Level B. B-suits include both encapsulated and non-encapsulated. Breathing apparatus consists of Scott Next Generation with 60 minute bottles. Towers and trucks carry 1 hour bottles and engine companies 30 minute bottles. In-suit communication is accomplished with OTTO throat mike push-to-talk units.

They also carry Chlorine A, B & C Kits, a mercury spill kit, Purple K fire extinguishers, and a bucket of Metal X for extinguishing Class D fires in flammable metals. Monitoring instruments include AREA RAE, Multi RAE Plus Meter, pH paper, HAZCAT Kit, Prime Alert Biodetector, Gamma Scinillator, Draeger, Draeger pump, PID, Ludlums for radioactive materials, Phosphine monitor, and smart strips. Additional support equipment includes laptop computers, fax/printer, digital camera, and radios.

Denver's Hazmat Team is very pro-active, and team members have a wide range of knowledge concerning hazardous materials. Much of their time outside of responses is spent on training and improving the team and resources.

Hazardous Materials Exposures

Denver is located near major Interstate transportation routes including I-25, I-70, I-270, and I-225. Additional major highways include U.S. Highway 6 and state routes 285 and 287. Currently the Burlington Northern Santa Fe and Union Pacific railroads have yards on the Northwest side of Denver near downtown. Plans are underway to move the rail yards out of the city. There are also numerous pipelines in and near the City of Denver. Fixed facility exposures in Denver include a major chemical distributing company, a chlorine manufacturer, anhydrous ammonia for cooling, and Air Products specialty gas company. Most of the nine Department of Transportation Hazard Classes of chemicals can be found in Denver's response area. Sulfuric acid is one of the most common chemicals present, and polymers and plastics are becoming a growing problem for Denver Hazmat Personnel. Denver uses the NFPA 704 fixed facility marking system for facilities that use or store hazardous materials. The system was adopted by amendment to the Uniformed Fire Code.

Incidents

Recent major incidents in Denver have included the removal of aging dangerous chemicals at Metro State College and multiple chemicals found in a private residence. As a result of the Metro College incident, Denver has developed a program to remove aging chemicals from schools before they create a problem. Back in April 1983 a leaking railcar of nitric acid created a significant hazmat incident in the Denver rail yards near downtown Denver. A noxious cloud of rust-colored fumes chased about 5,000 Denver residents from their homes for about 8 hours after a ruptured railroad tanker spilled at least 18,000 gallons of nitric acid at a near downtown rail siding. Most evacuated people were allowed to return later in the day after soda ash was used to neutralize the acid. Three firefighters suffered minor burns from contact with the acid and three dozen people were treated for eye irritation and breathing problems.

Denver Rail Yard Nitric Acid Spill

About 4:00 a.m. mountain standard time on April 3, 1983, a Denver and Rio Grande Western Railroad Company (D&RGW) switch crew was switching 17 cars in the D&RGW's North Yard at Denver, Colorado, when a coupler broke on the 4th car, leading to an undetected separation of 150 feet between the 3rd and 4th cars. The engineer, responding to a hand lamp signal from the foreman, accelerated the locomotive, with a caboose, an empty freight car, and a loaded tank car coupled ahead. The loaded tank car impacted a fourth car at a speed of about 10–12 mph. Upon impact, the end sill of the fourth car (empty boxcar) rode over the coupler of the loaded tank car and punctured the tank head. Nitric acid spilled from the car and formed a vapor cloud which dispersed over the area. As a result, 9,000 persons were evacuated from the area; 34 were injured. Damage to railroad property was estimated to be about $341,000 (*Firehouse Magazine*).

Durham, North Carolina Biological–Chemical Emergency Response Team (BCERT)

Durham, North Carolina, established in 1869 and known as the City of Medicine, is located in the North Central part of the state on Interstates 85 and 40 and is part of the "Triangle" or "Research Triangle" in an area encompassing three major universities, Duke in Durham, University of North Carolina in Chapel Hill, and North Carolina State in Raleigh. Because of the close proximity of Raleigh and Durham the term Raleigh-Durham is often used for the geographical area as well as they share the Raleigh-Durham Airport. Durham is the core of the four-county Durham-Chapel Hill Metropolitan Area. Durham has a population of approximately 282,737 in 2020 and an area of 69 miles^2. Durham's rich history includes tobacco companies, the first mill to produce denim and the world's largest hosiery maker. Today it is home to the world's largest university-related research park, Triangle Park. Durham is also home to the Durham Bulls AAA minor league farm club of the Tampa Bay Devil Rays and subject of the 1988 baseball movie Bull Durham, starring Kevin Costner and Susan Sarandon. Durham has a mild climate with an annual average temperature high of 69.8 and low of 47.3.

Durham Hazmat Overview

Within the City of Durham, both the police and fire departments provide hazardous materials and biological–chemical terrorism response and work together on many responses. They also work closely with the Durham County Emergency Management Agency and the Public Health Department. Durham Police and Fire Departments teams train together on a monthly basis. All non-criminal hazmat calls such as hydrocarbon fuel spills along with other types of hazardous materials incidents are dispatched solely to the Durham Fire Department. If a call is suspicious in nature such as white powder incidents or suspicious package, police respond for the initial investigation. Once the investigating officer determines that a Bio-Chemical Emergency Response Team (BCERT)/hazmat response is needed, the request is made through the Police Department watch commander and the commander notifies the BCERT Coordinator.

While in route typically BCERT requests a full hazardous materials response from the Durham Fire Department for assistance. Full hazmat response from the Durham Fire Department includes Hazmat Engines 3 and 13, Hazmat 13, the district engine, district ladder, a squad company, a paramedic ambulance, and a battalion chief. Once on scene, recon operations are conducted with one police officer and two firefighters and evidence collection is achieved with two police officers and one firefighter. Because of the requirements of the evidence chain of custody, police officers deal with evidence while the firefighter monitors environmental conditions.

Decontamination operations can be conducted by either team, however, since the fire department usually has more personnel on scene, they are usually called upon to conduct decon. According to BCERT Coordinator Tim Westcott, the splitting of duties has worked well for them and allows for the fire department and police department to work together both in training and on calls. The remainder of this article will focus on the Durham Police Department's.

Durham's Police Department is led by Chief Cerelyn "C.J." Davis with 512 uniformed personnel, 300 cruisers and approximately 50 personnel on duty per shift. Durham's Police Department has four major bureaus, the office of the chief; Administrative Services Bureau; the Operations Bureau which includes uniformed officers, K9, Traffic Enforcement, HEAT; and the Investigative Services Bureau. Within the Operations Bureau, in addition to the Bio-Chemical Team, the Durham Police also operates a SWAT Team, three "HEAT" (High Enforcement Abatement Teams), which handle prostitution, street level drugs, gangs, and other details as directed by the district captain. Special Operations Division handles Organized Crime, Interdiction, and the Major Crimes Unit. Durham Police implemented their body cam program in 2017. A total of 470 cameras were deployed. More than 12,000 videos were recorded, with an average length of more than 9 min each.

Hazmat Team History

Durham's Police Department started their Bio-Chemical Emergency Response Team in early 1999 in preparation for Y2K issues and the emergence of a growing threat of Weapons of Mass Destruction (WMDs) worldwide. Initially the team was composed of two technicians who also performed other functions within the department. White powder incidents, which peaked in October 2001, propelled the team to the forefront and resulted in the technicians becoming full time with the addition of a full time investigator to the team.

Hazmat Team

Currently the team is composed of a full-time team coordinator and eight reserve officers that have other primary jobs and respond to incidents as a secondary job function. Durham's Bio - Chemical Response Team covers the City of Durham and the Counties of Durham, Person, Grandville, Vance, and Wake. The team responded to 70 incidents in 2018. These included suspicious package calls, venue sweeps, and other agency assists. Their response vehicle is a 1999 International 4300, and they are located at the Police Headquarters, 602 East Main Street in downtown Durham (Figure 5.34). They are looking at replacing the unit in 2020. Additional vehicles assigned to the team are two Ford Crown Victoria's, and a Ford Explorer, all of which are unmarked and carry additional equipment.

BCERT only responds to criminal use of hazardous materials, all other types of calls are handled by the Durham Fire Department. Sweeps of major events citywide implemented since the Boston Marathon Bombing. They do provide assistance to the fire department team as requested. Patrol officers investigate all suspicious packages. If they determine the situation requires additional resources the request is made for BCERT through the duty Watch Commander who in turn calls the team leader for response. There are 1–3 personnel on per shift and all others are subject

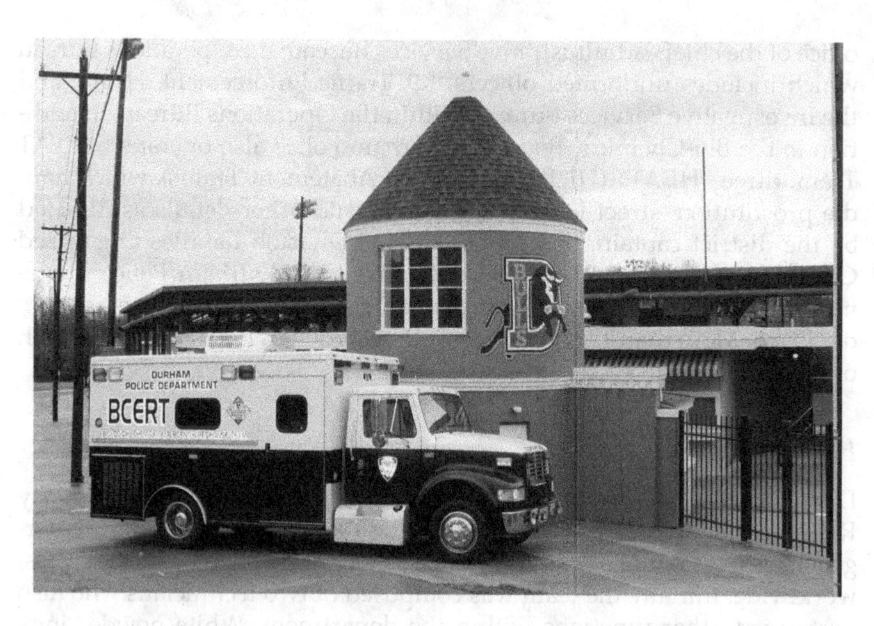

Figure 5.34 Durham Police response vehicle is a 1999 International 4300 and they are located at the Police Headquarters, 602 East Main Street in downtown Durham.

to recall and are issued vehicles to respond from home. All SWAT team members are hazmat technicians and part of the BCERT team response. Durham Fire Department has another 30–50 hazmat technicians. BCERT is the only law enforcement team of its kind in the State of North Carolina.

Mutual aid is available from the Durham Fire Department Team, Raleigh Fire Department Hazmat, North Carolina Regional Response Team (out of Raleigh, a state team asset staffed by the Raleigh Fire Department), and the National Guard Civil Support Team out of Greenville, NC.

PPE, Equipment and Training

Most equipment and personal protective clothing (PPE) has been purchased by grants or asset forfeiture money (money confiscated from criminal activity). The response vehicle was obtained from the University of North Carolina in Chapel Hill and was formerly a hospital to hospital transport vehicle (Figure 5.35). Interior work on the vehicle to convert to team needs was performed by team coordinator Tim Westcott and Mark Schell (Durham County Emergency Management). Durham County Emergency Management has been instrumental in obtaining grant funding for the response vehicle and new detection equipment.

Chemical protective clothing used by BCERT includes Level A, DuPont Tychem TK 650 and Level B, Blauer Manufacturing Multi Threat

Figure 5.35 The response vehicle was obtained from the University of North Carolina in Chapel Hill and was formerly a hospital to hospital transport vehicle. (Courtesy: Durham Police Department.)

Suit, Blauer XRT, Lakeland Tychem F and Blauer Manufacturing XRT (carried inside hazmat bags in every line patrol car). Breathing protection is provided with Survivair Panther SCBA with in-mask communications, North Safety escape hoods, Avon C50, and Avon FM53 APR's. In-suit communications, Survivair in-mask and Motorola headsets. Motorola APX6000XE Radios. Cooling vests are utilized for hot weather incidents.

Monitoring instruments carried on the response vehicle include Chemical and Biological, Smiths Detection Hazmat ID 360; Smiths Detection Responder Raman Chemical Identifier (RCI); Chemical only, Smiths Detection LCD 3.3; M8, M9 and 256 papers and kit, KI paper, pH paper and water paper. Biological only monitoring includes Advnt PS-5T Pro Strips, Alexeter RAID 8, and Biocheck 20/20. Trace explosives and narcotics are detected with Smiths Detection MMTD. Radiation instruments include Ludlum 2241 and Canberra ultra Radiacs. Air monitoring is accomplished with Rae Systems MultiRAE Pro for oxygen (O_2), carbon monoxide (CO), and hydrogen sulfide (H_2S); PID PPB 10.6; Lower Explosive Limit (LEL) and Gamma sensor; MSA Sirius for oxygen (O_2), carbon monoxide (CO), and hydrogen sulfide (H_2S); and MSA AltAir 4 for oxygen (O_2), carbon monoxide (CO), and hydrogen sulfide (H_2S) and LEL.

Other equipment carried on the response unit includes evidence collection kits for chemical and biological materials; decontamination shower, pool, hoses, wands brushes, and buckets; body recovery bags, laptop computers, portable weather station hydration coolers, research materials, Sony video camera, and Kodak digital still camera.

Team members are hazardous materials technicians and have extensive training at the Center for Domestic Preparedness in Anniston, Alabama, radiological training at the Nevada Test Site near Las Vegas, Nevada, explosives training at New Mexico Tech at Socorro, New Mexico, and additional WMD training at the Dugway Proving Grounds near Salt Lake City, Utah. Street officers in Durham receive 40 hour HAZWOPER training.

Team trains a minimum of 16 hours a month.

Hazardous Materials Exposures

Fixed facility exposures within the City of Durham include Duke University research labs, Research Triangle Park, North Carolina Central University, Federal facilities including Veterans Administration Hospital, and the Federal Building. Major transportation routes include Interstates 40, 85, and 540, US Highways 1, 15, 64, 70, 401, and 501 and State Routes 55, 147 and 751. Railroads in the Durham area are the Norfolk Southern and CSX. Incident response over the years has been composed of white powders, ethidium bromide and mercury spills, and clandestine drug labs. Mutual aid when assistance is needed is available from the Durham Fire Department (*Firehouse Magazine*).

Edmond, Oklahoma: Big Time Fire Department in a Small City Setting

> **Prolog:** *My purpose in traveling to Edmond was to teach the National Fire Academy 2 week class Chemistry for Emergency Response (CER) with former student and good friend Jack McCartt. You know you are getting old when you start teaching with former students! The class was scheduled at the fire training facility at the Edmond Fire Department. I also take the opportunity while traveling to find hazmat stories for* Firehouse Magazine. *From the moment I contacted Chief Hall about doing a story for Firehouse while I was in Edmond, to the day I left 2 weeks later, I enjoyed one of the most hospitable, friendly, professional, and educational visits of my 40 years in the fire service.*

Edmond, Oklahoma, is located in the Northwest suburbs of Oklahoma City west of Interstate 35. It has an approximate population of 93,849 in 2020 people and an area covering about 90 miles^2. Edmond had its beginnings as a coal and water station located at mile marker 103 of the Atchison, Topeka and Santa Fe Railroad, now the Burlington Northern Santa Fe (BNSF). During March of 1887 the station was officially named Edmond in honor of Santa Fe freight agent Edmond Burdick. Following the great Oklahoma Land Run of 1889 the first territorial public school was erected in Edmond between Broadway Boulevard and 2nd Street where it still stands today. Additionally, other Edmond firsts in the Oklahoma Territory included the first church building, St. John the Baptist Catholic Church, and the first continuous newspaper, the *Edmond Sun*. Historic U.S. Highway 66, known as the "Mother Road" and Route 66 runs through Edmond and is now known as Oklahoma State highway 66.

Fire Department History

During 1903 the first fire company was established in Edmond with bucket fire brigades and a horse drawn aerial ladder, which could carry a hose nozzle to the sixth floor of a building. Volunteers were summoned to Edmond's first fire station in 1904, located between 1st and 2nd streets on Broadway, by "Diane" the city's fire bell. During 1929, Edmond purchased its first motorized fire apparatus. In 1930 the fire department moved to a building which also housed the police department and city hall. In the 1950s the Edmond Fire Department consisted of a chief, four paid firefighters, six volunteers, and six part-time student volunteers. Until 1971, volunteers made up all or part of the firefighters in Edmond. Additional stations were constructed because of steady population growth. Stations 1 and 2 were constructed in 1976, Station 3 in 1983, and Station 4 in 1994. Station 5 was built in 2005. Continued growth in Edmond has created the need for station relocation and a new station to be built. Station 2 will be relocated and the new station completed in 2017. Land for Station 6 has been acquired and is projected to be completed by the end of the department's 5 year plan.

Today's Modern Department

Edmond's Fire Department consists of 116 uniformed personnel under the leadership of Chief Chris Goodwin. Edmond Fire operates out of 5 fire stations. Their first response area covers approximately 101 $miles^2$, which includes a small area outside the city limits. Firefighters operate 3 (ALS) engine companies, 2 (ALS) Quints, one aerial platform truck company (Figure 5.36), which is the department's only basic life support (BLS) company. Rescue 1 is housed at Station 1 near downtown and is equipped as a heavy rescue with an air cascade for filling self-contained breathing apparatus (SCBA), extrication equipment, air bags, rescue jacks, cribbing, and other rescue equipment.

Because some areas of Edmond have limited water supplies, the department has three tankers, all with pump and roll capability for use

Figure 5.36 Edmond firefighters operate 3 (ALS) engine companies, 2 (ALS) Quints, one aerial platform truck company.

in brush fire scenarios. Tanker 3 has a 2,500 gallon capacity and Tankers 4 and 5 carry 3,500 gallons each. Tankers are equipped with 6 inches dump valves for water shuttle operations and quick emptying of the tankers. Tankers are also equipped with pumps that create pressure inside the tanks to allow for emptying in 1–1.5 min. Each tanker carries a portable tank larger than the tankers water capacity for water shuttle operations. Engine 5 carries a portable tank as well. During high fire danger days Edmond has 4 brush pumpers that can be placed in service. Each pumper carries 300 gallons of water and is equipped with a single stage pump for firefighting.

CBRNE/Hazmat Region 8 response unit covers the region and responds to local hazmat and CBRNE incidents as well. Part of their hazmat/CBRNE response capability includes a Mass Decontamination/Casualty trailer. It is capable of handling a large-scale incident with two decontamination tents with built-in showers and portable hot water heaters. Tents are designed for quick deployment and stowing. Edmond Fire Department also houses and maintains a regional response heavy rescue trailer which is designed for structural collapse, high angle rescue, trench rescue, and confined space rescue.

The unit has the ability to self-sustain for 5 days. Other equipment includes dive team and the auxiliary rescue trailer that has a complete complement of extrication equipment, lights and 10 kW generator, breathing air compressor and air cascade system to fill SCBA bottles. Air bags and air driven tools can be operated from the air compressor as needed. Also built on board the unit is a compressed air foam system. Edmond Fire Department maintains two reserve engines and one reserve rescue. They provide fire suppression, non-transport ALS emergency medical services (EMS), technical rescue, hazmat mitigation, fire prevention, public education, and disaster preparation for the Edmond area. EMS transport is provided by EMS Authority (EMSA), which is a private firm funded by the cities that utilize their services.

Training Facility

Edmond has a state-of-the-art 40 acre training facility next to Station 5 along with a very unique Children's Safety Village life safety education center. Administrative offices for the fire department are located on the East wing of Station 5 along with spacious and well-equipped class rooms and meeting rooms. Edmond's training center has a wood construction house that can be burned and rebuilt; roof ventilation prop; flashover simulator; drill tower; driving course; and hazardous materials training area. Vehicles including school buses are used for extrication training evolutions.

Another vehicle prop is a propane fired vehicle fire simulator. Edmond's 40 acre training center includes hazmat props such as an Motor Carrier (MC)/Department of Transportation (DOT) 307/407 Kidde fire training prop and actual 306/406 tankers; propane props including tanks, trees, BBQ grill, stove for firefighting training; and confined space props. One very unique feature of the training facility is Skid truck driver training. At one time this was the only facility in the United States to have Skid truck training. Skid truck equipment is attached to an Edmond engine and used to train drivers in a very realistic setting. Skid truck can simulate driving conditions and situations. Edmond's training facility, one of the best in the metro area, is used by other fire, police, and public works departments in the metro area. The day I was touring the training facility, the Oklahoma City motorcycle police were doing driving evolutions on the driving course.

Fire Safety Village

Children's Safety Village is a complete city with child size buildings, streets, traffic lights, city park, and vehicles. Children learn fire, traffic, water, pedestrian, and other personal safety skills from uniformed Firefighters, Police Officers, and Public Works staff. In all my travels across the country I have never seen anything like it. In addition to their children's safety programs, Edmond provides a Citizen's Fire Academy. An

8 week program provides participants with a basic overview of fire, rescue, and EMS operations in the fire department. They will have the opportunity to participate in hands-on evolutions. Participants are equipped with appropriate protective equipment during attendance at the academy. Subjects introduced to students are:

- fire behavior
- water supplies
- fire hose
- fire investigations
- first aid and CPR
- forcible entry
- extrication and ventilation
- personal protective clothing and SCBA
- search and rescue
- thermal image cameras and fire extinguishers
- technical rescue

Attendance at the academy is free of charge and is open to Edmond residents or those who work in Edmond City limits. Participants must be a minimum of 18 years old.

Hazmat Team History

Edmond Fire Department started hazmat and CBRNE operations in 2002 following planning, equipment acquisition and training of personnel. Their first response vehicle consisted of a trailer and tow vehicle. Primarily the purpose for the formation of the team was the terrorist attacks on September 11, 2001, and the numerous white powder calls that followed. Prior to 911 Edmond Fire Department did not have a hazmat team.

Hazmat Team

Currently, Edmond's Hazmat and CBRNE Team operates a 38 feet duel axle trailer with a 20 kW generator, light tower, command center, air cascade system, and other equipment (Figure 5.37). The Region 8 trailer and other hazmat response equipment was provided to the Edmond Fire Department by the Oklahoma State Department of Homeland Security (OKDHS) utilizing federal grant money. Edmond in turn is required by an agreement with OKDHS to provide maintenance of the trailer and equipment, personnel and respond as part of the OKDHS Regional response team in Region 8 along with three other teams. Edmond's team can also be requested to assist anywhere in the State of Oklahoma they are needed by OKDHS. Edmond's Hazmat Station 5 is located at 5300 E. Covell Road.

Figure 5.37 Currently, Edmond's Hazmat and CBRNE Team operates a 38 foot duel axle trailer with a 20 kW generator, light tower, command center, air cascade system, and other equipment.

Also housed at Station 5 are Engine 5, Brush 5, Tanker 5, and the Auxiliary Rescue trailer.

Edmond's Region 8 CBRNE and Hazmat Team respond to an average of 100 hazardous materials calls per year. These included gas leaks or LPG, gasoline spills, chemical releases, and carbon monoxide responses. Engine companies carry 5 gallon pails of absorbent materials. Spills that cannot be handled with the 5 gallon pails trigger response of the hazmat team. Engine companies also carry four gas monitors and 5 gallon containers of Aqueous Film Forming Foam Alcohol Resistant (AFFFAR) with the eductor and nozzle to apply foam. There are 30 hazardous materials technicians on the Edmond Fire Department Team. On average 7–8 technicians are on duty each shift. Edmond firefighters work one 24 hour shift on with one shift off alternating for 5 days and then have 4 days off. Operations trained personnel assist with decontamination, freeing up technicians for incident stabilization. Edmond's Hazmat Team personnel are not dedicated to the Hazmat Unit. They operate all apparatus at Station 5 as needed. Personnel are switched around between other stations to free up technician personnel and utilize non-technician personnel to keep other apparatus in service. Mutual aid is available to Edmond for hazardous materials and CBRNE incidents from other Region 8 teams and other teams in the OKDHS response system. The 63rd Oklahoma

National Guard Civil Support Team is headquartered in Oklahoma City and available for mutual aid as well.

PPE, Equipment and Training

Personal Protective Equipment (PPE) for hazmat and CBRNE incidents for Level A is Kappler Zytron. Level B PPE is composed of Tyvek, Polycoated Tyvek, and Tychem Saranex. In-suit communications for PPE is accomplished with throat mikes with push-to-talk mounted on chest and is plugged into department portable radios. Equipment carried on the Region 8 CBRNE Unit includes Scott 4.5-1 hour SCBA, PAPR's, N95 HEPA, and Scott CBRNE Cartridges.

CBRNE specialized equipment includes a Hazmat ID; Draeger CNS; Amino Assay Kits; M-8 & M-9 test kits; and Reeves Mass Decontamination Tents, which are articulated frame tents.

Mitigation equipment carried includes:

- grounding and bonding
- air drill and chisels
- Chlorine A, B, and C Kits
- over pack drums
- dome covers
- transfer pumps
- neutralizing materials
- spill containment pools
- other typical hazmat equipment.

Basic hazardous materials and CBRNE training is conducted in house at Edmond's training facility. All personnel receive awareness and operations training. Operations personnel are also trained to do decontamination, which is conducted on scene under the supervision of a department hazmat technician. Technician training is provided by Oklahoma State University Fire Service Training and is conducted at various locations throughout the state. Some additional training taken includes National Fire Academy Chemistry for Emergency Response (CER), Chemical Agent Training at the Center for Domestic Preparedness in Anniston, Alabama, and radiological training at the Center for Radiological Nuclear Training in Nevada.

Hazardous Materials Exposures

Hazardous materials transportation routes going through Edmond include I-35, Oklahoma State Highway 66, U.S. Highway 77, and the Burlington Northern Santa Fe Railroad main line that travels right through downtown Edmond. Crude oil from North Dakota is regularly transported by rail through Edmond. In terms of potential regional response the Keystone Hub is located 52 miles from Edmond. This facility is the largest oil storage facility in the United States. Region 8 primarily covers the metro area of and including Oklahoma City, which is the State Capitol of Oklahoma.

Los Alamos National Labs Hazmat Challenge

Edmond's Hazardous Materials Team has competed for the past several years in the annual Hazmat Challenge at Los Alamos National Labs in New Mexico. Their team won the overall event in 2013 and came in second in the Technical category. Edmond won again in 2018 (Figure 5.38) (*Firehouse Magazine*).

Figure 5.38 Edmond won again in 2018.

Greater Cincinnati Hazmat Unit

Sharonville is a city in Butler and Hamilton counties in Ohio. The population is 13,560. The city is part of the Cincinnati metropolitan area on the North side of Cincinnati. Greater Cincinnati Hazmat Unit (GCHMU) coverage area has a population of approximately 1,470,000 people in approximately 2,638 miles².

Hazmat Team History

The GCHMU is a non-governmental, non-profit volunteer organization, except for Office Administrator Brooke Matzen. Annual operations budget is $170,000. Funding is provided by Public funds from Emergency Management Agencies in the covered entities. GCHMU was originally organized in 1990. Currently GCHMU contracts with ten entities in Ohio, Indiana, and Kentucky. These include:

Ohio

- City of Fairfield
- Hamilton County
- Clermont County
- Brown County
- Warren County

Kentucky

- Campbell County
- Fort Wright

Indiana

- Dearborn County
- Ohio County
- Franklin County

GCHMU is governed by a Board of Trustees formed by representatives of each of the entities served by the team. Board members can come from fire service, the public, local government, and in state businesses.

Hazmat Team

GCHMU is able to respond to both major and minor incidents. The team has over 20 years of experience and is capable of handling multiple incidents at one time, making multiple entries and sustaining a 24 hour operation. On average they respond to around 25 responses a year with 5 of them being full team responses. If additional help is needed, they request mutual aid from the City of Cincinnati.

Team members come from emergency response agencies and private industry throughout the response area. They can include firefighters, emergency medical technicians (EMTs), Paramedics, Law Enforcement and industry representatives from the tri-state response area. GCHMU operates out of their main station and office in Sharonville. Additional equipment is located at fire stations in their response area. Fire departments provide drivers for the equipment kept in their stations. The team is also a certified Type 1 team in the State of Ohio, which is the highest level and they can be called anywhere in the state if needed.

During normal business hours, Office Administrator Brook Matzen is a contact for the team as well. One Duty Officer is on call at all times year round. Only the Duty Officer can call for full team activation. Some calls are handled by duty officers without activating the full team and in some cases the Duty Officer responds and handles the request for assistance, again without the full team. Once the Duty Officer arrives on scene, he may also activate the full team, once he determines the need. Currently there are 129 members on the team and four chemists. GCHMU has three levels of response, which include:

Administrative Consultation:
The IC may request a Duty Officer for a phone consultation for resources and information such as contact with a Chemist, Clean-Up Contractor, etc.

Administrative Response:
A Duty Officer only responds to the scene and provides detection equipment and expertise in mitigation.

Full Team Response:
Full response includes Duty Officers, Chemists, Technicians, and appropriate apparatus.

GCHMU at their Sharonville headquarters, house one trailer and truck combination, the analytical response unit, with all monitoring and detection equipment and Unit 400, the Command and Communications Unit (Figure 5.39). Truck and trailer combination units are utilized for all response equipment and can be used as an environmentally controlled dress out area. Everything is in one area for transport, training, and sustainment. All three are equipped the same, except only two have Gators. Located at fire stations in their response area are two additional truck and trailer combinations; 2 foam trucks that contain AR-AFFF foam, 16 and 55 gallons of foam each, and 2 foam nozzles that flow 500 g.p.m.; 2 absorbent trailers that carry Spill X absorbent products (acid, base, caustic, formaldehyde), booms (spaghetti, oil) universal, Pads (Oil) universal, pillows, socks. Both trailers have the same materials; 3 duty officer trucks and two gators.

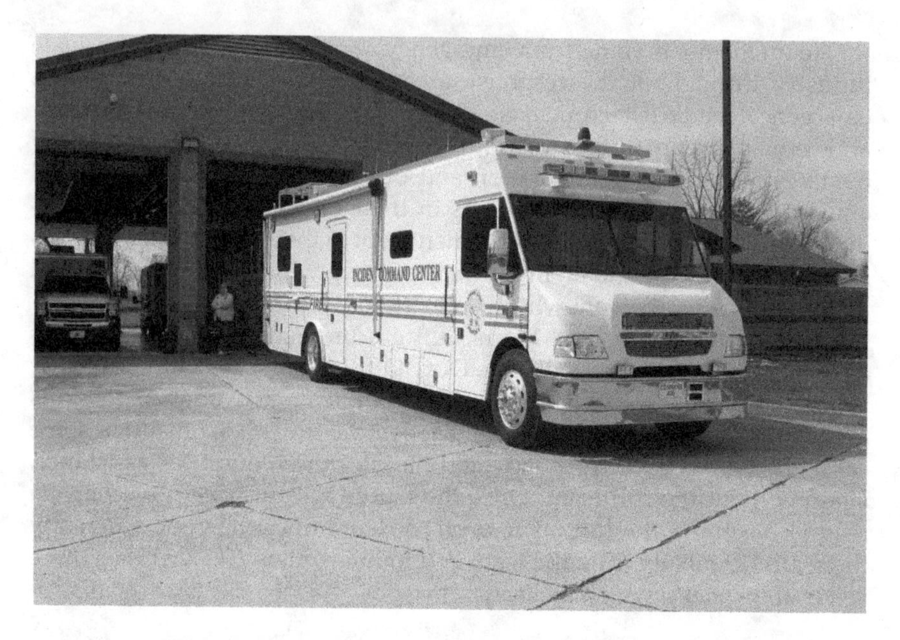

Figure 5.39 GCHMU at their Sharonville headquarters, house one trailer and truck combination, the analytical response unit, with all monitoring and detection equipment and Unit 400, the Command and Communications Unit.

Communications

Unit 400 Command and Communications vehicle is under the direction of the GCHMU and funded by GCHMU, Hamilton County Fire Chief's Association, and the Clermont County Fire Chiefs Alliance. Unit 400 was purchased in 2006 with Urban Area Security Initiative (UASI) funds. The unit can be requested by any public safety agency in UASI Region 6. An ACU1000 on board is one of the unique capabilities of Unit 400, which allows operators to patch into any radio frequency in any location for radio communications. This can be very useful in a multi-agency operations. According to Raytheon Communications, "The Raytheon JPS Communications ACU-1000 Modular Interconnect System is easily set up and controlled via software, and can be connected to a Wide Area Interoperability System using the WAIS controller". It can "Simultaneously cross-connect different radio networks, connect radio networks to telephone and SATCOM systems".

Mass Decontamination

Mass decontamination units are prepositioned throughout the response area. They have a capability of up to 50 persons per hour. Units are housed at fire stations and drivers are provided by those fire departments. Technicians and operations personnel can perform decontamination both gross and mass. Training is held monthly on the on the second Saturday. They also participate in local and regional Local Emergency Planning Committee (LEPC) exercises annually.

Drone Program

GCHMU is constantly looking to improve response operations and finding methods to make response personnel safer. Several tactics technician personnel are called upon to accomplish by making an entry into a Hot Zone place them in harm's way. Chief Bennett and Duty Officer B.J. Jetter have been looking into drones for application to hazmat response, with help from the University of Cincinnati. While they do not yet have on scene experience with drones, Chief Bennett is a licensed drone pilot and owns his own Phantom 4. Fire department and hazmat use drove his interest in purchasing a drone. He is concerned about sending personnel into the Hot Zone and would prefer to utilize a drone instead.

While Federal Aeronautics Administration (FAA) has a 400 feet height limit for drones, Chief Bennett says "you would be surprised at what you can see at 400 feet". Certainly a drone can provide a better view and send video and photographs that can be reviewed at the command post over and over to see all of the details. An added value is that a drone can also zoom in as close as required for a clear view. Using a drone places it in harm's way, but that is better than sending personnel in when a drone can accomplish the same thing. Chief Bennett also prefers to purchase a cheaper drone, if it will do the job, so if the drone is lost, it will cost less to replace it. Train derailments are one type of incident that drones are particularly suited for. Tracks do not always follow highways and may occur in areas where access is limited. Sending in a drone would provide valuable information in an area that accessing by foot in a chemical suit or even bunker gear would be difficult at best. With the frequency of ethanol and crude oil shipments, trains with multiple tank cars of the same material are not uncommon. If multiple cars derail, using a drone to size-up the situation would be invaluable.

Chief Jetter also wants to use a drone for air monitoring and sampling. This would require a second drone as most drones do not allow for multiple functions at the same time. Also, they do not allow for interchange of tools such as cameras and air monitoring equipment. This was Chief Bennett's concern as well. Being able to locate a leak at an incident scene and then monitor the air and or take a sample would not only reduce time on scene, but also keep responders out of danger. Because most departments do not do cleanup, with a drone, in some circumstances, entry by personnel might not be necessary.

There is the added problem of decontamination of the drone, which might not always be possible, if the drone is in a vapor, not much contamination will occur from vapors. On the other hand, vapor may damage the drone. Drones are not the replacement for robots or even hazmat personnel they are just another tool in the toolbox. However, having drones with interchangeability of equipment would certainly make them more valuable. Once drones are purchased, personnel need to be trained and certified to use them.

University of Cincinnati

The University of Cincinnati has provided seminars and credit courses for non-degree seeking students. One such course is Unmanned Aerial Vehicles (drones) for Emergency Responders. FAA has regulations for the ownership and use of drones (107 Certification). Drones are limited to flight 400 feet above the ground and can only be flown in the sight of the operator. Exemptions have been granted, however, getting one is not an easy task. Operating the drone requires additional insurance. According to Chief Bennett, "the industry has not caught up with the technology". Many carriers do not insure drones. Amounts of coverage that will be issued has increased from 1 to 2.5 million. Some insurers will provide insurance by the hour as low as $10.00 per hour (*Firehouse Magazine*).

Gwinnett County, Georgia Police and Fire Combine for Hazmat Response

Prolog: *Gwinnett County was my destination for a 2 week National Fire Academy Chemistry for Emergency Response Course. My original plan was to visit the Atlanta Fire Department while I was there to gather information and photos for a Firehouse Magazine article on their hazmat team. Those plans fell through, fortunately for me I found a hidden gem in Georgia in the form of the Gwinnett County, Georgia, Combination Police and Fire Department hazmat team. Theirs is an amazing story of police and fire pooling their resources and working together to provide an excellent hazmat team for the citizens of Gwinnett County. To me this was a completely seamless organization between two different agencies. It was also peach season while I was there and got to take home another of Georgia's best back to Nebraska. What a great double treat on my trip to Georgia.*

Part I Fire Department

Gwinnett County, Georgia, is located approximately 30 miles Northeast of Atlanta and covers 437 miles^2 with a population of 947,037 in 2020. It is the second most populated county in Georgia after Fulton, which is next to the city of Atlanta. It has also been one of the fastest growing counties in the United States over the past 20 years. There are 16 municipalities within Gwinnett County, and the county seat is located at Lawrenceville. It was created in 1818 by the state legislature from part of Jackson County and frontier land ceded to the state by the Creek Indians. Gwinnett County was named for Button Gwinnett a signer of the Declaration of Independence.

Fire Department History

Before a full-time, paid fire department existed in Gwinnett County, each county district had its own fire plan and protection. In 1967, seven lives were lost in the Norcross area of the county due to fires. This prompted the county commissioners to offer districts a countywide fire protection plan. A countywide referendum was held in 1970 and failed by a four-to-one margin except in the Pinckneyville District. Commissioners made another effort to consolidate fire protection by mandating that if one district wanted fire protection by the county, individual districts could vote it in. In 1970, the Pinckneyville District was the first area to pass a vote in favor of county fire protection, thus the beginning of the Gwinnett County Fire Department. On March 30, 1971, at 10:15 a.m. the Gwinnett County Fire Department responded to its first alarm. Under contract with the City of Norcross, the new department had 10 firefighters, one engine, and a budget of $89,000. Over the next 10 years, the county fire department's protection was so well received by Gwinnett citizens and businesses alike that other cities and districts voted to be included in the county coverage. Thirty years later, the department, now known as the Gwinnett County Department of Fire and Emergency Services, is the largest fire service district in the state of Georgia for the number of legal jurisdictions under one fire department.

Today's Modern Department

Gwinnett County Fire Department provides fire protection for the entire county except for Loganville which is partly located in Walton County. Russell Knick is the fire chief and has 850 uniformed and civilian personnel under his command. Over 700 Gwinnett County firefighters operate out of 31 stations. Additionally, Gwinnett County Fire Department has 30 engine companies, 10 truck companies, 23 Medic units, Hazmat, Bike Medics, Technical Rescue, Swift Water Rescue, and 2 manpower squads. Firefighters responded to over 83,172 calls for assistance in 2018. Of those calls approximately 4,121 were fires, 62,743 were Emergency Medical Service (EMS), and 16,306 other calls were handled. The department has an ISO Rating of 2. According to *Firehouse Magazine*'s National Run Survey the busiest companies in the county are Engine 15, Truck 5, Battalion 4, Medic 15, and TRT 24 the busiest squad. Firefighters work for an average of 54 hours per week with minimum staffing on engines and trucks of three personnel. Gwinnett County has their own fire academy for recruit and in-service training requirements. The Fire Academy's facilities include a burn building, training tower, apparatus building, drafting pit, low-speed driving course and classrooms. These facilities allow for a variety of training courses in both classroom settings and practical situations.

Hazmat Team History

Gwinnett County Fire Department started their hazardous materials team in the mid-1980s as a result of various hazmat calls that occurred in the county over several years. Originally the team was located in the western part of the county near several chemical and manufacturing facilities, a major interstate, and two rail lines. All personnel were trained to the technician level and equipped with response vehicles that included a quick response vehicle (QRV) and a hazmat unit. Additional equipment and vehicles were purchased over time as the need for the team grew. The team was eventually moved to Station 20 where it is presently located.

Hazmat Team

Hazmat Station 20 is located at 1801 Cruse Road in Lawrenceville (Figure 5.40), Gwinnett County, Georgia. Housed at the Hazmat Station are Hazmat Unit 20, a 2006 Spartan Chassis with an EVI Box (Figure 5.41); Engine 20, a 2008 KME 1,500 gpm pump with 5 inch hose; Medic 20; the "Pod" truck, a 2007 International with a mass decontamination pod (Figure 5.42) and bulk storage pod; Squad 20 the manpower squad,

Figure 5.40 Hazmat Station 20 is located at 1801 Cruse Road in Lawrenceville.

Figure 5.41 Housed at the Hazmat Station is Hazmat Unit 20, a 2006 Spartan Chassis with an EVI Box.

Figure 5.42 "Pod" truck, a 2007 International with a mass decontamination pod.

a 2002 Ford 550 which can carry 4 personnel; and the Water Tender 20 (tanker), a 2006 International 7400 with body by US Tanker, which carries 3,000 gallons of water and 100 gallons of foam concentrate. Two personnel are assigned to Hazmat 20 on each shift, and the remaining units are dispatched as needed for personnel and equipment. An average of 16 technician-level personnel are on duty each shift with a maximum of 22 at full staffing. Hazmat 20 also responds to medical and fire calls as needed.

Gwinnett County's Hazmat Team responds to between 300 and 400 calls per year. That total includes fuel spills, gas odors, and leaks. Engine companies carry absorbent materials and handle spills of 5–10 gallons. Greater volume spills require the response of the hazmat team. Mutual aid is provided by the DeKalb County Hazmat Team when assistance is required.

PPE, Equipment and Training

Chemical suits used by the Gwinnett County Hazmat Team include the Saint-Gobain ONE®Suit Flash Chemical Protective Hazmat Suit for Level A and Level B and DuPont Tychem. Ice vests are used for cooling in hot environments and a wire to mask earpiece button used for in-suit communications. Respiratory protection is provided by Scott with 60 minute bottles used by the hazmat team. They also carry Air Purifying Respirator (APR) masks for terrorist chemical agent protection.

Monitoring equipment carried on Gwinnett County Hazmat Units include Rae 4 gas with PID; Rae 4 gas; Sensit Gold 4 gas; Sensit Gold 2 gas; Honeywell EC-P2 with chlorine and ammonia chips; Thermo Scientific First Defender RMX; Thermo Scientific Tru Defender FT and FTI; Ludlum 14C Survey Radiation Meters; APD 2000; Smith's Detection LCD 3.2E; IdentiFINDER 2 Nuclide Radiation Meter; Camberra Ultra Radiac Plus; MCP RAM-R200; and Draeger Pump. Four Toughbook laptop computers with internet air cards are available to team members. Computer-based references used by the hazmat team include ERplan.net; WISER; National Weather Service Hazmat Response; CAMEO; ALOHA; and MARPLOT. Hard copy reference materials include Hawley's Condensed Chemical Dictionary; CHRIS Manuals, MERCK Index; Fire Protection Guide to Hazmat; the NIOSH Pocket Guide; and others.

Personnel assigned to Station 20 are trained to the technician level and all other firefighters on the department are trained to the operations level. International Fire Service Training Association (IFSTA) technician-level training is provided in house. Additional training is provided by National Fire Academy courses and Department of Homeland Security courses at Anniston, Alabama, the Nevada Test Site, and the New Mexico Tech training facilities. Personnel are granted school leave to take classes but receive no additional pay for attending the training courses.

Hazardous Materials Exposures
Major transportation exposures for potential hazardous materials incidents in Gwinnett County include Interstate 85, which runs north and south through the county, State Highways 20 and 985 as well as US Highways 23, 29, and 78. CSX and Norfolk Southern Railroads have lines into and through the county as well. Pipeline exposures are comprised major natural gas pipelines and the Plantation Pipeline Company. Fixed hazardous materials exposures consist of a Postal Facility which processes 50% of the mail in Georgia that has a biological agent detection system. The Centers for Disease Control (CDC) has a satellite facility in Gwinnett County with biological hazards. Specific chemicals that can be found in the county include chlorine, propane, ethanol, anhydrous ammonia, and others.

Police and Fire Joining Together
During 2003, Command Staff of Gwinnett Fire and Gwinnett Police began discussions on how the fire department's hazardous materials team and the police department's bomb team could begin working together. Following the events of September 11, 2001, and the increased potential of foreign and domestic terrorist events, both departments realized the need to share resources, knowledge, information, and experience. Today, Gwinnett County has one of the most comprehensive hazardous materials team and hazardous devices unit in the southeast. It is the only response model comprising fire and police teams in the Metro-Atlanta area and has become an example for other agencies. Through this collaboration, both teams have mitigated high-profile incidents including drug seizures and accidental hazardous materials releases along with potential and real explosive devices. During the 2012 Chamber Valor Awards, the combined unit was named "Public Safety Unit of the Year".

Part II Police Department

Under the leadership of Chief Tom Doran Gwinnett County has created a unique response to hazardous materials and hazardous devices incidents composed of fire and police teams is the only such system in the Metro-Atlanta area and has become a model for other agencies. Through this partnership, both teams have mitigated high-profile incidents including drug seizures, accidental hazardous materials releases along with potential and real explosive devices. One such incident occurred in December 2010 when a raid occurred on a methamphetamine "Super Lab" which turned out to be one of the largest in the United States. Officers removed 984 lb of meth with a value of approximately $44 million. Gwinnett County Police were assisted by the Drug Enforcement Administration (DEA) local police and the Gwinnett County Fire Department's Hazardous Materials Team.

Bomb Squad History

According to Major Bill Walsh, of the explosive ordnance disposal (EOD)/special weapons and tactics (SWAT) Unit, "in the late 1990s bomb squad response in Gwinnett County didn't amount to much more than a bomb suit, some x-ray film and imaging technology operated by a hand-crank". Now some 13 years later the Gwinnett County Police Hazardous Device Unit operates one of the premier hazardous device response vehicles in the State of Georgia. Gwinnett County's vehicle is a 2010 Pierce custom rescue body with a 400-HP Cummins ISL engine and Allison 3000 EVS transmission. The "Bomb Truck" as it is sometimes called is loaded with state-of-the-art equipment to handle a response to any type of terrorist or criminal incident involving hazardous devices (Figure 5.43).

A command center is located to the rear of the crew cab inside the vehicle. Equipment is located here to control remotely operated vehicles, view X-rays taken, along with communications and other equipment for response operations. Special features on the vehicle include a 30-foot telescopic camera; a powerful lighting system; flat-screen televisions that provide direct feeds from cameras remotely operated vehicles (robots) (Figure 5.44); storage areas; awnings to shade personnel; enough room to store three days worth of food in extreme emergencies, and a ramp for loading and unloading the remotely operated vehicles.

Figure 5.43 The "Bomb Truck" as it is sometimes called is loaded with state-of-the-art equipment to handle a response to any type of terrorist or criminal incident involving hazardous devices.

Figure 5.44 Special features on the vehicle include a 30 feet telescopic camera; a powerful lighting system; flat-screen televisions that provide direct feeds from cameras remotely operated vehicles (robots).

Remotely operated vehicles have the ability to X-ray suspicious objects or suspected explosives and return the film for development, thus not exposing personnel to hazardous objects. Response personnel can view developed X-rays on computer monitors inside the command center of the "Bomb Truck" or on large screen televisions on the exterior. Gwinnett County Fire Department played a key role in the design of the police response unit. The police solicited their fire department partners for their assistance. In particular, input from the hazardous materials team was sought because much of the equipment they have is utilized by both agencies. This support was fostered through years of cooperation between the two agencies.

Combined Team History

Several serious incidents have been averted over the past several years and most recently because of the partnership and organization of the two specialty teams from the Gwinnett County Police and Fire Departments. Both teams started training together in 2004. Both police officers and firefighters have volunteered to serve on both units (Figure 5.45). Personnel have received training far above the standard for law enforcement and firefighters in their individual fields. Firefighters learn the functions of the explosive technicians and are able to assist with preparing the explosive technician to enter a potential explosive device incident scene.

Figure 5.45 Both teams started training together in 2004. Both police officers and firefighters have volunteered to serve on both units.

Cross Training

Fire Department Hazardous Materials Technicians are cross trained on much of the equipment that is utilized by the Hazardous Devices Unit. This includes chemical, biological, and radiation detection and identification equipment. All Hazardous Device Technicians are required to be certified Hazardous Materials Technicians prior to attending the basic Hazardous Devices School. Patrol officers in Gwinnett County are trained to the Hazardous Materials Awareness Level. Some operations officers are also trained to the Hazardous Materials Operations Level.

Both units train together on a regular basis on the disciplines that are applicable to both units. They also train in supporting each other in their primary response roll. Training has been funded through the Department of Homeland Security with personnel traveling to the Center for Domestic Preparedness in Anniston, Alabama, The New Mexico Tech Explosives School, and the Radiation Training Site in Nevada. Efforts of both teams have proven beneficial in funding equipment. To date, grants have been received to purchase and maintain state-of-the-art equipment that exceeds six-figure dollar amounts. Because of the cooperative effort between police and fire departments, Gwinnett County has the ability to respond to multiple hazardous materials or explosive incidents occurring at the same time.

Police Department

Gwinnett County Police Department fields 758 sworn personnel under the leadership of Chief Brett West, which protect an area of 437 miles^2 and a population of 814,000 people in the unincorporated areas of the county and some small towns by contract. They responded to 898,954 calls for service in 2011. The Hazardous Devices Unit (Bomb Squad) consists of three full-time Hazardous Devices Technicians and an explosive detection K-9. Additionally, two officers are assigned to the unit part time as a collateral duty. The unit responded to 104 incidents in 2011 including Hazardous Device Unit, SWAT, and Hazardous Materials Support. Additional equipment for the police Hazardous Device Unit under consideration is an SUV for conducting on-site assessments of key resources and critical infrastructure. They are also looking at the purchase of a Bearcat that will enable to provide ballistic protection to both fire and police operators, as well as being able to operate the remotely operated vehicles from inside an armored vehicle. Currently the administration is looking at the future construction of a "Superstation" be built to house both the Fire Department Hazardous Materials and the Police Hazardous Devices Units.

Many times a response to a call requires the simultaneous dispatch of the Hazardous Materials Unit from the Fire Department and the Hazardous Devices Unit from the Police Department. Because these two units complement each other so well, they are used together quite often. When this does happen it depends on the type of call. For example, a call for an unknown suspicious substance would trigger a duel response. This ensures that all necessary resources will be at the scene. The Hazardous Devices Commander or the Fire Department Hazardous Materials Captain on scene will determine the need for additional resources.

Robots

Gwinnett County Hazardous Device Unit carries several remotely operated vehicles for use at Hazardous Materials as well as Hazardous Device scenes. These vehicles allow for the conduction of a preliminary evaluation remotely without having to send operators or other personnel down range into harm's way. In addition to monitoring instruments carried on the vehicles, they can deploy X-ray equipment to X-ray suspicious or actual explosive devices and bring the film back to personnel for development and viewing. The X-ray can also be utilized to look at pipes or valves on hazardous materials containers that may be wrapped in insulation and leaking. Video equipment on the vehicles can be used to view hazardous materials containers for markings and other identification information about the hazardous materials without personnel being exposed to danger.

Because the hazardous materials team members are familiar with much of the Hazardous Devices Unit equipment if additional manpower is needed at an incident scene they can be requested to respond. Training is also conducted with hazardous materials team members to provide emergency medical technician (EMT) and paramedic support. EMT's and paramedics are taught how to treat police personnel who might become injured during a hazardous device response. In particular, they are instructed on how to remove bomb suits during a medical emergency at an incident scene.

Firefighters and police officers on the Hazmat and Special Devices/SWAT Teams in Gwinnett County, Georgia, have one of the closest working relationships that I have ever encountered in between police and fire personnel in my career. You can tell by talking with them it's not just a professional relationship but extends to their personnel lives as well. It is much the same type of brotherhood that is shared by firefighters across the country. They seem to have developed a trust and confidence in each other's capabilities and expertise that allows them to work together seamlessly on the scenes of incidents that are either fire or police related.

Firefighters and EMS Taken Hostage

During April 2013, Gwinnett County Police Special Devices/SWAT personnel were called upon to rescue Station 10 firefighters Tim Hollingsworth, Josh Moss, Jason Schuon, Chip Echols, and Sidney Garner that had been taken hostage by a gunman hold up in his residence in the Suwannee section of the county. Firefighters were called to a medical emergency around 3:30 p.m. local time on April 11, 2013. Firefighters from Station 10 responded with Engine and Medic 10 having no reason to believe this call would be any different than any other they had responded to before. Upon arrival they took their equipment into the residence and began assessing the patient 55 year old Lauren Brown. Several minutes into the incident the patient took off the blood pressure cuff on his arm, displayed a hand gun and announced "it was time for the real reason they were there". He later displayed two other hand guns that were in the bed as well. After the initial shock to the firefighters, their training kicked in and they set about to lighten the mood and make Mr. Brown feel at ease. Firefighters were asked to remove their shirts to show they did not have any weapons.

Firefighters were told by Mr. Brown that he had been planning the incident for 4–6 weeks and chose the fire department personnel because he knew they would be unarmed and the police did not respond with them. Mr. Brown had been having financial problems and made demands that his power, cable television, and cell phone be turned back on and that windows on the residence be boarded up with insulation and plywood so that police could not view inside or get in. All demands were met as the day went on. He also told the firefighters he wanted the fire engine moved from the front of his residence.

Firefighter Jason Moss was released and allowed to move the engine. Before he left he made mental notes of the layout of the inside of the residence so he could pass the information on to the police. The incident went on for hours as the firefighters worked to gain Mr. Browns trust by telling him they were on the same team. He gradually began to allow the firefighters to move around freely in the residence, which gave them time to plan and communicate with police who were outside. Firefighter Tim Hollingsworth believed Mr. Brown planned to kill the firefighters, set the residence on fire and then kill himself. Firefighters did not try to disarm him because they believed he may have a trigger to set off explosives. This theory was formed because of the way he was positioned in bed and comments he had made.

Firefighters were alerted that SWAT was about to make a move to rescue them. Tim Hollingsworth and Chip Echols went to the kitchen on the pretense to get coffee. Firefighters Josh Schuon and Sidney Garner remained in the bedroom. They were the only firefighters in the bedroom when SWAT entered the residence. Tim Hollingsworth went to the front door on the pretense to get food that was being brought in by police and Chip Echols remained in the kitchen. After getting the go ahead, SWAT personnel entered the residence and set off an entry grenade to disorientate the gunman. Firefighters were successfully rescued from the incident with only minor injuries. One SWAT Team member Sergeant Jason Teague was shot in the arm. Mr. Brown was shot to death by SWAT Team members.

According to Major W.J. Walsh of the Gwinnett County Police Department Special Devices Team, "The firefighters being held hostage was of great concern to all personnel on scene. Particularly to Hazardous Devices Technicians at the scene who have developed a very close working relationship as well as a close friendship with many away from the office. The hazardous Devices Technicians deployed with the SWAT entry team". It is my opinion that the daily working relationship between the police and fire departments in Gwinnett County, Georgia facilitated the successful outcome of this incident. Agencies that train together and work together on a daily basis will also work well together on the scene of an emergency (*Firehouse Magazine*).

Hawaii Big Island Hazmat Team: Hazmat Response on the Island of Volcano's

Prolog: *Celebration of my 25th Wedding Anniversary brought me to Hawaii for a visit to three islands, Hawaii, Oahu, and Kauai. This visit to Hawaii could very well be a once in a lifetime opportunity, so I just had to visit the Hawaii Fire Department to do a story on their hazmat team.*

> *One of the things unique about the Hawaiian Islands is that each fire department is literally on an island. So, they have to be pretty much self-sufficient. No mutual aid is available. Evacuations on the island are also somewhat limited in the case of a hazardous materials release. When I was flying in for a landing in Kona on the Big Islands Southern Shore, I noticed what I thought were some large burned areas from a brush fire. Turns out it was just lava that has not yet been covered with vegetation. Apparently it takes 100 years or more for vegetation to start growing on the lava. Because of this lack of vegetation areas of the island look like they are on another planet! We had a wonderful time in the Big Island as it is also known and the visit to the hazmat team was an awesome part of it.*

The Hawaiian Islands are an archipelago of eight major islands and several atolls, numerous smaller islets, and seamounts in the North Pacific Ocean, extending some 1,500 miles from the island of Hawaii in the south to the northernmost Kure Atoll.

Hawaii County, sometimes referred to as the "Big Island" of Hawaii is the eastern most county in the State of Hawaii. It covers an area of 5,087 $miles^2$, but is still growing because of lava flow activity from the Kīlauea Volcano, which increased greatly during a major eruption in 2018. The Island of Hawaii is the youngest and largest of the Hawaiian Islands and many areas of land mass are lava flows without any vegetation. The Big Island has almost 63% of the land mass in the State of Hawaii with a population of 203,943 in 2020. There are no incorporated cities on the island and the government including the fire department covers the entire county. The county seat is in Hilo (pronounced Helo) on the Northern shore.

Fire Department History

Organized fire protection in Hawaii County began in 1888 with the formation of volunteer companies utilizing horse drawn apparatus. The first station was located at the corner of Kekaulike Street in Hilo. Jack Wilson was the first volunteer fire chief, and a steamer and hose wagon were located in the new station. Wood was used to fire the boiler, which produced steam to run the fire pump. By 1910 the Hilo Volunteer Fire Department had 60 members and responded to 11 fire calls that year. Additional equipment in service included a chemical wagon mounted on a car and a second car was used to tow the steam engine.

During 1919 the first motorized apparatus, a 750 gpm Seagraves engine was placed in service. The first career firefighters were hired in 1924 and included William Todd the first paid chief, an assistant chief, drivers, and hosemen; seven personnel total. The career service was supplemented by volunteers, which continue in service today. In 1927 the second motorized apparatus was purchased, which was a 1,000 gpm Seagraves. During 1931, 12 additional personnel were hired, and in 1937 a third engine was placed in service. Emergency medical service (EMS) was added to the department in 1972 with the first two firefighters trained as emergency medical technicians (EMTs).

Today's Modern Department

The present day Hawaii County Fire Department (HCFD), headquartered in Hilo, is a combination fire department with approximately 353 uniformed personnel and 230 volunteers under the command of Chief Darren Rosario. Hawaii County is the only fire department in the Hawaiian Islands with volunteer firefighters. HCFD provides fire suppression, emergency medical services, land and sea rescues, vehicle or other extractions, and hazardous materials mitigation for all Big Island residents and visitors. Hawaii County is divided into two operational areas: East and West. The Hawaii County Fire Department has 20 full-time fire/medic stations, and 24 volunteer fire stations.

There are over 60 apparatus and support vehicles, including 21 engines, 2 rescue companies, 15 ALS medic units, 2 hazardous materials response units, and a specialty truck for Waipio Valley incidents. The road into the Waipio Valley is only accessible by four-wheel drive vehicles. HCFD has two helicopters, Chopper 1: Rescue Chopper quartered at Waiakea Fire Station and Chopper 2: Medevac Chopper quartered at South Kohala Fire Station. Since they have no truck companies, engine companies are assigned truck operations as needed. The volunteer companies have additional apparatus for a variety of emergencies that can occur. Crash-Rescue at the airports on the island is provided by the State and is not part of the Hawaii County Fire Department.

Hazmat Team History

Hawaii County Fire Department organized their hazardous materials teams in 1995. The two hazmat units respond to an average of 190 calls per year including fuel spills and gas leaks. Most of the major calls involve hazmat-related fires and overturned petroleum tankers. Engine companies carry absorbent material and handle small fuel spills. Spills greater than 10 gallons require the response of the hazmat unit. Currently there are 102 trained hazmat technicians within the Hawaii County Fire Department with 30 assigned to the hazmat companies. There are 8 hazardous materials technicians on duty each shift. There is no mutual aid available since the entire fire department is on an island. If more personnel are needed, off-duty personnel are called back to duty. All firefighters on the island are trained to the Operations Level. Hazmat personnel are

Figure 5.46 Hilo Hazmat 1 is a 1997 HME on a Marion Body.

not dedicated to the hazmat units. They operate engine companies and respond to other emergencies as well.

Hazmat Team

Hazmat units are deployed to the East and West sides of the Big Island. The East station is located in Hilo at 310 Kaumana Drive and shares quarters with Engine 1, a hazmat equipment trailer for mass decontamination and mass casualty incidents and Ford F10 four-wheel drive utility pickup to tow the trailer. Hilo Hazmat 1 is a 1997 HME on a Marion Body (Figure 5.46). Engine 1 is a 1990 Seagraves 1,500 gpm with a 1,000 gallon water tank. Station 21 houses Engine 21 and Hazmat 21. Hazmat 21 is a 2005 Pierce and Engine 21 is a 2006 Pierce with a 1,500 gpm pump and 1,000 gallon water tank (Figure 5.47).

PPE, Equipment and Training

Equipment carried on Hawaii County hazmat units is typical of most hazmat units. Level A chemical suits are Lakeland and Saint Gobain ONEsuits®. Level B suits are encapsulated and non-encapsulated suits by Lakeland. In-suit communication is provided in an interface mounted directly to the MSA face mask and linked to a Motorola XTS 2500 via a push-to-talk (PTT). Respiratory protection is provided by MSA 60 min

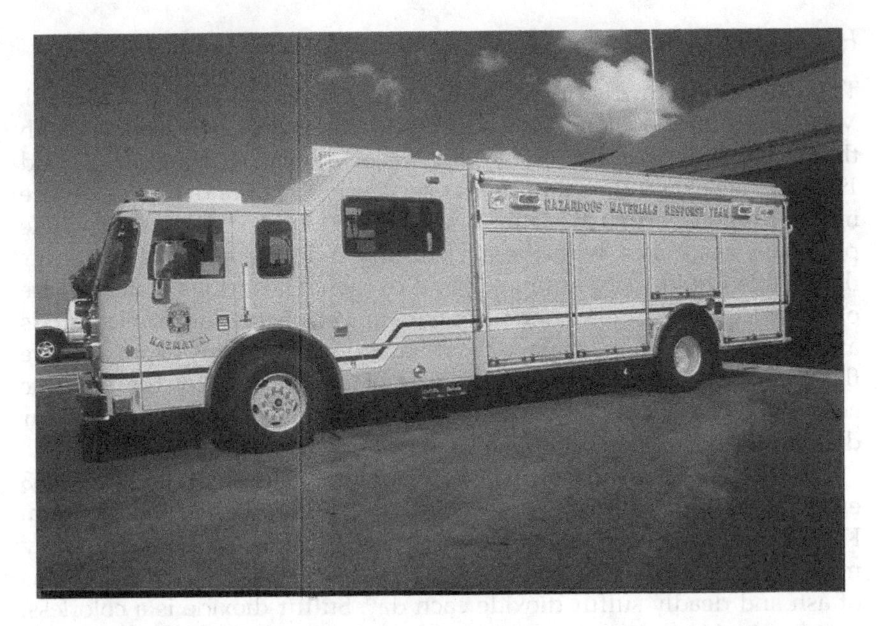

Figure 5.47 Hazmat 21 is a 2005 Pierce and Engine 21 is a 2006 Pierce with a 1,500 gpm pump and 1,000 gallon water tank.

SCBA's and MSA Air-Purified Respirators (APRs) used mainly for sulfur dioxide (SO_2) emissions from the Kilauea Volcano.

Monitoring instruments for hazardous materials carried in addition to sulfur dioxide monitors includes Canberra Dosimeters, Draeger Tubes, Thermal Imager, Ludlum 14C, Micro RAD, Draeger 4-gas, night vision binoculars, Sensor IR, and others. Terrorist agent monitors carried include APD 2000, Draeger CMS, and MSA PID.

Technicians receive their training on the island through the State Civil Defense Agency.

Reference Resources

Hazmat units are equipped with laptop computers with Internet access and printers. They have NIOSH, CAMEO, ALOHA, Marplot, COBRA, and Ex-PUB computer software programs to assist in research for hazardous materials characteristics and information. Hard copy reference materials carried on the units include the Coast Guard Chris Manuals, Condensed Chemical Dictionary, Farm Chemical Handbook, MERC Index, and the Association of American Railroads Explosives book.

Hazardous Materials Exposures

There are no railroads or interstate highways on the Island of Hawaii. Most cargo for the island comes through the islands, two seaports with the primary port located in Hilo. The major county highways are 11 and 19. There is currently no natural gas service on the island. Gas needs are met by propane. Primary hazardous materials found on the island include pesticides associated with the agricultural industry, propane, chlorine, flammable liquid fuels, ammonia, and oxygen. Hazardous materials are often transported in intermodal containers. Petroleum flammable liquids are shipped by barge. Primary fixed hazardous materials exposures are flammable fuel tank farms and power plants. Because of the volcanic activity on the island, there is always a potential for the release of sulfur dioxide (SO_2) emissions from the Kilauea Volcano.

Since 1952, there have been 34 eruptions, and since January 1983 eruptive activity has been continuous along the east rift zone. All told, Kilauea ranks among the world's most active volcanoes and is believed by many to be the most active in the world. Kilauea spews nearly 3,000 tons of ash and deadly sulfur dioxide each day. Sulfur dioxide is a colorless, non-flammable toxic and corrosive gas that is heavier than air with a very strong pungent odor. The primary route of entry into the body is through inhalation. It has an NFPA 704 classification of Health-3, Flammability-0 and Reactivity-0. NIOSH IDLH (immediately dangerous to life or health) is 100 ppm. AIHA ERPG-2 (maximum airborne concentration below which it is believed that nearly all persons could be exposed for up to 1 hour without experiencing or developing irreversible or other serious health effects or symptoms that could impair their abilities to take protective action) is 3 ppm.

Sulfur dioxide is primarily a respiratory hazard and can cause irritation to those with previous respiratory problems. Weather conditions on the island often cause the creation of "Vog" (fog from the volcano). While I was there in July 2008, I had to drive through areas of Vog with a distinct smell of sulfur in the air. Since the Vog formations are somewhat unpredictable and can happen quickly, the strategy for protecting the public is sheltering in place. Civil Defense personnel recommend that "safe rooms" be prepared in all homes exposed by the release of sulfur dioxide from Kilauea. They have also proposed the purchase of air purifiers for all public schools to create "safe zones" within the schools.

Hawaii County Fire Department has placed sulfur dioxide monitoring equipment at all fire stations on the island. Seven Area Rae sulfur dioxide monitors have been deployed to strategic locations on the island as well. Information from the fixed monitors is fed to a server at the RAE Company and can be viewed real time by fire department personnel for determining sulfur dioxide levels in the air at a given location.

2018 Kīlauea Eruption

Eruption Timeline

The 2018 Kīlauea eruption began between April 30, 2018, and May 4, 2018, with a series of earthquakes (Figure 5.48), the Pu'u O'o Crater collapse at the Kīlauea volcano summit, the movement of magma down rift toward the Lower East Rift Zone, followed by cracks opening on Mohala St. in the Leilani Estates subdivision of the Puna District.

- On May 4, 2018, lava broke through the surface in Leilani Estates, resulting in a 100 feet lava fountain spewing from the initial fissure.
- Over subsequent months, Hawaii Volcanoes Observatory (HVO) reported a total of 24 known fissures, 60,000 earthquakes, and an eruption equivalent to 8 years of Kīlauea's magma supply in just over 3 months.
- Given the volume of lava and associated hazards such as SO_2, ash, tephra, and laze, Island of Hawaii residents were severely impacted.
- Entire neighborhoods, such as Kapoho, schools, such as Kua O Ka Lā, and beach parks, such as Ahalanui Warm Ponds, were destroyed by Fissure 8, *soon to be renamed by the state Board of Geographic Names.*

HVO reduced Kīlauea's alert level from watch to advisory on Friday, October 5, 2018, after the passing of 30 days without seeing lava on the surface. There have been no active lava flows since August, though lava was seen inside Fissure 8 in Leilani Estates as of September 5, 2018.

Figure 5.48 The 2018 Kīlauea eruption began between April 30, 2018, and May 4, 2018, with a series of earthquakes.

Impacts

Verified impacts from the 2018 Kīlauea eruption include:

Natural

- 13.7 miles^2/35.5 km^2/8,488 acres inundated with lava.
- 875 acres new land created along shoreline.
- Kīlauea summit collapse.
- Erupted with a volume of 1 km^3 of lava; two-thirds from Fissure 8.

Housing/Structural

- 716 structures destroyed, including approx. 200 primary residences.
- 3,000 residents displaced.
- $296M community wealth loss.
- Estimated $236.5M in damages to roads, waterlines, and facilities (e.g. parks).

Economic

- Small businesses decreased revenues and closures.
- $27.9M farm losses resulting in decreased agriculture and floriculture production.
- Decreased tourism revenue and adjustments to marketing and products.
- Hawaii Volcanoes National Park 4-month closure, source of $222M/year economic influence (*Firehouse Magazine*).

Honolulu Hawaii Hazmat Team

Prolog: *Prior to my first visit to Hawaii I had been in 49 of the 50 states. So I was looking for an opportunity to visit Hawaii to complete my visits. It didn't dawn on me until I was actually there that the 50th state which was remaining to visit was actually the 50th state entering the union! I didn't plan things that way, it just happened. A teaching opportunity came up through the National Fire Academy in Honolulu at the Honolulu Fire Training Academy. Now, I knew there was no way I could bid on the class and have the low bid living in Maryland. I really wanted to visit Hawaii and complete my tour of the United States. So I knew it would require me to bid below cost if I really wanted to teach the class during January 2005. Let's see, Maryland, ice, snow, cold in January or Honolulu, Hula Girls and the beach in January, hmmmm? Yes, I really wanted to go! Well as luck would have it, I apparently wanted to go worse than anyone else and was awarded the class. While visiting Hawaii anyway I figured this would also give me the opportunity to do an article on the Honolulu Fire Department (HFD) Hazardous Materials Team as well.*

A year earlier I traveled to Anchorage, Alaska, and I thought that was a long flight. Well, Hawaii is one time zone further than Alaska! I didn't realize that the Pacific Ocean was that big, and I only flew over a small part of it. This was my first ever experience flying over an ocean. It was pretty awesome!

After many hours of flying, a couple of movies and lunch later, the plane started its decent to a 3:00 p.m. landing in Honolulu. The flight path took the plane down Waikiki Beach towards NASA's alternate shuttle landing runway for a smooth landing. The view was remarkable. Not only was it the longest flight I had ever taken, the taxi from the landing strip to the terminal was also seemed like the longest ever. In fact, I wondered if it was some kind of cool down period required to re-acclimate you to being on the ground again. Once at the terminal the 767 aircraft came to a stop at the gate and fortunately I had a seat towards the front of the plane and the exiting would not take long. The original seat I had selected on the internet was at the front of the coach section so I thought I would have a great view. However, when I sat down, there was no window! I told the flight attendant that it was my first trip over the ocean and to Hawaii and I would really like to look out. She found me a seat further up behind first class with wider aisles so it worked out great.

My journey began at 3:00 in the morning in Maryland, which was 11 p.m. Honolulu time the previous night, which meant it was now almost bedtime with 5h of daylight left in Honolulu. However, I was too excited to sleep. Having viewed many television shows and movies about Hawaii I was sure that a hula dancer or two would be waiting as I exited the plane to place a flower lei around my neck and plant a warm Hawaiian kiss on my cheek, so I was ready. As I entered the terminal, much to my disappointment, there were no hula dancers or flowered lei's waiting. Had we landed at the wrong airport? Of course once outside the terminal the 80+ degree temperatures in the middle of January told me at least I wasn't in Maryland anymore. While in Hawaii I learned many things that I did not already know. First of all, it snows on some of the islands! I thought how is that possible? Well there are mountains at a high enough elevation that it snows. I also got to taste Poi, a Hawaiian delicacy among natives which is a purple paste like concoction that you eat with two fingers from a bowl. Yuck, guess you have to stay around for awhile and acquire a taste for it!

Honolulu Oahu Hawaii

Honolulu is the capital and largest city of the State of Hawaii. It is also the county seat of the city and county of Honolulu and located along the southeast coast of the Island of Oahu (Oahu fittingly means "the gathering place" in native Hawaiian). It is the main gateway to Hawaii and a major portal into the United States. Honolulu is the most remote city of its size in the world and the western most U.S. city. The population of the City and County of Honolulu was estimated at 891,839 in 2020. Honolulu means "sheltered harbor" or "calm port". The city has been the capital of the Hawaiian Islands since 1845. Honolulu gained historical recognition following the attack on Pearl Harbor by Japan near the city on December 7, 1941.

Fire Department History

Honolulu's fire department was formed in 1850 by King Kamehameha III and was the first fire department in the Hawaiian Islands and the only one in the United States established by a ruling monarch. Firefighters were volunteers until 1893 when the legislature authorized funding for salaries for firefighters. During the attack on Pearl Harbor on December 7, 1941, three engine companies were dispatched from the Honolulu Fire Department to fight fires caused by the bombing. Engines 1, 4, and 6 responded and firefighting efforts resulted in the deaths of two fire captains and a hosemen and resulted in injuries to 6 other firefighters. The firefighters killed and wounded received Purple Hearts, the only firefighters in the United States to ever receive such an award.

Today's Modern Department

Honolulu's Fire Department today has over 1,100 uniformed personnel under the leadership of Chief Manuel P. Neves. When I was there in 2007, Kenneth G. Silva was the chief (Figure 5.49). Honolulu Fire Department covers the entire Island of Oahu except for the military installations and Honolulu International Airport. Honolulu Fire Department (HFD) boasts: 43 engine companies; 5 tankers, 15 truck companies (5 ladders, 8 quints, 2 towers); 2 rescue squads; 1 brush unit; 2 hazardous materials units; 3 helicopters; 1 tender; 1 mobile command center; 1 communications vehicle; 2 rescue boats and 10 rescue watercrafts. Companies are located in 44 fire stations and organized into 5 battalions. Honolulu Fire Department does not have medic units as they do not transport patients from emergency medical service (EMS) calls. That service is provided by another city department. Crews responded to 55,746 calls for service in 2017. Honolulu's Fire Department is the 17th largest in the United States.

There is no mutual aid between other Islands and Honolulu and they must rely on their own resources to handle any emergencies which arise. Limited mutual aid does occur between Honolulu and the Federal Fire Department Joint Base Pearl Harbor-Hickam and the 93rd Civil Support Team. State Fire Airport Aircraft Rescue & Fire (AARF) for hydrocarbon fluid spills. Honolulu does not provide EMS through the fire department; it

Figure 5.49 When I visited Honolulu in 2007, Kenneth G. Silva was the chief, and I had the pleasure of meeting him.

is a separate city department. When my class started in the HFD Training Center the familiar sound of a firehouse siren activated on the fire station next door. I was surprised to hear the siren since there are no volunteer fire companies on the island. As it turns out, fire stations located near major intersections are equipped with house sirens which sound when an alarm is dispatched to warn motorists of fire apparatus entering the street.

Hazmat Team

Honolulu's hazardous materials team was placed in service in 1990 and responds to all petroleum, radioactive materials, and toxic chemical incidents on the island. There are two hazardous materials units in separate companies, Hazmat 1 in the City of Honolulu at Station 32 located at 1861 Kam IV Road (Figure 5.50) and Hazmat 2 is located in Honolulu County at Station 40 Kapolei Fire Station 2020 Lauwiliwili Avenue (Figure 5.51). Both Hazmat 1 & 2 are 2,000 Pierce vehicles. Station 32 Kalihi Uka Fire Station houses Engine 32 as well as Hazmat 1. Station 40 houses Engine 40, Tower 40 and Brush 40 as in addition to Hazmat 2 and is Battalion 4 Headquarters. Engine 32 is staffed by 5 technicians and Tower 40 also has 5 technicians. Hazmat 1 and 2 each have 5 technicians. There are an additional 60 trained hazmat technicians stationed throughout the island. Firefighters work 24 hours on and 48 hours off. The shifts are called

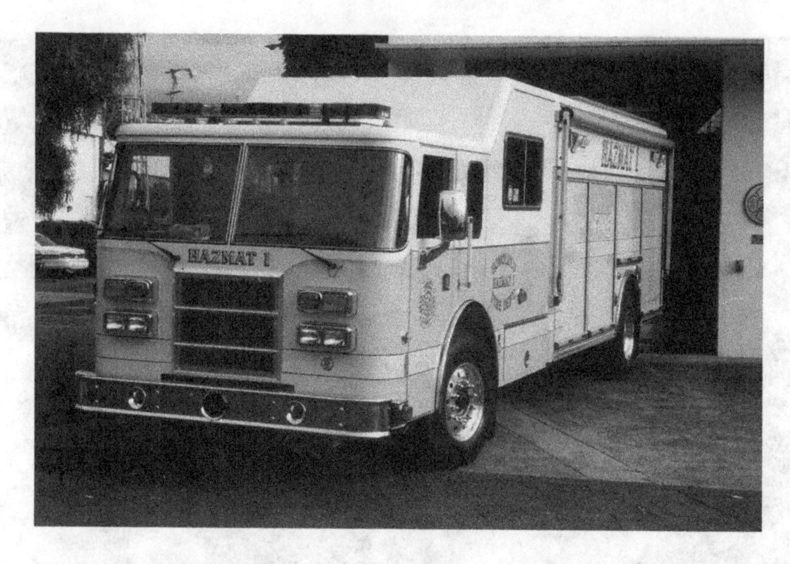

Figure 5.50 Hazmat 1 in the City of Honolulu at Station 32 located at 1861 Kam IV Road.

Figure 5.51 Hazmat 2 is located in Honolulu County at Station 40 *Kapolei Fire Station* 2020 Lauwiliwili Avenue.

"Watches", so they have 1st, 2nd, and 3rd Watch (like the old television show *Third Watch*!). Station uniforms are casual and the atmosphere is pretty laid back as you might imagine in such a warm climate; shorts, t-shirts, and flip flops.

Both hazmat units are equipped similarly with typical hazmat decontamination, patching and plugging tools and supplies, personal protective equipment (PPE), breathing apparatus, and miscellaneous tools. They also carry a portable weather station, an underwater digital camera, and a thermal imaging camera. The federal fire department at Pearl Harbor has a non-dedicated hazmat team available for mutual aid. Honolulu International Airport Fire Department is operated by the state and foam units are available for mutual aid. All engine and truck company personnel are trained to provide decontamination.

PPE, Equipment and Training

PPE used by the Honolulu Hazmat Team includes Trellchem VPS for Level A. Level B suits used are PSC-Beta, Tyvec Saranex Yellow, and Tyvec white. Breathing apparatus is manufactured by MSA G1 and 1 hour bottles are used for hazmat and 45 min quick entry on apparatus for firefighting. In-suit Communication is accomplished with OMNI Hazmat System, Cavcom Inc.

Monitoring instruments used by the Honolulu Fire Department include BW Multi Gas Monitors (4-gas 3, 3-gas 3), Draeger Tubes (18 gases or vapors), APD 2000, Hazcat, Guardian-Biocapture (Guardian Reader), Ludlums Micro Rem, Ammonia and Chlorine detectors, Pocket Dosimeters, M-8 paper, IRCAM, and others.

HFD Charles H. Thurston Training Center, Honolulu's training center, is located next to Station 8 Mokulele Fire Station located at 890 Valkenburgh Street near the Honolulu International Airport. All Honolulu firefighters are trained to a minimum of Awareness and First Responder Operations level. Captains assigned to the hazmat units provide hazmat training for all other department members. Hazmat technicians initially attend an 80 hour chemistry of hazardous materials class and 80 hours of hands-on training involving PPE, Decontamination, Monitors, Strategy and Tactics, and other equipment. Team members also attend National Fire Academy Terrorism training and are sent to the National Fire Academy in Emmitsburg, Maryland, for other hazardous materials classes.

Buildings in Honolulu do not have heat and many do not have air conditioning. Shopping centers are generally not enclosed and I was somewhat surprised when the outside door to the pool from the hotel corridor was just a metal gate and the corridor was open to the outside. Fire stations on the island of Oahu have only recently been air conditioned. Most dwellings on the island have limited insulation because it is not needed. There are no snakes, crocodiles, or large cats on Hawaii. In fact, Hawaii boasts only two native mammals: the **Hawaiian Monk Seal** and the **hoary bat**, both of which are common to the islands. The hoary bat is Hawaii's only native land mammal.

Hazardous Materials Exposures

Because Oahu is an island, there are no interstate highways or railroads. Everything brought onto the island is brought in by air or water. Trucks then transport hazardous materials to locations of storage and use on the island. Honolulu has a large port facility with intermodal containers of hazardous materials. Hazardous materials exposures include two refineries (with their own fire departments), sulfuric acid, propane, anhydrous ammonia used in cold storage facilities, pipelines, petroleum storage, and pesticides. They have the largest synthetic natural gas plant in the United States and a 12 million gallon petroleum storage facility that was built into a mountain during World War II. Hawaii Gas is the primary synthetic natural gas (SNG) and gas utility began bringing liquefied natural gas (LNG) into Oahu in 2016. Hawaiian Electric Company will begin bring in LNG in 2019.

Incidents

Kahuku Wind Farm

August 3, 2012 – Battery room fire at Kahuku wind-energy storage farm.

Downtown Explosion

August 14, 2009 – Gas explosion in State Office Tower in downtown Honolulu. Gas leak in the street pipeline followed the electrical conduits up into the fiber optics room.

Fork Lift Propane Fire

December 19, 2016 – A heavy duty forklift caught fire lifting a 5,000 gallon propane tank with direct flame impingement. Fire extinguished without BLEVE.

Ammonia Leak

April 24, 2017 – Major ammonia leak in a refrigeration system at an ice-making facility in Honolulu (*Firehouse Magazine*).

Epilog: *Throughout my visit to Oahu I was continually impressed by the hospitality of the members of the Honolulu Fire Department from the Chief to the individual firefighters. Firefighters everywhere are great, but the Honolulu Firefighters went above and beyond the call of duty to make sure I had a good time while I was there. Just in case anyone was wondering, I did finally get my flowered lei, in fact three of them! When I mentioned my disappointment to the students in the class that I had not been greeted at the airport they were appalled (not really), but they did provide one when I threatened not to pass anyone in the class). Anyway, following the lunch break the first day a firefighter came forward with a flowered lei and placed it around my neck followed by a warm Hawaiian kiss on the cheek, just as I had imagined (Figure 5.52).*

Figure 5.52 Following the lunch break the first day a firefighter came forward with a flowered lei and placed it around my neck followed by a warm Hawaiian kiss on the cheek, just as I had imagined.

By the way, it was a girl firefighter! The next day one of the guy firefighters brought in a flowered lei he had made from some flowers in his yard (he got the girl to give it to me with another kiss). The flowers had a very fragrant scent like nothing I had ever smelled. Finally, when I checked into my hotel on Waikiki the door man greeted me and placed a beautiful orchid lay around my neck. No, he did not kiss me.

Houston, Texas Hazmat Team: "Petrochemical Capitol of the World"

Prolog: *During my trip to Houston in October 2018, I had the honor and privilege to ride with Hazmat 22 on October 25th (Figure 5.53). Among the responses I rode on, included a transformer knocked down by a semi-tractor trailer, which was leaking with live wires down, and a truck on fire with possible hazardous materials. Also went to a daycare center for a visit. Additionally I was also given a tour of the Houston Training Academy and the extensive Hazmat Training Props located there. Then on October 27, I spent a 24h shift with my cousin Dustin Schroeder who is the Senior Captain at Station 68 (Figure 5.54). He is officer on Truck 68, but they had a reserve vehicle that day with only four seats, so I rode with Engine 68 throughout the shift except for an apartment building fire in the evening and I rode with District Chief 8. This was a working fire*

Figure 5.53 During my trip to Houston in October 2018, I had the honor and privilege to ride with on Hazmat 22 on October 25.

Figure 5.54 October 27, I spent a 24 hour shift with my cousin Dustin Schroeder who is the Senior Captain at Station 68.

in a second floor apartment of a very densely populated complex and the first in crew made a quick stop. My experience during these visits was awesome and I thank all of those responsible for making me feel at home and sharing their on duty time with me.

Fire Department History

The Houston Fire Department was established in 1838 with one station, Protection Company No. 1. It grew to a volunteer fire department status with three stations by 1859. After having provided volunteer firefighting services for 57 years, the Houston Fire Department began paying its firefighters in 1895.

Today's Modern Department

Houston, Texas, is the 4th largest city in the United States covering 654 $miles^2$ with a population of 2,340,890 in 2020, which swells to over 4.2 million during the daytime. Houston's Fire Department is the third largest in the United States. Samuel Pena is the fire chief leading 3,900 uniformed personnel, including 21 District Chiefs, 10 emergency medical service (EMS) Supervisors, 2 Shift Commanders, and 3 Safety Officers. They operate with 87 engine companies, 32 truck companies, 3 Tower Ladders, 11 booster trucks (for small fires, brush, trash, etc.), 36 medic units, 56 ambulances, 2 hazmat units, 1 Hazmat Foam Engine, 3 technical rescue units, 3 air cascades, and a rehab truck. Aircraft fire rescue equipment is located at Houston's George Bush Intercontinental Airport (IAH) and William P. Hobby Airport (HOU). These include ARFF Crash Trucks, 4 ARFF Trucks, 2 ARFF Medic Units, 2 ARFF Rapid Intervention Units, 1 Triage Vehicle and 2 ARFF Triage Trailers. The department is organized into 21 districts with 93 fire stations and several new ones in various stages of funding, planning, construction, and development. Houston personnel responded to over 347,662 alarms in 2019 including 45,482 fires and 302,180 EMS calls.

Hazmat Team History

During the 1970s the idea of organized hazardous materials response by fire departments and other organizations began to develop. Houston has been referred to as the "Petrochemical Capital of the World". It is reported that there is a greater concentration of chemical plants and refineries in Houston than anywhere else in the world. Most of the facilities are located along the shipping canal in southeast Houston. Several significant high-profile incidents involving railroad tank cars in Crescent City, Illinois, Waverly, Tennessee, and Kingman, Arizona, resulted in changes in tank car safety, hazmat transportation markings, and regulation at the federal level. Organization of hazardous materials response teams soon followed with many established by fire departments. Starting a hazardous materials team in the beginning was difficult because there wasn't much guidance available. There were no guidelines for training from National Fire Protection Association (NFPA) or Occupational Safety and Health Agency (OSHA) at the time and little information about SOPs and proper equipment. Industry was heavily relied upon for expertise, equipment, and training. Firefighters were recruited for the team from department ranks, some reluctantly. From the original recruits, only Firefighter William T. Hand has been with the Houston hazmat team since its inception.

Despite a lack of guidance, much of the subject matter for training provided to Houston's first hazmat team was not that different from today's lesson plans. Training began with the NFPA's "Handling Hazardous Materials Transportation Emergencies," a slide-and-tape program designed for the U.S. Department of Transportation. Visits were made to local chemical facilities, refineries, and tank farms for live chemical demonstrations and tours. A class was held on cryogens, along with a show-and-tell with a MC-338 cryogenic tanker truck. Industrial instructors conducted a class covering the use and care of monitoring instruments and "acid suits" (terminology of the time for today's chemical suits).

Training also was conducted at a chlorine plant covering the characteristics of chlorine, chlorine containers, and hands-on training for Chlorine A, B, and C Kits. Team members were then exposed to several chemistry class sessions. A field trip took team members to the Richmond Tank Car Co. to learn how tank cars were constructed. Scott Aviation provided training on new positive-pressure self-contained breathing apparatus (SCBA).

Interestingly, some of the recommended procedures at the time are still in use today, such as establishment of the "hot zone" and use of the buddy system and backup teams. Pesticide training was accomplished using the National Fire Academy's "Pesticide Spills & Fire Control" slide program. Hands-on training was provided by Dow Chemical Co. representatives on mitigating leaks from drums, tanker trucks, and rail cars.

Teams which developed as the brain children of pioneer leaders, like Chief Russell Yarbrough, Ron Gore, William T. "Bill" Hand and Chief Max McRae formed in Jacksonville, FL, Houston, TX, and other locations. These team formations preceded the Bhopal, India, industrial accident in 1984 which was the high water mark in getting regulatory attention of the United States Congress. Legislation followed by regulations from the OSHA, the Environmental Protection agency, and Department of Transportation revolutionized hazardous materials response.

The first hazmat vehicle used by the Houston Hazardous Materials Response Team was an old 1967 International step van rescue truck retrieved from the salvage yard (Figure 5.55). Following several months

Figure 5.55 The first hazmat vehicle used by the Houston Hazardous Materials Response Team was an old 1967 International step van rescue truck retrieved from the salvage yard.

of training and equipping of the response vehicle and facing an uncertain future, the team was activated at 06:30 hours on October 5, 1979. Designated as HM-1, the unit carried Chlorine A, B, and C Kits, a high-expansion foam generator and foam concentrate, SCBA, proximity suits, lifting bags, acid suits, and chemical gloves. Monitoring equipment included a combustible gas indicator with an oxygen cell, carbon monoxide monitor, and Civil Defense Geiger counters. Following several months of training and equipping of the response vehicle and facing an uncertain future, the team was activated at 6:30 a.m. on October 5, 1979. The cost of training and equipping the new team was approximately $19,000.

The new hazardous materials response team in Houston was assigned to Fire Station 1 and would also run rescue calls with a new rescue unit as one of eight rescue teams in the city. On their first day of service they were called upon to mitigate a small chlorine leak. An additional 29 hazmat alarms were answered that first year. More hazmat alarms had occurred during the first year and through the years to come, but the dispatchers did not always remember they had a hazmat team and other companies were dispatched sometimes instead.

> ***Author's Note:*** *In spite of their reluctance and apprehensions in the beginning, Chief Max McRae and Bill Hand went on to do an outstanding job for the Houston Hazmat Team and had the respect and admiration of hazmat personnel throughout the United States and Beyond.*

Hazmat Team

Despite early organization difficulties and growing pains, Houston's Hazardous Materials Response Team became the busiest and one of the best teams in the country. They were used as a model for other fire departments starting their own teams during the 1980s and 1990s. Hazardous materials personnel from around the world have rode with the team in its Ride Along Program. Many of the visitors would experience more hazmat response in a week in Houston than they would in months in their own departments.

Houston's Hazardous Materials Team is currently a standalone unit (since 1983) that is dedicated to hazmat response and other hazmat duties. Team quarters moved from Station 1 (now a restaurant) to Station 17, to Engine 22, Ladder 22, and Medic 22 and then relocated 20 miles to its current quarters. Engine 78 is located at the former Engine 22 Station and Engine 22 and Ladder 22 were taken out of service. Hazmat 22 currently operates out of relocated Station 22 at 7825 Harrisburg Street. Station 22 is close to the cities petrochemical area and hazardous materials units have a 20 to 25 minute response time anywhere in the city.

Station 22 is home to two hazardous materials response units. Hazmat Unit 1 is a 2015 Spartan Gladiator and Unit 2 is a 1995 Super Vac (Figure 5.56). Also located at Station 22 is foam Engine 22, a 2016 Spartan

Figure 5.56 Station 22 is home to two hazardous materials response units. Hazmat Unit 1 is a 2015 Spartan Gladiator and Unit 2 is a 1995 Super Vac.

Gladiator with a 2,000 gpm pump 500 gallon water tank, 750 gallon foam tank, 1,000 inches of 4″ hose, and a Williams Hot Shot II 90 gpm. Foam System. Foam Engine 22 also carries decontamination equipment and can respond alone on some types of hazardous materials alarms. The hazardous materials station is a dedicated station, and there are normally eleven hazmat techs on duty each shift. Approximately 69 hazmat techs are in reserve. Station alarms are received through an animated voice dispatch system generated by a computer.

Houston firefighters work a 46.7h work week with 24 on, 24 off, 24 on, and 5 days off. Hazmat 22 responded to 1,448 hazmat calls in 2017. Of those, 766 were turnarounds where they were canceled before arrival and 39 turned out not to be hazmat related. Below are the statistics on types of incidents:

- Fire/Explosion 30
- Spill/Leak 161
- Gaseous Release 363
- Sampling 7

Transportation incidents included 116 to highway and only 1 rail incident. There were no hazmat team injuries. Civilian injures totaled 11 and there was 1 civilian fatality. Incidents involved 32 different chemicals. They resulted in 15 instances requiring sheltering in place or evacuation from buildings or areas. Freeway incidents totaled 39 with 8 of those requiring closure of the freeway to traffic. During 2017, 1 civilian died as a result of exposure to hazardous materials and 22 were injured. There were no hazmat team deaths or injuries.

I thought it might be interesting to get a list of chemicals Houston Hazmat responded to, being one of the busiest teams in the country and having a large volume of hazardous materials exposures. Things absent from the list are the "exotic" chemicals, which seems to be the case across the country and reinforces the basic premise that most of the calls we respond to in hazmat are common chemicals.

The list follows:

Natural Gas	Suspicious Substance	Motor Oil	Diesel
Gasoline	Corrosive Liquids	Mineral Oil	Mercury
Hydraulic Oil	Propane	Methanol	Carbon Monoxide
Nitric Acid, Fuming	Hydrochloric Acid	Acetone	Paint Thinner
Carbon Dioxide	Nitrogen Gas	Phosphoric Acid	Sodium Hydroxide
Suspicious Powder	Sulfur	Propylene	Sulfur
Non-hazardous	Hydrogen Fluoride		

Houston Hazmat response area includes Harris County, and they have responded as far as Austin, Montgomery County, and Galveston. Huston relies heavily on the Channel Industries Mutual Aid Group (CIMA) for backup during incidents. Engine companies throughout the city carry 10 gallons of hydrocarbon dispersant for use on small fuel spills. Booster trucks are used to apply the dispersant materials. Anything larger requires the response of the hazardous materials unit. Houston's hazmat personnel operate as a team with a great deal of confidence in their abilities and each other. Captains function like quarterbacks and not coaches.

PPE, Equipment and Training

Chemical suits are Kappler Responder for Level A including 4 flash suits and Kappler CPF 2 & 3 for Level B. They have both encapsulated and non-encapsulated Level B suits. Team members are trying Blue Tooth Technology on in-suit communications. Station alarms are received through an animated voice dispatch system generated by a computer. Houston uses Scott 4.5 SCBA's with 1 hour bottles and they received new Scott 3000 masks in 2018 and had to re-design their hoods to fit the new masks. A new policy was implemented which prohibited wearing hoods around the neck when not in use. This was done to help prevent cancer due to recent cancer studies involving firefighters. No other type of respiratory protection is used.

Monitoring equipment carried on the hazmat unit includes four-gas meters, Mini RAEs, Sol Mini-CAD, radiological instruments, and APD 2000. Reference materials consist of electronic with a laptop computer with CAMEO, Aloha, Wiser, Peta 2, Gem's Transtar, and Internet access. Hard copy includes Emergency Response Guidebook (ERG), National Institutes for Occupational Safety and Health (NIOSH), Bureau of Explosives Book, Condensed Chemical Dictionary, pesticide manual, and various military agent books. They also have 10,000 Tier II Reports.

Firefighters who want to be on the hazmat team must have 5 years on the department and submit an application. Personnel must take the Houston Tech Course to get on the hazmat team. Hazardous materials team physicals are conducted annually on all personnel. All firefighters are trained to the Operations Level.

Incidents

RIMS Incident

On July 2016, 8:15 a.m., an 18 wheeler which included an MC 331 Propane Tanker went around a curve and hit a concrete embankment at 55–60 mph and skidded 200 feet on the concrete roadway. Friction between the steel tank and the concrete left scrape marks on the tanker's side.

Considering the mechanism of injury (damage) to the tank, hazmat team members on arrival did a visual inspection of the tank and air monitoring to determine if there was a leak. Liquid propane is transported in a non-insulated tank so the temperature of the liquid is much the same as ambient temperature. When the tank was loaded in Arkansas the ambient temperature was cool. As ambient temperature increases, so does the pressure inside the tank. Increases in heat of any kind when chemicals are involved are dangerous. Increase in heat from any source is the worst-case scenario when dealing with a hazardous material inside a container.

MC 331 tankers are equipped with temperature, pressure, and liquid level gauges. These can be helpful in determining what is going on in a tank that is not leaking. If a tank is leaking or on fire, it is too dangerous to worry about these gauges. This situation was ideal for using the gauges for important tank information. Gauges give you a visual indication there is a leak. At the time of the incident, the temperature in the container was 85°F and the tank pressure was 130. Weather reports indicated that summer temperatures in Houston would reach 110°F–115° F during the day. That temperature increase would raise the pressure inside the tank. The decision was made to keep the tank cool with a Ventura using a hose stream, which would also provide a secondary cooling effect.

This accident occurred on an elevated section of the expressway and there was no water supply. 1,000 feet of 5 " hose for a supply line was used to supply a ladder pipe that was used to hook the hose to, like an artificial standpipe. A tent was fashioned from the ever innovative firefighting tool, the salvage cover. The tank was tented and hose line placed to begin the cooling process.

The entire well-planned process worked as intended and kept the tank cool until it could be safely offloaded and the accident scene investigated and cleared of debris. No further damage or injury occurred.

Houston Distribution Warehouse Complex

One of the largest fires to ever challenge the Houston Fire Department occurred on June 24, 1995 (Figure 5.57). Seven alarms were transmitted for the Houston Distribution Warehouse complex located at 8550 Market St., starting with the first at 8:33 a.m. Two-thirds of the city's on-duty force was called on to fight the stubborn fire over the next 31 hours. Thirty-one engine companies and 14 ladder companies fought the fire. Warehouses in the complex were loaded with organic and inorganic chemicals, including corrosives, motor oils, plastics, solvents, cleaning compounds, lubricants, organophosphorus pesticides, flame retardants, and metallic compounds.

Hazmat units responded on the second alarm. The fire overcame the building's sprinkler system and firefighter hose streams, and by 11 a.m.

Figure 5.57 One of the largest fires to ever challenge the Houston Fire Department occurred on June 24, 1995. (Courtesy: Houston Fire Department.)

the building was totally involved. Exposures included five tractor-trailers loaded with organic peroxides and a liquefied petroleum gas rail car. Lines were placed in service to protect exposures and protective booms were placed to keep runoff water from entering storm drains. Mutual aid was requested from Channel Industries Mutual Aid (CIMA) at 7:30 p.m., which brought 10 foam tankers, four 2,000 gpm monitors mounted on trailers and a 6,700 gallon foam tanker to the scene. After pouring foam on the fire all night, firefighters brought it under control at around 3 o'clock the following morning. Fire units remained on scene for several days putting out flair ups. Despite the magnitude of the fire and the chemicals involved, only three firefighters sustained minor injuries.

I-610 at Southwest Freeway Ammonia Tanker Incident 44 Years Ago

The worst transportation incidents to occur in Houston happened on May 11, 1976, at approximately 11:08 hours (Figure 5.58). This incident occurred a little more than 2 years before the hazardous materials team was formed. An MC 331 tanker truck hauling 7,509 gallons of liquid anhydrous ammonia struck and penetrated a bridge rail, then struck a support

Figure 5.58 The worst transportation incidents to occur in Houston happened on May 11, 1976, at approximately 11:08 hours. (Courtesy: Houston Fire Department.)

column and fell approximately 15 feet to the Southwest Freeway below. All of the ammonia was released from the tank resulting in 6 fatalities, 78 hospitalizations, and approximately 100 other persons treated for exposure to the ammonia. Had this incident occurred at another time or location the results could have been catastrophic. If the location had been in downtown Houston next to office buildings, apartments, or a congested residential area, there could have been many more injuries and fatalities. As a result, Houston created designated routes for hazardous materials transported within the city to avoid such vulnerable areas.

Mykawa Train Yard Vinyl Chloride BLEVE

The Houston Fire Department experienced a tank car incident on Mykawa Road resulting from a train derailment on October 19, 1971. Two tank cars of vinyl chloride monomer ignited. There were six cars total of vinyl chloride and one each of acetone, caustic soda, formaldehyde, plasticine, and butadiene. The butadiene tank car experienced an explosion. Photographs and video of this incident have been circulating through the fire service for years showing a firefighter (Andy Nelson) on an aerial ladder engulfed in flames and smoke following the explosion. There was some reservation about the contents of the derailed cars but officials were slow to identify the cargo. A second explosion claimed the life of Inspector Truxton J. Hathaway and severely burned Firefighter Andy Nelson. The blast also burned 37 firefighters and civilians. Two foam trucks arrived at the scene and put out the fire in 40 minutes. The railcar derailment was believed to have been caused by poor track conditions combined with too few railroad cross ties.

It was thought that it was the Mykawa incident that motivated Chief V.E. Rogers (a district chief at the time of the incident), who was also burned at the incident, to form a hazardous materials team in Houston. However, the team was formed as a result of Rogers attending a fire chiefs' conference at which a presentation was made by Ron Gore about the new Jacksonville, FL, Hazardous Materials Response Team. Rogers returned home and directed McRae, then a district chief, to organize Houston's hazmat team.

Borden's Ice Cream Explosion

On Sunday, February 2, 1983, a call reported an ammonia leak at the Bordens Ice Cream plant on the corner of Milam and Calhoun (now St Joseph Parkway). The two-story brick building with a basement occupied the whole downtown city block. Fire companies from Fire Houses 1, 7, and 8 responded. On arrival, the crew knew that they would have to go in the building to turn off a valve to control the leaking anhydrous ammonia vapor. Several events delayed firefighters from entering the building before the explosion occurred that certainly would have caused firefighter injuries if not death. Before they entered the side door the on-duty plant maintenance engineer informed the firefighters that he knew where the valve was and would be willing to go in with them. The firefighters put an air pack on the engineer and started through the door when he had second thoughts about his safety.

The firefighters then backed out of the doorway and went around the corner to the other side of the building. Once they reassured him he would be safe, they started back toward the side door. Before they reached the door they felt the ground shake and a fire ball shot out of the doorway in which they had just been standing. Hazmat personnel had not yet arrived on scene when the explosion took place, and that may have saved them from injury and death as well. The force of the blast was so strong that it picked up a nearby manhole cover and propelled it into the windshield of a nearby parked car.

At the time, there was a great deal of surprise among firefighters that ammonia would burn or explode. Previous procedures were to enter the building with chemical protection and stop the leaks. Following the Borden plant incident, procedures were changed to ventilate the building thoroughly before entering to plug leaks. Ironically, 25 years later, Fire House 8 was built on the same block.

Houston's Hazmat Team Marks 25 Years of Service

Houston, TX, Fire Department Hazardous Materials Response Team personnel from past and present teams gathered on October 16, 2004, to commemorate the first 25 years of service to Houston and South Texas. A dinner was held at the Marriott Houston, where 250 people gathered for the celebration. The "father" of Houston's hazmat program, Max H. McRae, and former Fire Chief Vernon E. Rogers were recognized for their efforts in getting hazardous materials response started. Representatives of the Houston chemical industry and railroads were also present at the dinner (*Firehouse Magazine*).

40th Anniversary Houston Hazmat Team

On October 5, 2019, the Houston Fire Department Hazardous Materials Team, Company 22, celebrated 40 years of organized hazardous materials response. A celebration was held at the Houston Fire Department Pension Center with retired team members returning along with current team members and other fire department personnel.

Imperial County, CA Hazardous Emergency Assistance Team (IV-HEAT)

Established in 1907, it was the last county to be formed in California covers 4,482 miles^2 with a population of 182,333 in 2020. El Centro is the county seat. It is the smallest but most economically diverse region in the state. It is located in the Southern California Border Region and the Imperial Valley. The County borders Arizona, Mexico, and Riverside County in California. The county is home to the El Centro Naval Air Station, Imperial Sand Dunes Recreation Area, and the Painted Gorge.

Today's Modern Department

Imperial County Fire Department (ICFD) under the leadership of Alfredo Estrada Jr. operates from eight stations and six contracting agencies. Contracting agencies include Brawley, Calipatria, Holtville, Westmorland (Volunteer), Salton City (Volunteer), and Salton Sea Beach run county engines. It has a number of 75 uniformed personnel, and they operate 8 Engines, 2 Trucks, 2 Rescue Trucks. Other apparatus includes a Crash unit, two

Water Tenders, one Hazmat Unit, and two Brush Trucks. Imperial County Fire Department provides services to unincorporated communities of the county, townships, and the City of Imperial. Services include Fire Protection, Medical Basic Life Support (BLS)/Advanced Life Support (ALS), Aircraft Rescue Firefighting, Technical Rescue, and Hazardous Materials responses. Ambulance service is provided by Gold Cross contracted by the county and provides ALS services with a paramedic and emergency medical technician (EMT). All firefighters are required to be EMTs and provide BLS service until an ambulance arrives. By contract they provide fire protection for the City of Imperial. They are required to provide ALS service within the city. Boley Field (Imperial County Airport) has protection from the Imperial County Fire Department providing a Crash Unit. It is housed at Station 1 in the City of Imperial.

County fire stations are located in the communities of Heber, Seeley, Ocotillo, Palo Verde, Niland, Winterhaven, and City of Imperial. Each station is staffed with a captain, firefighter, and reserve firefighter except for the Palo Verde station that is staffed with a firefighter and reserve firefighter. Firefighters work a shift of 2 days on and 4 days off (48/96). Each station is equipped with a Type I Engine (Structural Firefighting). The City of Imperial and Heber Stations also house a Ladder Truck. Seeley and Heber Stations also house Type III Engines (Wildland firefighting).

Hazmat Team History

The Imperial Valley Hazardous Emergency Assistance Team was created in 1995 through the formation of a Joint Powers Agreement (JPA). Formation of the JPA established the first joint countywide emergency response for hazardous incidents. The following member agencies make up the JPA: Cities of Brawley, Calexico, Calipatria, El Centro, Holtville, Imperial, and Westmorland, the County of Imperial (Fire and Environmental Health Departments). The response role of the IV-HEAT Team is to Isolate, Contain, Identify, Mitigate, and Stabilize.

Hazmat Team

Station 2 located in Heber is the hazardous materials stations. The team averages 3 hazmat calls each year. That total does not include fuel spills, gas odors, and leaks or odor complaints. Four hazmat technicians are on duty each shift and operations personnel handle decontamination and other assistance. There are 20 hazmat technicians in the county. Mutual aid is available from El Centro Naval Air Station, Riverside County, and Yuma, AZ. The hazardous materials unit is housed at the Heber Station. Their hazmat response vehicle is a Type 1, 2017 Freightliner (Figure 5.59).

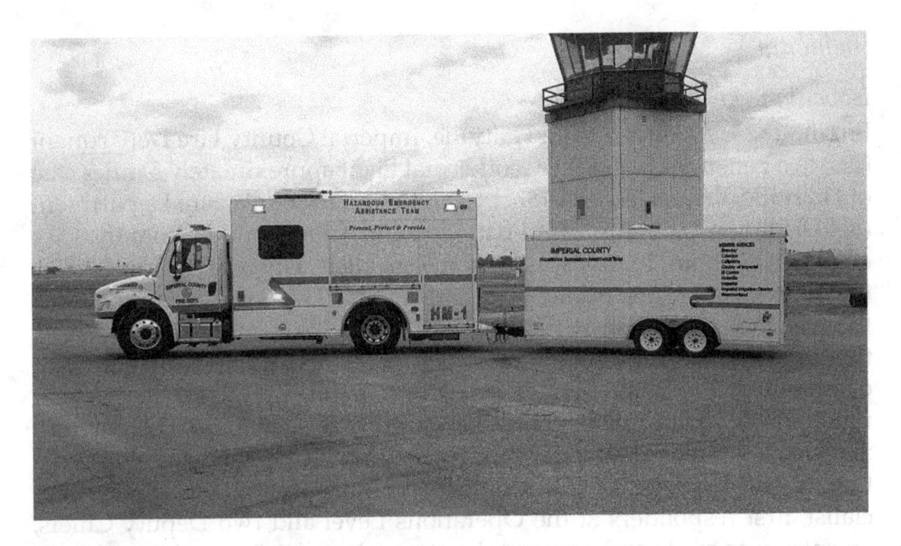

Figure 5.59 Hazmat 1 and decon trailer are housed at Station 3 in the township of Seeley, CA. (Courtesy: Imperial County Fire Department Captain Oscar Robles.)

PPE, Equipment and Training

Level A, B, and C personal protective equipment (PPE) is used as needed. Respiratory protection is the use of self-contained breathing apparatus (SCBA). Handheld radios are used for in-suit communication. Monitoring equipment includes QRAE34 gas monitor, PID, Hazcat System, and radiation detection equipment. Laptop and internet are the primary resources for product information.

Hazmat Techs are required to First Responder Awareness (FRA), First Responder Operations (FRO), Hazmat Technician of 4 weeks through the state and Hazmat Specialist, two additional weeks of training. All firefighters in the county are trained to the Awareness and Operations levels. Annual refresher training includes 12 h of hands on and 12 h of didactic each year.

Hazardous Materials Exposures

Being largely an agricultural county, common hazards include fertilizers, liquid and solid, and pesticides. Fuel spills also make up a large percentage of responses. Railroads transport hazardous materials through the county and there are fuel pipelines as well. Kindermorgan Co. in the City of Imperial has a fuel tank farm with millions of gallons of gasoline, diesel, and jet fuel.

Incidents

Isopentane Leak

Engine 2823 responded from Holtville, Imperial County Fire Department (FD) to a generator fire at the Geothermal Plant approximately 5 miles east of Holtville. Requested assistance form Imperial County Fire Department (ICFD) Station 1, Engine 1, and Battalion Chief 1 responded from the City of Imperial about 20 miles west of the incident. Upon arrival of E-2833, they encountered a specialized generator fire. They proceeded to attack the fire and extinguished the fire with about 1,000 gallons of water. At this time Engine 1 and Battalion Chief (BC) 1 arrived and assisted with overhauling and command. BC took command and activated Hazmat unit to respond due to Isopentane leak around the generator. Hazmat unit arrived on scene with hazmat coordinator and immediately established a hot zone and evacuated it. With the help of three hazmat techs, one specialist, first responders at the Operations Level and two Deputy Chiefs, an entry was made and scanned the atmosphere for flammable and toxic gases. With levels being normal including oxygen and samples of puddles and dirt were taken to test the area and incident was secured and the incident turned over to the Geothermal Health and Safety Manager (Imperial County Fire Department).

Jacksonville, FL First Hazmat Team in the United States

Jacksonville, Florida, is located in the northeast corner of the state. Jacksonville City and County government are combined into one jurisdiction and cover a geographic area of over 850 miles^2 with a population of over 926,371 in 2020.

Fire Department History

Bucket Brigade

Jacksonville's first organized firefighting efforts took place about 1850 and involved digging wells and stockpiling ladders and other equipment at three intersections. A fire bell was hung from a five-story tower built over a well at Newnan and Adams streets. If a fire broke out, someone would run for the bell while yelling "fire". The yelling and clanging would bring out the town's able-bodied men who would form lines to pass water buckets from the wells to the fire.

Hand Pumper

The city soon acquired its first fire engine, an apparatus with handles on each side for pumping, but lost it to the flames in a big fire in 1854.

First Organized Fire Protection
The city's first organized fire fighting force formed on January 10, 1868, when the Friendship Hook and Ladder Co., a group of volunteers, began providing fire protection. Several other volunteer companies formed by 1870 to became part of the Jacksonville Volunteer Fire Department. The city provided little support to the volunteers, who often battled in the streets over access to fire hydrants while buildings they were supposed to save burned, and the organizations gradually lost members and began to disband. In December 1885, a disastrous fire that lead to the death of a volunteer firefighter may have been the final event to convince the City Council that it needed a paid department.

Career Department Established
On April 20, 1886, the City Council passed an ordinance establishing a paid fire department, and on July 15, the city hired its first 20 paid firefighters. That year three staffed stations and a fire chief established a legacy protecting a core city area of approximately 39 $miles^2$. Station 1 was located on Forsyth Street just off of Pine Street, which is now known as Main Street. Station 2 was located at Pine and Ashley streets. Station 3 was located in the 500 block of East Bay Street. All three stations burned down in the Great Fire of 1901 and were later rebuilt. Station 1 was rebuilt at Adams and Ocean streets.

Station 2 was rebuilt at the same location. Station 3 was rebuilt on Catherine Street near its original location. Bricks salvaged from buildings destroyed in the fire were used to construct the stations. Station 3 became known as the Catherine Street station. It housed a new 1902 American LaFrance Steam Engine and was staffed by four African-American fire-fighters. One of the first Jacksonville firefighters to die in the line of duty was an African-American by the name Henry J. Bradley.

The Catherine Street station is registered with the Jacksonville Historical Society and with the Northeast Florida African-American Historical Society. Today the station has been relocated to Metropolitan Park, where it now serves as the department's Fire Museum.

The Great Fire of 1901
The Great Fire of 1901 started at 12:30 p.m. on May 3, when a cinder from a nearby chimney landed on moss and fiber drying in the sun at the Cleveland Fiber Factory at Union and Davis streets. Eighteen-mile-per-hour westerly winds fanned the flames, and the fire and destruction did not stop until it reached what is now known as the St. Johns River.

By the time it was over, the fire had destroyed 146 city blocks and 2,368 buildings. Property damage totaled $15 million. The official death count tallied two people from burns, three from drowning, and two others from fright. Some survivors questioned those numbers, saying there was no way of knowing how many people actually died in the smoke and flames or drowned fleeing in the river.

Beginning Ambulance Service

In 1962, Jacksonville Fire Department Asst. Fire Chief James Dowling, Jr., began a push to end the practice of funeral homes using hearses to provide ambulance service to the city. He argued that the funeral homes were more concerned with having funerals than with providing transportation of the sick and injured to the hospital. In November 1967, Mayor Hans Tanzler placed emergency ambulance service permanently in the care of the Jacksonville Fire Department.

Rescue Division

The Rescue Division began with six station wagons, each staffed by a chief and two firefighters, equipped with first aid kits and folding Army cots for stretchers. Within a few months the department equipped and staffed six new modular transport vehicles for continuous 24 hour service. Crews soon became aware that they were in over their heads due to the nature of the calls, a large proportion of which were cardiac related, so the department connected with area doctors eager to provide better training. With advanced medical training and better equipment, the Jacksonville Fire Department saved more lives, and Jacksonville became known as the "safest city in the world to have a heart attack".

City County Consolidation

In 1968, the Jacksonville City and Duval County governments consolidated. Today, the Jacksonville Fire and Rescue Department provides fire and emergency medical services to a metropolitan, suburban, and rural areas.

Notable accomplishments of the Jacksonville Fire and Rescue Department (JFRD) are:

1. Establishing one of the first Advanced Life Support (ALS) service in the nation;
2. Establishing the first hazardous materials team in 1977;
3. Becoming the first fire department to successfully extinguish a fully involved petroleum tank fire.

Today's Modern Department

Jacksonville's Fire Department and Hazmat Team cover the entire jurisdiction. Under the leadership of Chief Keith Powers, 1,450 uniformed employees operate 58 engine companies, 13 truck companies, 1 rescue units, and 58 medic units from 62 stations. Jacksonville also operates a technical rescue team, Urban Search and Rescue Team Florida Task Force 5, air cascade, 3 fireboats, 2 hazmat units, and a mass casualty decontamination unit. Firefighters in Jacksonville respond to an average of 160,000 alarms per year of which about 70% are emergency medical service (EMS) related.

Hazmat Team History

Chief Yarbrough's Vision

Jacksonville, Florida, Fire Department is the home of the first hazardous materials response team organized in the United States. During the late 1970s, Chief Yarborough of the Jacksonville Fire Department envisioned the need to deal with hazardous materials response in a trained and organized manner. He realized that Jacksonville was the "Great Shipping Center of the South". Railroads had major operations in the city and there was significant military activity in the area involving nuclear materials. No one had ever taken an organized approach to hazardous materials response before. Being the first at anything is a difficult undertaking because there is little if any guidance available to assist. Chief Yarborough and Captain Gore believed a specialized team was the best approach and they needed 15–20 volunteers from the firefighting ranks to form the nucleus of the new hazardous materials response team.

This came directly from the man so many of us credit as the "Godfather of Hazmat". "It wasn't my idea", said retired Capt. Ron Gore during a recent visit to Fire Station 7. In town for a reunion of JFRD's original Hazmat Team members in late June, Gore shared how the specialty team's concept originated with Fire Chief Russell Yarbrough in the 1970s. Tanker car derailments, shipments of nuclear weapons to local military installations, dangerous and volatile cargo passing through our city's main thoroughfares and Jacksonville's growing chemical industry got the chief's attention. If something went awry, Yarbrough wanted JFRD to be as prepared as possible.

> ***Author's Note:*** *While Ron Gore doesn't want to take credit, even though it was Chief Yarbrough's idea, it was Captain Ron Gore who made it happen. Not just in Jacksonville, but all across the United States. Ron Gore is one of the most influential people in the world of hazmat and he has touched thousands of people during his training sessions and Ron, you are the undisputed "Godfather of Hazmat" in the American Fire Service. I for one do not think you get enough credit for what you have contributed to all of us in Hazmat Response.*

The "Godfather of Hazmat"
Engine 18's Gore was about to become Hazmat's Number One Son. As a new lieutenant in the early 1970s, Gore was assigned to the Training Academy for a year. This was common practice back then, but Gore's experience would be far from common after he met Training Chief Simon Joseph King, Jr., Gore was still getting used to pinning bugles on his collar when King assigned him the task of teaching hazardous materials as part of a fire science course at Florida Junior College now Florida State College at Jacksonville. There was just one problem. "I didn't know the subject matter", said Gore. Neither did anyone else on the job. Initially, Gore and his students, many of them firefighters, spent time with chemistry teachers. But lectures on molecular structures and chemical reactions didn't translate well to emergency response.

A frustrated Gore soon found relief in practicality. He convinced Florida Junior College (FJC)'s administration to allow him to train his students in the field. "We trained at the rail yard, ship yard, the tank farm", he said, adding that private industry was very cooperative in providing access and institutional knowledge. Gore's class became an on-site assessment of industrial hazards, the chemicals, production methods, and the containers. Through that familiarity, Gore and his students began to understand the risks, and they devised response strategies.

America's First Emergency Services Hazmat Team
Together, the members focused on making America's first hazmat team as capable as it could be, often at their own expense and while off duty. They looked to military surplus outlets for protective gear, they collaborated with private industry and JFRD's machine shop to design the most effective tools, and they traveled the country to understand the volatility of new chemical products. Back then, Jacksonville provided very limited funding to the team, Gore said. That didn't stop them; they willingly bore the costs of travel, training, and even some specialized equipment purchases. Jacksonville's first hazmat units were Engine 9 and Engine 9A (Figure 5.59A).

The initial request for volunteers produced more than 50 firefighters wishing to become team members. There was no additional pay for being on the hazmat team. Captain Gore looked for personnel for the team who had some previous experience with chemicals. There were firefighters who had worked for the gas company, some with military experience and others who had experience with various chemicals. As with anything new, there was a lot of distrust and resistance to the new hazardous materials team from officers and firefighters alike.

"We had the attitude that when we were working together, there was no event we couldn't handle" Phil Eddies, JFRD Retiree, and Original

Figure 5.59A Jacksonville's first hazmat units were Engine 9 and Engine 9A. (Courtesy: Jacksonville Fire Department.)

Hazmat Team Member. Theirs was an extensive and unrelenting discovery process. There were no precise OSHA standards for hazardous materials response to guide them, Gore said. Although they lacked technical knowledge and defined procedures, the team had confidence. "When we formed that team, it was the best group of people you could have hoped for", said JFRD Retiree Phil Eddins original Hazmat Team member. And handle they did: propane tank fires, derailed train cars leaking hydrogen chloride or muriatic acid, petroleum tank farm fires. There were exposures, but no serious injuries. Each incident was a response as well as a lesson for the team (Figure 5.59B).

On January 1977 the Jacksonville Hazmat Team became operational. There were skeptics of the team early on, and some who poked fun by labeling the hazmat members as the "Clorox Team" or "bleach drinkers". Eddies said he eventually took the monikers as a compliment, and JFRD Retiree Jim Croft embraced the identity as well. On occasion, when he got transferred to another station, Croft would show up with his gear in one hand and a genuine bottle of Clorox in the other. "I'd come inside the station and set it down on the table", he said. As the team continued to prove itself, the nicknames subsided, and the interest from departments across the country increased.

Personnel from the Jacksonville Fire Department pretty much had to make things up as they moved forward. There were no response procedures to follow so they had to be developed as the team gained experience. Attempts at tactics were developed as they responded to incidents.

Figure 5.59B Jacksonville's original hazardous materials team members. (Courtesy: Jacksonville Fire Department.)

Sometimes they worked and sometimes they did not. They developed procedures and equipment following incidents where they saw the need for something that would have helped them during an incident.

There wasn't much hazmat equipment available commercially so much of what they used was created in their maintenance shop by the chief of maintenance. They literally invented things as they went along. Incidents that occurred were discussed on all three shifts, which is the way many of the early team members learned to deal with hazardous materials, basically through trial and error. It would be difficult to train every firefighter to deal with hazardous materials. When the team was organized, little was known about personnel protective clothing for chemicals outside of the military. Early on they would use plastic trash bags over their firefighter turnouts for chemical protection.

They would soon find out that suits obtained from the military were good for some chemicals but not for all. During a tank car leak incident at Union Camp that occurred in July 1978 a technician's suit was breached during operations on top of a hydrogen chloride tank car dome. Little was known in the beginning about suit compatibility and the suit being worn was not compatible with hydrogen chloride. Decontamination didn't exist at the time and there were no hot or warm zones like we use today. Most personnel in protective clothing or not were right up in what would be the Hot Zone. Much was learned about personal protective equipment (PPE) and suit failure at this incident. It was also the incident that in spite of the issues helped to establish the credibility of the new hazardous materials response team.

Many of the procedures and equipment mentioned above would seem highly unusual in today's world of hazmat response. However, keep in mind that it was the early efforts of the Jacksonville Fire Departments trial and error flying by the seat of your pants operations that have evolved into today's organized and structured hazardous materials response. Hazmat response was new to city government as well. The city did not provide money for equipment or training. Much of the training and travel expense was paid out of pocket by team members. In spite of the uncertainty and lack of procedural guidance available at the time the team was organized, there was no loss of life or serious injury among team members.

Gore's travel included trips with Chief Yarbrough, who was the President of the International Association of Fire Chiefs. This gave Gore numerous opportunities to present to large audiences that were hungry for information that only he and JFRD's team could offer. Gore said he doesn't recall any other fire department challenging Jacksonville's position as the first to forge a municipal hazardous materials response team. "It makes me proud", he said.

JFRD Retiree Andy Graham joined the team in 1980 and remained a member until retiring in 2004, longer than anyone else. During that time, he interacted with a lot of visiting fire departments. "We had people from Canada, the Virgin Islands, New York", Graham said. "They wanted to ride with us. A lot of times, we didn't have anything noteworthy, but they would ask questions. We would just talk about hazmat. They knew that Jacksonville was the leader in this, and they wanted to pick our brains". Graham said he joined the hazmat team to try something different. "After a couple of weeks, I thought 'Why didn't I do this sooner?'" he recalled. "They were so eager to learn. That's what impressed me".

Graham also recalled how JFRD members initially had their doubts about the team, but after a few years, the field took notice of their effectiveness and cautious work habits. "We'd show up decked out in air packs", Graham said. "People began to understand if we had our stuff on, they better have theirs on". Although the team faced plenty of unknowns, Graham said that Gore trained him and everybody so well that he "can't ever remember fearing" for his life. "Capt. Gore loved the job, and he taught me how to love it", Graham said. "He gave me the push to love what I did, and I love it. When I retired, it took me about a year to get over leaving".

My Visits to Jacksonville

Not only was Jacksonville the first hazardous materials team in the country, they also know how to make a fellow firefighter and hazmat responder feel very welcome. My trip to Jacksonville was the first time I was met at the airport and escorted to my destination, many thanks to District Chief Randy Wyse and Lieutenant Todd Smith for all their assistance while I was in Jacksonville and during my first visit to Jacksonville. During the trip to Hazmat Station 7, we took a short detour to visit old Station 9 (now a museum) at 24th and Perry Streets, where the first hazmat team in the country was originally housed (Figure 5.60). Station 9 was chosen because of its strategic location near I-95 and the 20th Street expressway which allowed them quick access to all parts of the city including industry to the East and rail yards to the West.

The trip to Station 7 was followed by a barbeque attended by former hazmat team members including Ron Gore, Bob Masculine, Neil McCormack, Richard Morphew, Phil Eddins, Jim Croft, and Davis Love who together represent over 60 years of hazardous materials experience (Figure 5.61). Jacksonville Fire Chief Richard Barrett, current hazmat team members, and other Jacksonville fire department personnel were also attended. Over the next several hours I experienced some of the most enjoyable conversations about hazardous materials of my career.

Figure 5.60 Station 9 (now a museum) at 24th and Perry Streets, where the first hazmat team in the country was originally housed. (Courtesy: Jacksonville Fire Department.)

Figure 5.61 The trip to Station 7 was followed by a barbeque attended by former hazmat team members including Ron Gore, Bob Masculine, Neil McCormack, Richard Morphew, Phil Eddins, Jim Croft and Davis Love who together represent over 60 years of hazardous materials experience.

Captain Ron Gore Retires

Gore retired from JFRD in 1989, but he continues to teach the subject through his hazardous materials company, Safety Systems, which he founded in the late 1970s. Gore offers training and emergency response services. His clients include the military, private industry, and other fire and rescue departments. Gore estimates that he and his team of instructors have trained "hundreds of thousands of people" nationwide (Figure 5.62).

Gore is now the hazmat subject matter expert that he was desperately seeking in the early 1970s, when he was tasked with arguably one of the most challenging assignments ever entrusted to a firefighter in Jacksonville or elsewhere. He may not accept the "Father of Hazmat" label, but just Google "Ron Gore" and you'll learn how his place in the hazardous materials circle is renowned. At 78, Gore is still eager to teach what he's learned over the last 40 years, and what he's learned from the intrepid firefighters who formed America's first Hazmat Team. He has no plans to retire either. After being forced to leave his comfort zone that first year he became a JFRD lieutenant, Gore believes his career path was forged by Chiefs Yarbrough and King and, most importantly, God. "I was created to do this", he said.

Figure 5.62 Gore estimates that he and his team of instructors have trained "hundreds of thousands of people" nationwide. (Courtesy Jacksonville, FL Fire Department.

In addition to Ron Gore's efforts in Jacksonville, he traveled throughout the United States conducting hazardous materials training classes through Safety Systems Incorporated a business he founded. As I have talked to members of hazmat teams across the country, many have referenced Jacksonville and particularly Ron Gore as having influenced the formation of their teams.

> ***Author's Note:*** *My own early interest in hazardous materials was fueled by a Safety Systems training course conducted by Ron Gore at the Fire Museum of Texas in Arlington during 1981. Following that class in Texas my Chief Bob Cox at the Verdigris Fire Protection District in Claremore, Oklahoma, and I arranged for Ron Gore to bring his training class to our department. During this class I had the opportunity to get to know Ron Gore and realized that Ron was likely the most knowledgeable person in the field of hazardous materials response in his time. Ron's impact on the formations of countless other hazmat teams in the United States is incalculable. During my travels across the United States to visit fire departments and hazmat teams I heard countless stories of how Ron Gore influenced other hazmat teams. His legacy, Captain Ron Gore is the "Godfather of Hazmat in the United States."*

Hazmat Team

Jacksonville's Hazardous Materials Teams are not totally dedicated. Hazmat 7 has two personnel assigned and Hazmat 21 has one (Figure 5.63). Personnel from Engine and Truck Companies at both stations provide the

Figure 5.63 Jacksonville's Hazardous Materials Teams are not totally dedicated. Hazmat 7 has two personnel assigned and Hazmat 21 has one. (Courtesy: Jacksonville Fire Department.)

remainder of the manning for the hazardous materials units when an alarm comes in. There are 20 hazardous materials technicians including the battalion chief on each shift at Stations 7 & 21. There are an additional 200 Technicians city wide. Jacksonville's Hazardous Materials Team responds to an average of 1,100+ alarms per year including fuel spills. Engine companies carry absorbents and respond to small fuel spills. The hazmat team is called in for spills larger than 10 gallons. Squads, Ladders, and Hazmat Companies carry monitoring equipment.

PPE, Equipment and Training

Jacksonville Hazardous Materials Units carry typical hazardous materials response equipment. Level A Chemical Protective Suits are Trelborg and Trelchem 6 HPU Disposable. Level B Chemical Protective Suites are SP100 Tychem and Saranex. In-suit communications are provided through a bone mike system. Breathing apparatus consists of MSA 45 min Firehawks for firefighting and 60 min bottles for hazmat. They also use Positive Pressure Air Purifying Respirators (PAPRs) and cartridge respirators for weapons of mass destruction (WMD) incidents.

Hazmat technicians attend the 160 h Florida Certification, technician level along with the International Association of Fire Fighters hazmat certification, Team task book, monthly training, and individual daily company training. All other companies in the city are trained to the Operations Level. There are no special requirements for Jacksonville Fire Department Personnel to transfer to the hazardous materials stations.

Hazardous Materials Exposures

Transportation hazardous materials exposures in Jacksonville include Interstates 95 and 10 and US Highway 1. CSX and Norfolk Southern Railroads have tracks and rail yards in Jacksonville. They also have major natural gas and petroleum pipelines. Portions of Jacksonville border the Atlantic Ocean and there are Port facilities and ocean going vessels and barges that carry hazardous materials in intermodal containers and others traveling to and through the city. Fixed facility exposures include 2-Navy bases that they respond to through mutual aid, FMC Corporation and Millennium Chemical Companies, Van Waters and Rogers Pesticide Company, and petroleum storage tank farms for Coastal, Chevron, and BP oil companies. Almost any kind of chemical could be found transported through or used in Jacksonville. Major chemical hazards include chlorine, propane, sulfuric acid, and styrene.

Incidents

Dave & Busters

During my visit to Jacksonville the hazmat team received a call for a chemical odor at a popular sports bar/restaurant on the East side of town in Station 21's area. Of course Station 21's crews were at Station 7 on the West side of town so the response from Station 1 resembled a small parade of fire apparatus responding to the call with all of the resources of Stations 7 & 21 in route. District Chief Randy Wyse (Fire 4 B-shift) asked me if I wanted to go, so I rode along with him. The initial response was handled by Engine and Ladder 44. The call turned out to be a small fire caused by spontaneous combustion of grease laden grill rags that had been washed and dried and placed in a pile on a shelve in the laundry area. Washing cloth that has been used to clean areas covered with animal/vegetable oils does not completely remove all the oil from the cloth. When washed and dried, the dryer heats up the animal/vegetable oil.

Piling the clothes together on a shelve causes concentration of the heat. Oxygen reacts with the double chemical bonds in the animal/vegetable oil creating additional heat and breaking more bonds, which results in ignition of the cloth material. Each of the clothes in the pile had a uniform brown pattern on the cloth indicating an internal rather than external heat source. Firefighters and hazmat personnel from Stations 7, 21, and 44 quickly removed the clothes from the room and ventilated the area. Patrons were allowed back into the restaurant in about an hour.

Stewart Petroleum Fire

On January 1993, JFRD and Hazmat responded to a gasoline storage tank fire at the Stewart Petroleum Facility. The tank contained more than 2,000,000 gallons of gasoline. It took 6 days to extinguish, and most of the time aqueous film-forming foam (AFFF) was used. This was the first time in the United States that a fire department successfully extinguish a fully involved petroleum tank fire without any outside help from industry.

Faye Road Incident T2 Laboratories

At 1:33 p.m. on December 19, 2007, a powerful explosion and subsequent chemical fire killed four employees and destroyed T2 Laboratories, Inc. (T2), a chemical manufacturer in Jacksonville, Florida (Figure 5.64). It injured 32, including 4 employees and 28 members of the public who were working in surrounding businesses. Debris from the reactor was found up to 1 mile away, and the explosion damaged buildings within one quarter mile of the facility.

On December 19, T2 was producing its 175th batch of methylcyclopentadienyl manganese tricarbonyl (MCMT).

MCMT is an organomanganese compound used as an octane-increasing gasoline additive. The Ethyl Corporation originally developed MCMT in the late 1950s. T2 manufactured and sold MCMT under the trade name Ecotane. MCMT is a combustible liquid and is very toxic by inhalation or skin contact. Both the National Institute for Occupational Safety and Health (NIOSH) and OSHA set exposure limits for MCMT. Although

Figure 5.64 At 1:33 p.m. on December 19, 2007, a powerful explosion and subsequent chemical fire killed four employees and destroyed T2 Laboratories, Inc. (Courtesy: Jacksonville Fire Department.)

MCMT decomposes quickly when exposed to light, the Environmental Protection Agency (EPA) designates MCMT as an extremely hazardous substance (EHS) (Section 6.2.2).

At 1:23 p.m., the process operator had an outside operator call the owners to report a cooling problem and request them return to the site. Upon their return, one of the two owners went to the control room to assist. A few minutes later, at 1:33 p.m., the reactor burst and its contents exploded, killing the owner and process operator who were in the control room and two outside operators who were exiting the reactor area.

The Chemical Safety Board (CSB) found that a runaway exothermic reaction occurred during the first (metalation) step of the MCMT process. The CSB tested the T2 batch recipe to determine the most likely failure scenario. A loss of sufficient cooling during the process likely resulted in the runaway reaction, leading to an uncontrollable pressure and temperature rise in the reactor. The pressure burst the reactor; the reactor's contents ignited, creating an explosion equivalent to 1,400 lb of TNT.

Responding to the explosion and subsequent fire were the JFRD; U.S. Naval Air Station Mayport Fire Department; Jacksonville International Airport Fire Department; Jacksonville Sheriff's Office (JSO); City of Jacksonville Environmental Resource Management Division; City of Jacksonville Planning and Development Department; Florida State Fire Marshal; Florida Department of Environmental Protection (FDEP); U.S. Bureau of Alcohol, Tobacco, Firearms, and Explosives (ATF); U.S. Environmental Protection Agency (EPA); U.S. Department of Homeland Security, Chemical Security Compliance Division (DHS-CSCD); U.S. Department of Labor (*Firehouse Magazine*).

> ***Hazmatology Point:*** *Like the Phillips 66 explosion and fire in Houston, this was a unique facility of a type that would not be found in many other places in the country. Fixed facilities that manufacture or use "exotic" chemicals should be known to local fire departments and hazmat teams. Plans should be made including the facility operators to pre-determine what actions should be taken by plant officials and emergency responders in the event of an accident like the one at T2. These are extremely dangerous facilities and pre-planning is critical so that no responders are killed or injured in responding to these once in a career events.*

Kansas City, Kansas Hazmat Team

Kansas City, KS, is located at Kaw Point, the junction point of the Kansas and Missouri rivers. It is the third-largest city in Kansas and the third-largest city in the Kansas City metropolitan area, which has a population of more than 2 million. Kansas City, KS, an area of 127 miles^2 of land and 3.5 miles^2 of water, is situated on the western shore of the Missouri and Kansas rivers and has a population estimated at 153,600 in 2020.

Today's Modern Department

The Kansas City, KS, Fire Department has approximately 436 uniformed personnel under the leadership of Chief Michael Callahan. Firefighters operate out of 18 fire stations with 15 pumpers (an apparatus with a pump is called a pumper in the Kansas City metro area), 3 aerial ladders, four quints, one heavy rescue, 9 medic units with three backups, 4 fire-rescue boats, a foam tender, three brush trucks, two hazardous materials response vehicles, and a hazmat support unit. The department also operates a special Urban Search and Rescue (USAR) heavy rescue vehicle. Specialized units include hazmat, high-angle rescue, trench rescue, dive, and confined-space rescue teams. The Fire Communications Division operates out of the Unified Government's Public Safety Dispatch Center and is responsible for processing, dispatching and coordinating emergency, and non-emergency calls. Kansas City Fire Department responds to over 26,000 calls annually.

Hazmat Team

Kansas City, KS, formed its hazmat team in 1979. The team was originally based at Station 6 at 1103 Osage Ave. along with Pumper 6. The first hazmat unit was a converted International army truck. Today's Hazmat Units 2 and 3 are identical 2007 Freightliner step-van-type response vehicles with independent-entry capability (Figure 5.65). Hazmat 2 has

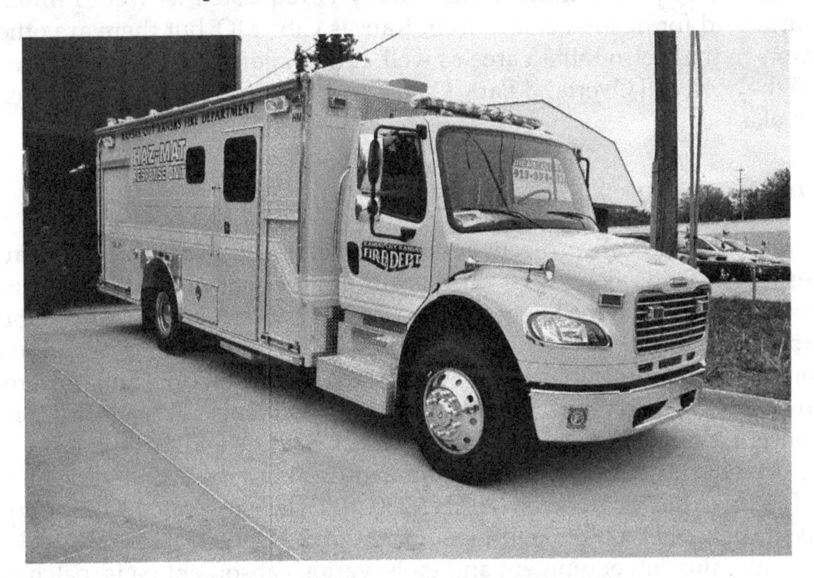

Figure 5.65 Hazmat Units 2 and 3 are identical 2007 Freightliner step-van-type response vehicles with independent-entry capability.

reduced monitoring capability when responding alone. Hazmat 3 personnel specialize in monitoring and entry. Hazmat 2 carries extra radios not available in Hazmat 3. Hazmat 2 team members specialize in library and research functions. Kansas City Hazmat has gone to all computerized and internet reference materials.

Kansas City's hazmat units are not dedicated. The crews of Stations 2 and 3 provide personnel for the hazmat units during a response. Station 3 personnel include two medics and four firefighters on Pumper 3 and Truck 3, all hazmat technicians. There are always six personnel in Station 2 who are technicians including two medics and four firefighters. Each shift has 11 assigned hazmat technicians on duty with nine or more others at additional companies around the city. Station 2, at 6241 State Ave. in West Kansas City, houses Quint 2 and EMS 2, which is an advanced life support (ALS) ambulance, in addition to Hazmat 2. Quint 2 is a 1997 E-One with a 75-foot ladder and 1,250-gpm pump. Station 3, at 420 Kansas Ave. in East Kansas City, houses Hazmat 3, Pumper 3 and Medic 3. Station 16, at 1437 South 55th St., houses the decontamination team (Pumper 16) that responds on hazmat calls to provide support for Hazmat 2 and 3. The hazmat team responded to 213 incidents during 2017.

Full hazmat assignments in Kansas City would include the department safety officer, a battalion chief, a medic unit, both hazmat units, the first-due pumper company, and the hazmat coordinator. If entry is required, Pumper 16 is dispatched for decontamination. When Level A entry is required, mutual aid usually is requested as well. Primary mutual aid for hazmat comes from Kansas City, MO, but there are other teams in the metropolitan area as well. If they require additional mutual aid they can call Overland Park, Olathe and the Civil Support Team from Topeka.

PPE, Equipment and Training

Level A chemical protective clothing includes Kappler Lifeguard Responder Suits, along with various gloves and boots. Level B protective clothing carried on the hazmat units also is made by Kappler. Respiratory protection carried on the hazmat units includes MSA G1–60-minute self-contained breathing apparatus (SCBA), cartridge respirators and powered air-purifying respirators (PAPRs). For WMD incidents a Rescue Taskforce Concept is utilized. Each company carries two vests and to helmets.

Monitoring and detection equipment includes the M9 and 561 kits, pH paper, Radiac meters, AIM 3,250 gas detectors, Draeger pumps and tubes, MSA Orion multi-gas meter, and HazCat chemical test kit. Decontamination equipment and tools, various absorbent pads, patching and plugging tools, and assorted other equipment and supplies common to hazmat units are carried on Hazmat 2 and 3.

Hazmat team members volunteer to be on the team. They receive training from Environmental Protection Agency (EPA) Region VII, local community colleges, and equipment manufacturers. Team members also train with other metropolitan departments. All other response personnel on the Kansas City, KS, department are trained to the hazmat operations level. Firefighters are trained to the Awareness and Operations Level. Operations personnel are trained to do decontamination. Primarily dry decon is utilized. Engine and Squad Companies carry Peatsorb in 5 gallon containers and any spill over 25 gallons requires the hazmat team response. The company officer can also make the decision to summon hazmat.

Kansas City, KS, also operates one of nine Metropolitan Medical Strike Team trailers that are strategically located throughout the metro area. Trailers are set up differently, depending on tactical objectives. The Kansas City team specializes in decontamination, Mark I Auto-Injector Kits, tents, water heater, showers, pools, and heaters for winter and misting fans for summer.

Hazardous Materials Exposures

Major transportation routes include Interstates 35, 70, 435, 635, and 670. U.S. Highways 24 and 40 also pass through Kansas City. State routes include K5, K7, and K32. Railroads that service Kansas City include Union Pacific, Burlington Northern Santa Fe, Southern Pacific, and KCT. However, responses to railroad related incidents have gone down.

Major hazardous materials fixed facilities include Procter & Gamble, Barsol Solvents, ConocoPhillips and Magellan pipelines, and Kansas University of Kansas medical laboratories. Common hazardous materials found in the city include bulk petroleum, chlorine, anhydrous ammonia, propane, and refrigerants. There has been an increase in response to incidents involving fentanyl.

Incidents

South West Boulevard Fire

It was a beautiful but very hot summer day in the Kansas City metropolitan area, sunny with temperatures in the 90s and a south wind of 13 mph. Before the day would end, five Kansas City, MO, firefighters and one civilian would die in an inferno of burning gasoline referred to by KMBC-TV reporter Charles Gray as "when all hell broke loose". Gray, always a strong supporter of the Kansas City, Fire Department, called it "one of the darkest days in modern history of Kansas City firefighting".

The fire started on a loading rack at the combination bulk plant and service station in Kansas City, KS, near the Kansas-Missouri state line.

On the initial alarm, Kansas City, KS, dispatched three pumpers, two ladder trucks, and two district chiefs (a fire apparatus equipped with a pump is referred to as a pumper in the Kansas City area).

Even before the fatal tank rupture, dozens of firefighters had been treated for heat exhaustion from the combination heat, humidity, and radiant heat created by the burning flammable liquids. Ambulances dispatched from across the city stood by in line in case they were needed to transport injured firefighters to a hospital.

The tanks began to fail at approximately 10 a.m., about 90 min after the fire started. Tank 4 was the last to fail and when it did, it moved 94 feet from its cradle into Southwest Boulevard through a 13 inch brick wall, spreading burning gasoline and flying bricks in its path. Firefighters with 2½ inch hose lines were just 74 feet from the tank when it ruptured, so their positions were over-run by the tank and burning gasoline that completely crossed Southwest Boulevard. Chief officers had ordered personnel back from the fire lines when Tank 4 began to roar like a jet engine. It was during their retreat that the tank failure occurred. The rear of Tank 4 failed, and the force coming out of the tank contributed to its forward movement. Two pumpers were destroyed and three damaged by the fire.

Following the failure of Tank 4, Kansas City, MO put out a call for six reserve companies and recalled one shift of firefighters. All available ambulances were requested from the metropolitan area, and station wagons were placed into service as make-shift ambulances. Area hospitals put their disaster plans into effect and prepared to receive injured firefighters, police officers, and civilians. Twenty-two firefighters were admitted to hospitals, five in critical condition, along with civilian "firefighter" Francis J. "Rocky" Toomes.

All five critically injured firefighters and Toomes died, the first at 2:45 p.m. the day of the fire and the last on August 24th. An additional 35 firefighters were treated at hospitals and released. Approximately 40 firefighters were given first-aid at the scene. All suffered from burns caused by contact with the burning gasoline from Tank 4. The cause of the tank ruptures, including the fatal rupture of Tank 4, was determined to be over pressurization of the tanks on fire because of inadequate venting of the tanks. Uninjured firefighters picked themselves up following the rupture of Tank 4 and continued to fight the fire with more determination than before. By 11 a.m., the fire had been extinguished.

It was the second-largest loss of life in Kansas City, MO Fire Department history. The fire changed the way flammable liquids were stored at automotive service stations and how flammable-liquid fires were fought involving horizontal storage tanks. National Fire Protection Association (NFPA) codes were changed to require flammable-liquid storage tanks at automotive service stations to be placed underground following the Southwest Boulevard fire.

Magellan Distribution Terminal

One of the city's major hazmat responses occurred at the Magellan distribution terminal at 401 Donovan Road on June 6, 2008. Lightning struck a 1.2 million-gallon unleaded gas storage tank, setting it on fire (Figure 5.66). Hazmat units conducted air monitoring at 60 locations throughout the metro area affected by smoke from the fire. Air monitoring was conducted in partnership with other teams from around the Kansas City metropolitan area, including EPA Region 7, using Area RAE detection equipment, as the incident affected a large portion of the metro area on both the Missouri and Kansas sides of the state line. Fire units decided to let the fire in the tank burn itself out while protecting surrounding tanks and other exposures. The fire was contained to the original tank and a berm area designed to contain leaking fuel from the tank. The facility also contained tanks of jet fuel, ethanol, diesel, and other fuel oils that were not involved in the fire (*Firehouse Magazine*).

Figure 5.66 One of the city's major hazmat responses occurred at the Magellan distribution terminal at 401 Donovan Road on June 6, 2008. Lightning struck a 1.2 million-gallon unleaded gas storage tank, setting it on fire. (Courtesy: Chris Phillips.)

Kansas City, Missouri Hazmat Team

Kansas City, Missouri is the largest city in the State of Missouri and is located in the Northwest corner of the state at the confluence of the Missouri and Kansas rivers. The city encompasses 319 miles^2 with parts located in four counties. Population estimates for 2020 include 505,102 within the city and over 2 million in the Kansas City, Missouri and Kansas City, Kansas Metropolitan Area.

Fire Department History

Kansas City's early fire department was organized around 1858 with the formation of volunteer bucket brigades. Church bells would signal a fire alarm and everyone would assemble at the scene of the fire to help. In 1867, the city abandoned the volunteer bucket brigade system and replaced it with a fully paid fire department. Initial firefighting equipment included a Silsby Rotary Engine with hose and two wheeled hose wagons. Col. Frank Foster was elected as the first paid fire chief. Late in 1869, the first ladder company was organized, and named McGee Hook and Ladder #1, in honor of ex-mayor E. Milton McGee. By 1872, the department consisted of three steamers, one hook and ladder, one chemical engine, and thirty-six paid firefighters. By the 1920s, the fire department had grown to 30 stations, with 40 companies. In 1928, the first training school was opened, and the department was fully motorized. The Kansas City Missouri Fire Department celebrated its 150th Birthday on March 14, 2018.

Today's Modern Department

Today the Kansas City, Missouri Fire Department operates from 34 fire stations with 1,200 uniformed personnel under the leadership of New Chief Donna Maize hired March 14, 2018. Chief Maize is the first woman to lead the Kansas City Fire Department. Kansas City Fire Department refers to its apparatus with a pump as a "Pumper". There are 34 stations housing 34 pumpers, 13 trucks, 3 rescue units, 54 medic units, 1 hazmat, 2 foam pumpers, and 3 air boats. The city is divided into 7 battalions, and each battalion chief has their own driver. Each battalion driver is a captain and functions as the safety officer on incident scenes. Kansas City Fire Department has taken over EMS service for the city which was previously a separate city agency. Firefighters trained to the basic EMT level respond as first responders to medical calls. Fire protection for Kansas City International Airport is provided by the Kansas City Fire Department from Station 5 located at the airport. Currently, the Airport Division is operates 4 frontline apparatus with a staff of 26 personnel including an administrative battalion chief, a division training officer, and 24 uniformed personnel working on three 24-hour shifts. Firefighters in Kansas City are on duty for 24 hours, off for 48 hours and get an extra day off following each on duty shifts.

Hazmat Team History

On September 24, 1989, the Kansas City Hazardous Materials Team was placed in service. The team was formed following an explosion at a construction site in 1988 that killed 6 Kansas City Firefighters. Pumper companies 30 and 41 along with their personnel were lost during the explosion. Pumper company numbers 30 and 41 were added together to form the number for Hazmat Unit 71 which was originally located at Station 47. The first hazmat vehicles included an E-One 1989 Pumper (P71), a 1980s model Dr. Pepper roll-up delivery truck donated to the fire department (S71), and a pick-up truck (U-71). The Dr. Pepper truck was replaced by a 1993 Super-Vac Hazmat Response Vehicle, the first hazmat vehicle built by Super-Vac. During 2007, a motor vehicle accident (MVA) placed the Super-Vac out of service and a reserve rescue unit was used until the new HM71 arrived. Initially, the hazmat team was not dedicated; personnel from Station 47 manned the hazmat vehicles when a hazmat call came in.

Hazmat Team

Currently, HM71 has 4 dedicated personnel assigned to it each shift. An additional 4 technicians are assigned to Pumper 27. The crews of HM 71 and Pumper 27 change units each month to give them experience on each unit. Foam 71 and Utility 71 are unmanned unless a call for them comes in. When that happens, HM 71 and Pumper 27 crew members place them in service. All rescue companies in the city are trained to the Hazardous Materials Operations level. All other firefighters in the city are trained to the Hazardous Materials Awareness level. Station 27 was built in 1911 and housed a pumper until the late 1980s. A squad replaced the pumper until the early 1990s when the station was closed. In the mid-1990s, the original Station 27 was torn down to make way for a new dedicated hazmat station.

The new Station 27 located at 6600 Truman Road, was constructed in 1996 and houses Foam Pumper 71, a 2007 Pierce with a 2,000 gpm pump, 1,000 gallon water tank, and a 50 gallon foam tank and an additional 35 gallons carried in containers on top; Hazmat 71, 2017 Pierce Custom with an internal walk-thru, filtered fume hood, elevated camera, and lights (Figure 5.67); Utility 71, a 1997 GMC with a Hackney body; and Pumper 27, a 2016 Pierce, 1,500 gpm pump. Pumper 71 responds to fire, EMS, MVA, and general non-emergency calls in addition to hazmat runs. Utility 71 is a secondary hazmat unit that carries supplies needed for extended hazmat incidents. Car 301 is the Hazmat Division Chief, and Car 310 the Hazmat Chemist.

Additional hazmat equipment deployed around the Kansas City Metropolitan area includes six identically equipped Metropolitan Medical Response System (MMRS) Mass Casualty Trailers and a

Figure 5.67 Hazmat 71, 2017 Pierce Custom with an internal walk-through, filtered fume hood, elevated camera and lights.

decontamination trailer. The MMRS trailer contains 100 Level A suits, 500 chemical agent antidote kits, 500 redress suits, communications equipment, and equipment for inclement weather.

Kansas City, Missouri's Fire Department responds to approximately 1,900 Hazmat calls per year including fuel spills, odors, leaks, and carbon monoxide (CO) alarms. All pumper companies in the city carry a bag of absorbent material and 5 gallons of foam to handle small fuel spills. Truck companies carry 4 gas meters. Personnel from Pumper 27 respond with Hazmat 71 to make up the initial response. There are 33 total hazmat technicians in the Kansas City Fire Department, all assigned to Station 27. Each pumper company is staffed with 4 firefighters, each truck 4 firefighters and each rescue 6 firefighters. If the Kansas City Fire Department requires assistance through mutual aid there are 8 regional hazmat teams available in the Kansas City Metropolitan Area. Additionally, the National Guard 7th Civil Support Team from Fort Leonard Wood, Missouri and the 73rd Civil Support Team from Topeka, Kansas can provide assistance for hazmat and terrorist incidents.

PPE, Equipment and Training

Personal protective equipment for the hazmat team consists of Kappler and DuPont for Level A and Kappler CPF-2, CPE-4 for Level B. Respiratory protection is provided with Scott 4500 with 1 hour bottles for hazmat and 30 minute bottles for firefighting purposes. In-suit communications are supplied with chest mounted radio harnesses and the Motorola HT 1,000 radios with the Cav Com system, which works with a bone mic that is placed in the ear. Kansas City Hazmat Personnel have found these radio units work better than typical bone mikes. Technicians are provided with an 80 hour Operations class that includes hazmat chemistry and the State of Missouri 40 hour Technician Course along with an 8 hour recertification annually. An additional 40–60 hours of refresher training is held each year for team members. Technical Rescue personnel are trained to the operations level and conduct decon operations when necessary.

Reference Resources

Resource materials are primarily electronic through computers or cell phones.

Hazardous Materials Exposures

Hazardous materials transportation exposures in Kansas City include the number 2 railroad hub in the United States with rail yards that run through the city in some places 46 tracks wide. Railroads that serve the city include Burlington Santa Fe, Union Pacific, and Kansas City Southern. Highway exposures occur with Interstates 35, 70, and 435 which run through the heart of the downtown section of the city. State Highways 24, 40, 69, 71, and 169 are also hazardous materials exposures. Major pipeline systems cross the city in several locations. Fixed facilities with major hazardous materials exposure include Bayer Chemical, Honeywell, MRI, and hazardous waste facilities. Major chemicals found in the city include propane, chlorine, anhydrous ammonia, pesticides, and nerve agents.

Incidents

Ammonium Nitrate Explosion

Several major hazmat incidents have occurred over the years in Kansas City. On November 29, 1988, explosions at a construction site killed six firefighters instantly as they fought fires in a trailer/magazine containing blasting agents (Figure 5.68). The fire also involved two vehicles and a second trailer/magazine, which exploded as well. The trailers were loaded with explosives. The blast left two large craters in the ground and destroyed two fire department pumpers. The force of the explosions, about half an hour apart, was so great that people were jarred from their beds as far as 10 miles away. Many windows in surrounding neighborhoods were broken. A seventh firefighter, arriving on the scene as a driver for a battalion chief, was injured when one of the explosions blew his Fire Department car 25 feet and shattered its windows. The fires at the construction site were believed to have been set by an arsonist. This incident

Figure 5.68 On November 29, 1988, explosions at a construction site killed six firefighters instantly as they fought fires in a trailer/magazine containing blasting agents.

was the worst loss of life fire/hazmat incident in terms of firefighters in Kansas City, Missouri history.

ChemCentral Company

February 8, 2007, a fire and series of explosions occurred at the ChemCentral Company in Kansas City. Firefighters reported that the building was full of 55-gallon barrels and that smaller explosions occurred up to two hours after the initial blast, which began at about 2:20 p.m. local time. Fire Chief Richard Dyer indicated the blaze started when employees were off-loading a petroleum-based product. Several 55-gallon drums ignited, touching off the explosions that injured two workers inside the plant. The ChemCentral Corporation is a chemical distribution facility that blends and repackages chemicals such as waxes, resins, solvents, acids, and other chemicals. The explosion generated flames approximately 100 feet in the air and covered a section of downtown Kansas City in thick black smoke. A company official previously said the chemical may have been polybutene, which is used in a variety of products, including liners for cereal boxes. Fire fighters worked from a defensive mode because of the chemical fire and chances of additional explosions. The fire burned for approximately 9 h before firefighters used foam to bring it under control. Police evacuated homes and businesses within 1 mile of the plant. Approximately 500 people and 6 schools were evacuated from the area. The fire destroyed the 5-acre plant, with losses estimated at $1 million (*Firehouse Magazine*).

Kingman, AZ Hazmat Response

Kingman is located in Northwest Arizona 26 miles east of the Nevada border and about 98 miles from Las Vegas. It is the county seat of Mohave County with a estimated population of 28,068 and an area of 34.82 miles^2. Kingman is located on Historic Route 66 also known as Andy Devine Boulevard, a native of Kingman.

Fire Department History

Kingman's Fire Department was formed after World War I by a group of returning soldiers even before the town was incorporated. American Legion Post #14 was established with a primary goal of creating fire protection for the community. Approved by the Board of County Commissioners at a meeting held on September 22, 1921, planning for the fire department began. The Kingman Fire Protection District was formed and 25 men chosen and elected to become members of the department. The first meeting was held on October 6, 1921, and an operational budget of $10,000 was established. A new 1922 American La France pumper and eleven fire hydrants were ordered. A fire station was built at the corner of 5th and Beale Street. The building and the American La France Pumper are still in the community today as a reminder of the fire departments beginnings.

Today's Modern Department

Today's modern Kingman Fire Department is part of municipal government and provides fire protection within the city limits. Under the leadership of Chief Jake Rhoades 45 full time and 13 part time personnel staff Kingman's Fire Department. Additionally, there are 2 civilian support and 3 hydrant maintenance personnel as well. The stations are staffed with three shifts (A, B, C) 15 firefighters per shift who work 48 hours on and 4 days off. Each shift is managed by a battalion chief. Firefighters operate from 5 stations with 4 Engines, 1 Truck, 1 Rescue, 1 Brush Truck and the Hazmat Unit. Kingman FD does not provide EMS transport. Each engine has a paramedic and provides ALS first response. AMR (River Medical) does ALS transport. Station 1 houses Engine 211, Rescue 215, Brush 246, and Battalion 2. Station 2 houses Engine 221 and Squad 2. Station 3 houses Engine 231, Ladder 234, and Reserve Engine 206. Station 4 houses Engine 234 and Reserve Engine 242. During 2017, Kingman Fire Department responded to 7,511 calls for service. EMS composed the largest number of calls at 6,114. There were 1,313 fire calls. Special Operations which includes Hazmat, Rope, and Water Rescue made 84 calls for service.

Hazmat Team History

Kingman's Hazmat Team was established in the 1980s. Hazardous materials known at fixed facilities and transported through the community. Kingman formed a hazmat team to deal with any incidents that might arise from those exposures. In the beginning, personnel were trained to the operations level. Later as needed, members of the team completed the Arizona Department of Emergency Management Technician program. They receive 40 hours of refresher training each year to help maintain their skills and knowledge. Additional equipment was purchased as the community grew and situations dictated in order to provide response to the hazards in Kingman.

Hazmat Team

Station 22, Kingman's 5th Station is located at 1605 Harrison Street houses the hazmat team, apparatus and equipment (Figure 5.69). The hazmat trailer is towed by a 2011 Dodge Ram 2500. Total annual hazmat call volume includes fuel spills, gas odors, and leaks. Engine companies carry absorbent materials. Any spill greater than 55 gallons triggers response of the hazmat team. Eight hazmat technicians usually are on duty each shift. The hazmat team is not dedicated. The department has 24 technicians

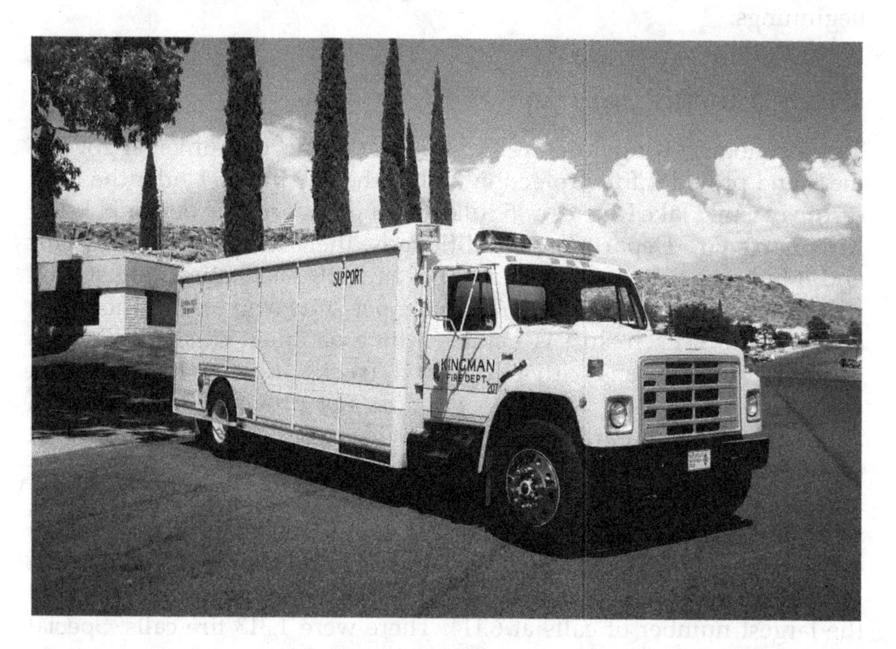

Figure 5.69 Station 22, Kingman's 5th Station is located at 1605 Harrison Street houses the hazmat team, apparatus, and equipment.

available. If mutual aid is required Lake Havasu Fire Department, National Guard Civil Support Team, Clark County NV Fire Department, Las Vegas NV Fire & Rescue, and Henderson NV Fire Department are available.

PPE, Equipment and Training

Levels A, B, and C are carried in the hazmat trailer. In-suit communication is provided by Motorola XTS 2,500 radios (Intrinsically Safe).

Monitoring and Detection Instruments

- Haz Mat ID 360
- Identi Finder 2.0
- Ludlum Radiological
- Haz Kat Kit, 3 Multi Rae Pros
- Draeger tubes and CMS Chips
- Mercury Meter
- Various monitoring papers

Reference Resources

The hard copy, computer, and Internet are available for product research. CAMEO, ALOHA computer resources. ERG, NIOSH Triage Desk, and Smiths Detection Reach Back.

Hazardous Materials Exposures

Burlington Northern Santa Fe Railroad (BNSF) and I 40 are major transportation routes for hazardous materials through the city and county. BNSF transports over 150 million gross tons annually through the area. The BNSF mainline carries approximately 120 trains daily, with 90% being intermodal containers. I 40 passes through the center of Kingman. Arizona Department of Transportation estimates 33,000 vehicles utilize the route each day. Many of those vehicles carry hazardous materials. There are two Tier II facilities in Kingman with various hazardous materials.

Incidents

Propane Rail Car Explosion

Kingman, Arizona with a population of 7,500 in 1973, is a desert community located approximately 80 miles southeast of Las Vegas, Nevada, 184 miles northwest of Phoenix and 147 miles west of Flagstaff. At the time of the incident, the Kingman Fire Department was a part career and part volunteer force operating out of two fire stations. There were 6 career firefighters with one on duty in each station at all times and 36 volunteers. Kingman's equipment in service at the time of the explosion included 4 engines and 1 rescue vehicle. Station #2 was located just a half mile west of the site of the explosion.

On July 5, 1973, a propane tank car being off-loaded in Kingman, Arizona, caught fire, which resulted in a Boiling Liquid Expanding Vapor Explosion (BLEVE), that killed 11 Kingman firefighters and 1 civilian (Figure 5.70). Another 95 persons were injured by the blast, and over $1,000,000 in property damage occurred to surrounding exposures. Except for one career firefighter/engineer who was severely burned but survived, those injured were mostly spectators that had gathered along Historic U.S. Highway 66 to watch the incident. Most of those injured were approximately 1,000 feet from the explosion and ignored police warnings to stay back. Photographs of the spectacular BLEVE incident have appeared in countless articles, books, and training programs over the years. Instructors have often referenced the Kingman incident when warning emergency responders of the dangers of flame impingement on the vapor space of propane tanks.

Figure 5.70 On July 5, 1973, a propane tank car being off-loaded in Kingman, Arizona, caught fire, which resulted in a Boiling Liquid Expanding Vapor Explosion (BLEVE), that killed 11 Kingman firefighters and 1 civilian. (Courtesy: Hank Graham.)

Yet as recently as September 8, 2018, Firefighters in Tilford, S.D. found a single-family residence fully engulfed in flames. Initial attempts to fight the fire were hampered by the neighborhood's narrow gravel streets, congestion of nearby homes, sheds and garages, parked vehicles, and overgrowth of summer grasses and underbrush. There was a large amount of combustible fuel for the fire. Downed utility lines and a number of propane tanks also threatened, as firefighters attempted to evacuate other homes and search for a reported missing person. A 500-gallon propane tank located on the south side of the burning home exploded in a massive ball of flame when a "boiling liquid, expanding vapor explosion", or BLEVE occurred. Fischer was attempting to move a Sturgis VFD command vehicle, a Chevrolet Suburban, parked in a driveway north of the burning home. A large portion of the exploding tank cleared the burning home, a fire engine and the Suburban and struck Fischer, killing him instantly. The tank embedded in the walls of the well house. Another part of the tank was found about 350 yards to the south, near Interstate 90.

"He never knew (what hit him)," said Sturgis Fire Chief Shawn Barrows, appearing in dress uniform at Saturday's news conference, his badge wrapped in black in honor of Fischer, who served as assistant fire chief and was a 22-year veteran of the department. Fischer was also an active member of the Sturgis Ambulance Service and a member of the 82nd Civil Support team with the South Dakota Army National Guard. After Fischer's death, Sturgis firefighters continued to battle the blaze until the fire was essentially out, Barrows said. Once relieved by another department, they escorted the body of their fallen comrade to a Sturgis funeral home. "Through the process, we stayed with Dave", Barrows said. "We will have a last watch, at least one person with Dave, until he is laid to rest".

Louisville, Kentucky Hazmat Team

Louisville, KY is located in the northern part of the state on the Ohio River and bordering the State of Indiana. It is the largest city in the Commonwealth of Kentucky. The city was founded in 1778 making it one of the oldest west of the Appalachian Mountains. Louisville is home of legendary boxer Muhammad Ali, the Kentucky Derby, Kentucky Fried Chicken, Louisville Slugger baseball bats, and the University of Louisville. It is also home to the Son's of The American Revolution and the Honorable Order of Kentucky Colonel's. Louisville and Jefferson County merged in 2003 and its borders have been the same as the former Jefferson County since that merger. The official name of the consolidated city and county government is the Louisville/Jefferson County Metro Government and is abbreviated to Louisville Metro. Its population is estimated to be 624,890 in 2020, covering an area of 1,924 $miles^2$.

Fire Department History

Louisville Fire Department is the third oldest all-paid fire department in the United States. In 1798, the Kentucky Legislature granted townspeople the right to form their own fire companies, which Louisville quickly organized. Five companies were originally formed, each consisting of 40 men. The number of units would eventually grow to 11, before the disbanding of volunteer units in 1858. On the evening of May 27, 1858, the General Council organized the Steam Engine Fire Department of Louisville, to be effective June 1, 1858. The Division of Fire consisted of three fire stations: #1 at Preston and Jefferson; #2 on Jefferson Street between Sixth and Seventh; and #3 on Main above Shelby Street.

A.Y. Johnson (a member of the mechanics company) was appointed Louisville's First Fire chief, and his responsibility, with the aid of 65 men, 23 horses, and 5 newly purchased steam engines, was to provide fire protection for the 70,000 inhabitants of the city. The first official fire run was on July 2, 1858 to the home of a Mr. Waters, on Campbell Street between Main and Creek Street. The fire damage was estimated at $500.00 and the cause was incendiary. The newly formed Fire Department answered 6 alarms the first month and 2 of them were false. The first "General Alarm" fire occurred on July 1, 1864. It involved ten buildings on the north side of Main Street, between Eighth and Ninth Streets. The loss of the first was a total of $1,197,800.00. The largest of the one-owner buildings was the Federal Government and their loss was $800,000.00.

The 20th century was ushered in with a roar of a motorized fire engine, although horse-drawn engines remained a common sight as well. In 1907, the division obtained its first motor apparatus, a gasoline electric vehicle for Truck Company #1. While horse-drawn engines remained in the majority until 1925, it was nevertheless becoming obvious that very soon the days would disappear when citizens, having heard a fire alarm, could dash to the nearest engine house, catch hold of a rope attached to the engine, and run in front of the engine to the scene of the fire. The first engine fully operated by gasoline was purchased and put into service in November of 1917.

In the late 1950s, mobile radios were installed on the fire apparatus. Prior to this, once a Company was dispatched on a fire run, their only contact with the Communications Bureau was by a telegraph key inside the fire alarm box or by telephone. The radios gave the Department the flexibility for the Companies to move about their fire run districts for familiarization, driving practice, and inspections. Also, it allowed for multiple company training. But the greatest advantage was constant contact with the Communications Bureau and the availability of the Companies to respond to fire. A new Training Center was constructed in 1957. It consisted of a five story brick building, with a classroom on the first floor, a model of a wet and dry sprinkler system and a concrete pad around the outside of the building. In 1973 the Insurance Services Office evaluation gave the Fire Department a Class 1 Rating. This rating was held by only 5 other Fire Departments in the country.

Today's Modern Department

Under the leadership of Fire Chief Gregory Frederick, the Louisville Fire Department has 409 uniformed personnel operating out of 21 fire stations. Louisville fire personnel operate 16 engine companies, 16 regular, and 3 chemical with 250 gallons of foam, 7 truck companies, Quints (3), Aerials (5), Tower Ladder, 14 medic units, 3 hazardous materials companies, and 2 technical rescue companies. Louisville does not provide EMS transport. All personnel are trained as EMTs. Rescue companies provide Swift Water Rescue, Dive Rescue, Rope Rescue, Confined Space Rescue, Trench Rescue, and Structural Collapse Rescue. Louisville does not provide paramedic engine companies, however, there are some firefighter/paramedics assigned to selected suppression companies.

Scalar Rank Structure:
- Fire Chief (Colonel)
- Fire Executive Assistant Chief (Colonel)
- Fire Assistant Chief (Lieutenant Colonel)
- Fire District or Battalion Chief (Major)
- Fire Company Commander (Captain)
- Fire Apparatus Operator (Sergeant)

Typical company staffing includes four personnel, 1 Captain, 1 Sergeant and 2 Firefighters. The city is divided into four Battalions or Districts, each with a chief.

Louisville responded to over 42,000 calls for service during 2018. Of total responses 21,460 were EMS related and 18,984 were fire related. Louisville is surrounded by 14 fire protection districts and provides automatic aid to the Shively FPD.

Hazmat Team

Hazmat 1 responded to 1,451 calls in 2019 most of which were fuel spills and natural gas leaks. Incidents are calculated on a per call basis and some incidents involved multiple hazmat company response. Statistics for hazmat responses within Louisville do not include local engine runs for hydrocarbon fuel spills. Each engine carries approximately 5 lb of Hazorb material for cleaning up small fuel spills without calling a hazmat company. Larger spills would require a hazmat company response. Louisville relies on Jefferson County Fire Departments Hazmat Team for mutual aid when needed.

Louisville has a primary Hazmat Company which is dedicated Hazmat 1 and Engine 1. Engine/Hazmat 19 and Engine/Hazmat 21 are secondary companies. Secondary companies are utilized as support for the primary hazmat company. They are trained and equipped for detection and mitigation of incidents. Secondary hazardous materials companies are equipped with matching 2,002 Seagraves Pumpers with on board foam storage and foam pumping capabilities. Engines 19 and 21 carry 25 gallons of Class A and 300 gallons of Class B foam in on board tanks. Two hazmat units both are 2006 Pierce Lance Hazmat Trucks. Hazmat 1 is a 1995 Marque/GMC Hazmat Truck with roll up compartment doors and a command area carry the hazardous materials equipment (Figure 5.71). It is located at 1100 Grade Lane. Engine/Hazmat 19 are located at 3401 Bohne. Engine/Hazmat 21 is located at 300N. Spring Street. A boom trailer is located at Truck2/Engine5.

PPE, Equipment and Training

Personnel Protective Equipment used by Louisville for Level A is the Kappler Tychem Responder with Flash protection and for Level B Tyvec. Respiratory Protection is provided with Scott 30 minute Self-Contained

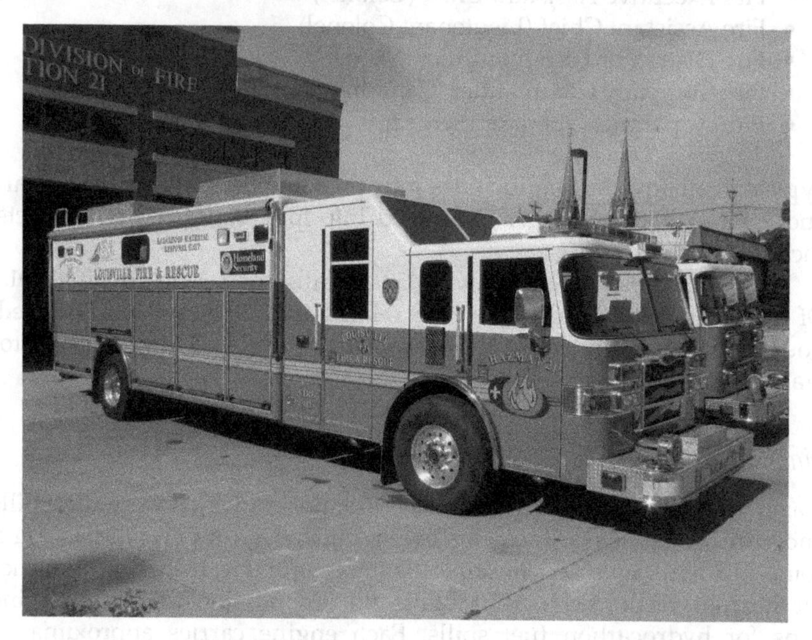

Figure 5.71 Hazmat 1 is a 1995 Marque/GMC Hazmat Truck with roll up compartment doors and a command area carry the hazardous materials equipment.

Breathing Apparatus (SCBA) for suppression companies and 60 minute bottles for the hazardous materials companies. In suit communication is provided by Scott's voice communications, part of the Scott 60 minute SCBA's used by the hazmat team. Monitoring instruments and identification equipment used by the Louisville Hazmat Team includes 290 Victoreen Geiger-Muller Counter, APD-2000 detects chemical warfare agents, pepper spray, and mace, pH Paper, Phd Biosystem, monitors for oxygen, carbon monoxide, LEL, and hydrogen sulfide, MultiRAE Plus combines a PID (Photo Ionization Detector) with the standard four gases of a confined space monitor (O_2, LEL, and two toxic gas sensors) in one compact monitor with sampling pump and Draeger tubes.

Research Resources that are computer based include CAMEO (Computer Aided Management of Emergency Operations) and various other hard copy reference books are available in the command section of the hazardous materials units.

Hazardous Materials Exposures

Transportation exposures include interstate highways, railroad, pipelines, and barges on the Ohio River. Louisville's bounded by Interstate highways 64, 65, and 71 which all pass through the center of the city. The Norfolk and Southern Railroad has a major hub in Louisville. Barge traffic passes the city on the North side on the Ohio River. Fixed facilities in Louisville include the United Parcel Service (UPS) Center, refineries, chemical companies such as DuPont and Rohm & Hass, the Navel Ordinance Facility, and others. Known hazardous materials in the city include cyanide, plastics, petroleum products, Oleum (fuming sulfuric acid), paint, varnish, anhydrous ammonia, and liquid hydrogen. Louisville has available a mass decontamination trailer with two tents, water heater, generator, and built in showers. They also have a negative pressure fan for the tent to keep vapors and other materials out of the tent during decontamination (*Firehouse Magazine*).

Martin County, Florida: First Volunteer Hazmat Team in United States

Martin County is located along Florida's Treasure Coast Region North of Palm Springs. It is the fifth largest county in Florida covering an area of 753 miles^2 (1,950) of which 543 miles^2 is land composed of a mix of suburban, agricultural/rural lands, and waterfront. The fire department covers all of Martin County, except for the Town of Stuart which is the county seat. Estimated population was 163,152 in 2020. During the winter season (November–April), the population can increase to over 200,000.

Fire Department History

District-2 (Station 18) is the original Martin County Fire Department. During the 1970s, there were 12 volunteer fire companies, 1 city department that was career and part volunteer. The Grumman Fire Department at the airport. The county career fire department began in 1976 with the hiring of the first career firefighter assigned to Station 21 in Palm City. Additionally, there were 4 volunteer BLS ambulances and 1 career ambulance operated by the City of Stuart Police Department. Chief Wolfberg was hired in 1981 to organize and place in service a career community wide ALS Ambulance service. One year later the ALS service was operational. During this transition the county took over ambulance response in the City of Stuart. One of the volunteer ambulances closed and many members began a new career with the county EMS service.

Around 1981/1982 a career fire marshal was added to the department. Within the next 2 years the job transitioned into Chief of Operations/Fire Marshal for the entire county. His responsibilities included oversight of all volunteer fire departments in Martin County. In 1994 the Martin County Fire Department and Martin County EMS merged and became today's Martin County Fire-Rescue.

Today's Modern Department

The Fire Rescue Department consists of three divisions: Operations, which oversees the Rescue Bureau and includes the Air Medical Transport Unit and Ocean Rescue lifeguards; the Division of Emergency Management, which oversees Radiological, Emergency Planning and Special Needs and 9-1-1 Communications; and the Administration Division, which oversees the Fire Prevention Bureau, the Training and Safety Bureau as well as the Fleet Services Section and maintenance of all Fire Rescue facilities.

Martin Counties Fire Department operates with 352 employees under the leadership of Chief William Schobel. He entered the fire service as a volunteer in Pinellas County, Florida in 1984 and became a fulltime firefighter later that same year. He spent the next 26 years ascending through the ranks, while taking some time off to serve his country in the US Army during Desert Storm. In 2010, he relocated to Martin County, accepting the position of Bureau Chief of Training & Safety for Martin County Fire Rescue.

Schobel was named Interim Fire Chief in May 2016 and in October 2016 the Martin County Board of County Commissioners appointed Bill Schobel to Chief of the Fire Rescue Department. Martin County Fire & Rescue Department operates out of 11 Fire-Rescue Stations strategically located throughout the county. Apparatus consists of 8 Engines, 3 Trucks, 14 Medic Units, 1 Air Cascade Unit, 3 Hazardous Materials Units,

1 Decontamination Unit, 1 Medical Helicopter, and 7 Wildland Brush Units. Special teams include a Dive Team and Technical Rescue that combines rope rescue, confined space rescue, trench rescue, and structural collapse. Martin County Fire Rescue protects 528 miles^2 of a mix of suburban, agricultural/rural lands, and waterfront with a population of approximately 140,000. During the winter season (November through April), Martin County's population can swell to nearly 200,000.

Responses in 2018:
- 18,013 calls were medical emergencies
- 13,247 patients were transported to area hospitals
- 4,283 calls were fire related

Special Teams Responses:
- LifeStar - EVAC Helicopter = 210
- Hazmat Team – Chemical or Biological specialized response = 118
- Special Response Team – Technical Rescue Operation Responses = 103

Life Guard Service

Martin County Ocean Rescue provides protection and response to emergencies at Martin County's four guarded beaches. Beaches are staffed 365 days a year with EMT/Ocean Lifeguards. Beaches guarded include: Bathtub Beach; Hobe Sound Beach; Jensen Beach and Stuart Beach.

Hazmat Team History

Chief Edward B. Smith became chief of the Martin County Fire Department in 1976. Trade magazines were routinely publishing stories of hazmat incidents occurring across the United States. Outside his office window trains traveled through Martin County with tank cars carrying hazardous materials. Hazardous materials response was a new and growing addition to the fire service. Chief Smith, recognizing the potential in his community for chemical incidents began preparing himself and his firefighters for hazmat emergencies taking classes and learning as much as they could. Also in 1976, young captain Ron Gore of Jacksonville FD was tasked with starting the first emergency services hazmat team in the United States.

Captain Gore became the "Godfather" of hazardous materials response. Captain Gore also created his own company Safety Systems, Inc. to train emergency responders not only in Florida, but across the country. His company was one of the first, if not the first to provide hazmat training for emergency response personnel. Chief Smith attended a Safety Systems Class and became friends with Mr. Gore and a partnership developed between Safety Systems and the District-2 Fire Department. District-2 would routinely host Safety Systems classes and in turn District-2 members could also attend the classes.

On December 6, 1978, District-2 Fire Department placed their hazardous materials team in service specifically for responses to Hazardous Materials Incidents. Their response vehicle had been donated by Chief John Caserta of the Grumman Fire Department located at Withham Field in Stuart, who was also a volunteer firefighter with District-2 and formally the City of Stuart (Figure 5.72).

Hazmat Team

Stations 16 and 18 are home for the hazmat team and equipment. Station 16 is located at 2710 NE Savannah Road in Jensen Beach and houses Hazmat Engine 16, and Hazmat 16. Station 18 located at 1995 NW Britt Road, Stuart, FL houses Hazmat Engine 18, Hazmat Unit 18, and the Decontamination Trailer. Each hazmat station is not dedicated and has a minimum staffing of 4, they also run suppression and rescue and hazmat as needed. Hazmat response consists of the Duty Station, Hazmat Unit, and Squad which has an air cascade unit on board. Hazmat 16 handles all odor, gas leaks, and CO type calls.

No monitoring instruments are located on any other companies. On average, Hazmat responds to around 100 hazardous materials calls

Figure **5.72** Their first response vehicle had been donated by Chief John Caserta of the Grumman Fire Department. (Courtesy: Martin County FL Fire Department.)

each year. Statistics do not include engine runs for hydrocarbon fuel spills. Engine companies carry 5 lb of Sphag-sorb for cleaning up small fuel spills without calling Hazmat. Larger spills would require calling out a Hazmat Company. Hazmat 16 is a 2002 Freightliner tractor pulling a 30 ft Hackney trailer with rollup doors and a command center (Figure 5.73). The unit has an onboard generator, light tower, 200 gallons of Class B foam in an on board tank. Within the command center are located computers, printer, fax, and computer and hard copy reference materials for product identification.

Other equipment carried includes decontamination; PPE; respiratory protection; Chlorine Kits A, B, and C; patching and plugging; LP flare off; dome clamps; grounding and bonding; and miscellaneous tools. MCFD also has a mass decontamination trailer with two tents, water heater, generator, and built in showers. Also carried are Pt. IDECON kits, additional PAPR's, WMD filters, HEPA vacs, and Tyvek suits. Law enforcement and EMS do not carry hazmat or WMD PPE. In the event MCFD needs additional assistance, they can call Palm Beach and Lt. Lucie County for assistance. A & B foams are available from the airport and Palm Beach.

PPE, Equipment, and Training

Hazmat PPE is composed of Trellchem HPS, VPS for Level A, Tychem for Level B, and DuPont Tyvec for Level C. Respiratory protection used is

Figure 5.73 Hazmat 16 is a 2002 Freightliner tractor pulling a 30 ft Hackney trailer with rollup doors and a command center.

Scott with 60 min. bottles, Scott & 3M PAPR's, and cartridge respirators. In-suit communication is provided by Scott Voice communications.

Monitoring and Detection Capabilities For Product Identification

- **Ludlum Personnel Radiation Monitor.** Displays in mR/h, uSv/h, dpm, Bq, cpm, or cps. Dead-Time Correct (DTC) allows Gamma Measurements up to 500 mR/h or up to 1,999 uSv/h. Count rate, Dose/Exposure, and Counting Alarms. Bright Red Flashing Alarm LED.
- **Multi-RAE Lite.** Monitors for LEL, CO, H_2S, VOC, and Chlorine. The wireless connection sends threat and alarm data in real time to a central command, providing awareness for fast incident response.
- **MSA Sirius** Multigas Detector. LEL, Co, H2S, VOC, Chlorine, and Methane.
- **APD 2000** Hand-held monitor and detector designed to emergency first responders to hazardous chemical releases and Chemical Warfare Agents (CWA). Simultaneously detects nerve and blister agents and recognizes pepper spray and mace.
- **Gas ID System** Identifies unknown gases and vapors in minutes. Each chemical has a unique infrared fingerprint. It measures how gases and vapors interact with infrared light and can quickly tell if a gas or chemical is present or not. Analysis takes 10 min and compares the results against database libraries that include nerve and blister agents.

Terrorist ID

First Defender; True Defender; M-8 Paper; M-9 Paper; M-256A Kits; HAZMATCAD; RAMP-ricin, anthrax, small pox botulinum;

Corrosive

pH paper and Sulfur sticks.

Qualitative Detector Tubes

Draeger Tubes

There are a total of 30 technicians on the team. All have received the State of Florida technician certification. Members go through 160 hours of training and pass a test at the end. Forty hours of continuing education is required each year, in addition to in-service training and off-duty training. Personnel are sent to National Fire Academy training and the LEPC also funds additional training.

Reference Resources

Computer Programs: CAMEO; Marplot/Aloha; Wiser. Hard copy reference books: *ACGIH Guide of Occupational Exposures; Chem-Bio Handbook; CHRIS (Coast Guard Manual); Emergency Care for Hazardous Materials Exposure;*

Emergency Response Guide Book; Fire Protection Guide to Hazardous Materials; Guidelines for Selection of Chemical Protective Clothing; Hawley's Condensed Chemical Dictionary; Hazardous Materials Desk Reference; Hazardous Materials Response Handbook; Medical Management of Biological Casualties; Merck Index and NIOSH Pocket Guide.

Hazardous Materials Exposures

Transportation hazmat exposures are two railroads, CSX and FEC, the Atlantic Ocean Intracoastal Waterway and Lake Okeechobee, and two natural gas pipelines. Major highways in Martin County are Interstate 95; SR 710, U.S. 1 and the Florida Turnpike. Martin County also has fixed hazardous materials facilities. They are within the EPZ for Hutchison Island Nuclear Power Plant. Spent fuel rods are stored on site. At times they are moved off site by train and they notify the fire department. Others include an ice plant that utilizes anhydrous ammonia, propane gas stations, commercial fishing industry, and explosives for building demos. Chlorine is utilized in water treatment by the county (*Firehouse Magazine*).

Memphis, Tennessee Hazmat: Evolution of Hazmat to All Hazards Rescue

Memphis, Tennessee is located on the Mississippi River in the Southwest corner of the state, covering an area of 324 miles^2 and is the county seat of Shelby County. The Memphis population is approximately 647,374 in 2020 making it the 23rd largest city in the United States and the largest city in Tennessee and largest on the Mississippi River. If you include the metro area which encompasses counties and cities in Tennessee, Arkansas, and Mississippi, the population is 1,316,100. Memphis is known for Blues and Barbecue, and is the home of founders and pioneers of various other American music genres, including Memphis soul, gospel, rock n' roll, Buck, crunk, and "sharecropper" country music. Memphis is home to Sam Phillip's Sun Records, Saint Jude Children's Hospital, Elvis's Graceland, FedEx's primary hub, Stax Records, and Beale Street.

Fire Department History

Memphis had its first organized fire response formed in 1846 with the advent of the first Independent Fire Company Number 1. Within the next 2 years a second company was formed, the Eagle Fire Company Number 2. That was followed by the Liberty Number 3 in 1849. Memphis promoted its first fire chief A.B. Jewell in 1859, a volunteer firefighter who received $1,200 per year for his services. Two assistant chiefs were employed part time at $150 per year. In the first chief's report, 589 active volunteer firefighters were shown. There were five good engines and two out of order with 2,900 feet of hose in the department. Each fire company received $125 quarterly for maintaining equipment and the first company getting water on a fire received $15.

In 1864 on Christmas Eve, the most disastrous fire on record, known as the Specht fire occurred. Fourteen lives were lost. In 1872 the Chief Engineer was in command of an Assistant Engineer, Fire Captains, four Fire Steamer Engineers and thirty-two firefighters. Firefighters were divided into four engine companies and one hook and ladder company. Each firefighter on the department received $3.00 per day. A board of Fire and Police Commissioners was created in 1879. During 1912 the first motor driven fire apparatus was placed into service. House to house fire safety inspections began in 1933 resulting in the lowest number of losses due to fire in 23 years. Emergency ambulance service was established in 1966 in Memphis and was one of the first fire department based services in the country.

Today's Modern Department

Fire Department has employs 1,685 uniformed personnel operating from 57 stations located throughout the city under the command of Director of Fire Services Gina Sweat, who is the first female Director of the Memphis Fire Department. Housed in those stations are 56 engine companies and 21 truck companies, 3 rescues (squads) that handle hazmat, 34 medic units, 8 squads, 6 brush trucks, 1 collapse trailer, and 2 decon trailers. In addition, Memphis has several specialized units including 3 heavy rescues (Squads) which handle hazmat response and all hazards rescue, 2 Zumro Fire Boats and crash rescue equipment at Memphis/Shelby County International Airport. The airport is the busiest air cargo hub in the world. Fire companies in Memphis are composed of an officer, driver, firefighter, and a firefighter/paramedic.

Memphis Fire Departments $1.7 million "Fire Barge" a 120 foot barge retrofitted with fire suppression pumps and equipment to combat fires on the Mississippi River and on shore facilities in and around the Port of Memphis and Presidents Island (Figure 5.74). The barge has two main water cannons that each can shoot 6,000 gallons of water per minute roughly 500 feet. Also on board are two portable cannons that can flow 1,500 gpm of water each. Mainly the barge will be used for catastrophic events as it needs to be towed by a tug boat to where it is needed. Firefighters told me it could also be used for evacuations on Presidents Island in the event the one road in and out was blocked. When needed, the barge will be staffed by a 20 member special operations team.

Firefighters work a schedule of 24 on 24 off, 24 on 24 off, 24 on followed by 4 days off. Memphis Firefighters responded to 123,405 EMS calls and

Figure 5.74 Memphis Fire Departments $1.7 million "Fire Barge" a 120 foot barge retrofitted with fire suppression pumps and equipment to combat fires on the Mississippi River and on shore facilities in and around the Port of Memphis and Presidents Island. (Courtesy: Memphis Fire Department.)

21,621 fires in 2018. The Memphis Fire Department operates 34 ALS ambulances referred to as "Units", 56 ALS engine companies, and 21 BLS truck companies. Training for all disciplines in the fire department is conducted at the impressive Chester Anderson Fire Training Campus in Memphis, which was constructed in 2001. Chester Anderson was the Director of Memphis Fire Department at the time the center was built.

Hazmat Team History

Following the Waverly Tennessee derailment and explosion Memphis began looking at ideas for the formation of a hazardous materials response team in Memphis. Captain James Covington, and Chief Adelman Chief of Training spearheaded the Memphis hazmat team formation in the spring of 1978. Following the firefighter strikes in the summer of 1978, the hazmat started. By late 1978 the entire Memphis Fire Department had been trained in hazardous materials awareness. Chief Adelman and several others went to Texas A&M for hazmat training. When they returned firefighters to use AFFF foam and prepared them for fighting fires involving propane and hydrocarbons. The hazmat team was phased in over 1978 and 1979 beginning qualifications and training for team members.

> ***Author's Note:*** *Captain Jim Covington retired and went on to the National Fire Academy to teach in the Hazmat Curriculum. He was my instructor for Hazardous Materials Operating Site Practices in 1982 and we remained friends.*

All Hazards Rescue (Special Operations)

All Hazards Rescue (Special Operations) in the Memphis Fire Department began in the 1970s with the formation of two hazardous materials teams. Rope rescue was added in the 1980s; structural collapse, confined space, and trench rescue in 1997; and water rescue in 2007. In 1995, Fire Squads 2 and 6 were hazmat certified and each covered approximately half of the city. Fire Squads evolved into today's rescue companies 1, 2, and 3. Rescue 1, a 1997 E-One, is located at Station 56 at 7445 Reese Road along with Engine 56, Truck 25, and Unit 26. Rescue 2, a 2000 E-One, is housed at Station 36 at 3215 South 3rd street, which also houses Engine 36, a rescue boat and Unit 17. Rescue 3 is a 2007 Pierce and is stationed with Engine 27, Unit 22 and a Zumro boat at 2530 Whitney.

Station 9 is located at 2785 Rudder Road and houses A-1 a 1997 F-350, A-2 a 2001 Oshkosh TI 3000, and A-3 a 2000 Oshkosh TI 3000 S. Station 33 is located at 2555 Winchester Road and houses Engine 33; Truck 23; Unit 9; and Reserve A-2, A-3, and A-5. Engine 32 is located on Presidents Island, at 1670 Channel Drive, along with the largest concentration of chemical facilities in Memphis.

Engine companies carry adsorbent and dispersal materials (tow trucks carry absorbents) and respond to fuel spills of 10 gallons or less. Larger spills require hazmat team response. Approximately 30 hazmat personnel are on duty each shift with 130 technicians city wide. These numbers drop when Tennessee Task Force 1 Urban Search & Rescue Team is activated. Task Force 1 is stationed in Memphis. Rescue 1, 2, and 3 are dedicated units. All Hazards Rescue personnel receive cross training on all types of rescue, not just hazardous materials (Figure 5.75). Initially, 132 hours of training is required for personnel for certification. During 2018 the All Hazards Units responded to 378 hazardous materials incidents. Hazmat response statistics include fuel spills, gas odors, and leaks. If mutual aid is required, Shelby County Fire and Refinery firefighting crews are available. A warehouse of 55 gallon drums of foam is located at the training academy. Also foam is deployed on the fire barge. Memphis Hazmat relies almost totally on electronic research on computer, Internet, and smart phone use.

Training

Personnel on Engine 32 are trained to the Awareness, Operations and Technician Levels. Next due companies to Presidents Island, Engine 10, and Truck 9 personnel are also trained to the Awareness, Operations and Technician Levels. Engine 10 and Truck 9 are located at Station 10 at 148 S. Parkway along with Unit 21. Personnel from Station 10 are also trained to the Hazardous Materials Awareness, Operations, and Technician Levels. Airport Stations 9 and 33 personnel have been trained to the Awareness,

Figure 5.75 Rescue 1, 2, and 3 are dedicated units. All Hazards Rescue personnel receive cross training on all types of rescue, not just hazardous materials.

Operations, and Technician Levels. All personnel who are assigned to Rescue Stations are trained to the hazardous materials technician level. Firefighters throughout the rest of the city are trained to the hazardous materials operations level.

Hazardous Materials Exposures

Highway routes for hazardous materials include I 40, I 55, and U.S. Highway 51. Railroads that operate in the Memphis area are the Burlington Northern Santa Fe and Canadian North. Barge traffic on the Mississippi often carry hazardous materials. A pipeline is located at President's Island on the Eastern Shore of the Mississippi River. Fixed facility hazardous materials exposures include Du Pont Chemical, Praxair, Drexel Chemical, USZINC, Valero Refinery, Cargill, and numerous other chemical facilities in the city limits. Chemical exposures include solvents, fertilizers, hydrogen peroxide, anhydrous ammonia, cyanide, sulfuric acid fuming, sodium hydroxide, propane, chlorine, and pesticides. The greatest concentration of hazardous materials exposures in the city are located on Presidents Island. The island is also considered the largest commercial and manufacturing area in the South. Presidents Island is part of the International Port of Memphis.

This sea port is the second largest inland port on the shallow draft portion of the Mississippi River, and the 4th largest inland Port in the United States. International Port of Memphis covers the Tennessee and Arkansas sides of the Mississippi River from river Mile 725 to mile 740.

Within this 15 mile reach, there are 68 water fronted facilities, 37 of which are terminal facilities moving products such as petroleum, tar, asphalt, cement, steel, coal, salt, fertilizers, rock & gravel, and coarse grains. The International Port of Memphis is 400 river miles from St. Louis and 600 River miles from New Orleans and is ice free year round.

Incidents

Drexel Chemical Company Fire & Explosion

In the heat of summer 9:25 a.m. on July 5, 1979, about a year after the formation of the hazmat team Memphis Fire and Hazmat faced a third alarm fire at the Drexel Chemical Company 155 W. Bodley Avenue near the intersection of Pennsylvania and Mallory Streets (Figure 5.76).

Figure 5.76 In the heat of summer 9:25 a.m. on July 5, 1979, about a year after the formation of the hazmat team, Memphis Fire and Hazmat faced a third alarm fire at the Drexel Chemical Company. (Courtesy: Dick Adelman Collection.)

An 8,000 gallon tank of parathion, a pesticide, caught fire. In an interview with the Commercial Appeal maintenance mechanic David Trumble said of the initial explosion "I heard one hell of an explosion and ran like hell". Another employee Robert Belden in a nearby warehouse who "heard an explosion. It was like a dull boom and I saw the doors of the warehouse coming off. Then it knocked me down, and the flash burned my hands in a couple of places and I went outside".

Richard Arwood, who 25 years later would become Director of Fire Services (Fire Chief), was assigned to Engine 10. Earlier that day he had driven by Drexel while doing a familiarization of the area and noting locations of hydrants. Back at the station he was looking out the window talking on the phone, when he saw the explosion occur and a white smoke plume rising above. He told me during an interview "it was a straight shot to Drexel, the wind was blowing North to South so he could approach the fire from upwind, the wind was blowing in the best direction it could have been" because of the toxic nature of the chemical involved and the toxic smoke blowing downwind.

Richard drove the first apparatus to arrive on scene. Upon arrival he saw fire blowing out of all the windows. After dropping off a master stream device, he laid dual 2½ lines and caught the hydrant on the SE corner of Pennsylvania and Bodley. He pumped the master stream through most of the early part of the fire. When 55 gallon drums started rocketing from the fire building, he thought if one came close he would dive under Engine 10. Some shot over his head 60–75 feet, but never got any closer to his position.

> ***Author's Note:*** *Richard is a friend and former instructor of mine in a course I took at the National Fire Academy in the 1980s. I had the chance to do a phone interview about the Drexel fire and the development of the Memphis Hazmat Team in preparation for this Volume.*

Drexel Chemical fire was in a concrete block and flat roof building 100 feet × 200 inch long. Firefighters were told by dispatch that organic phosphate pesticide was involved in the fire. When firefighters arrived on scene they all put on SCBA. Master streams devices were deployed and at the height of the fire 4,000 gpm of water was being played on the burning chemicals. Numerous explosions occurred sending 55 gallon drums flying through the air. Captain Jim Covington, who helped start up the hazmat team, figured out early on that they would just let the fire burn. Following dangerous explosions, rocketing 55 gallon drums and massive fire, Drexel Chemical Company was allowed to burned to the ground.

Water from the firefighting efforts spread the fire through a cache of methyl parathion which they poured into the Mississippi River because the fire could not be contained. Sulfuric acid was also dumped into the river as a toxic cloud spread beyond the plant, endangering the lives of area residents. Over 3,000 residents were evacuated including 400 who were taken by transit authority buses to an evacuation center set up at Whitehaven High School. Over 200 people visited hospitals complaining of bleary eyes, headaches, and vomiting. Treatment was hampered by a lack of medical familiar with treatment of chemical exposures. There were no firefighter injures as a result of the fire.

In 1984, Memphis local government took steps to help prevent another disaster like the Drexel Chemical fire. A hazardous materials advisory committee was formed with representatives from emergency services, government, and industry who secured a $100,000 federal grant to create a hazardous materials plan. As a result of the plan and extensive training, the Chemical Manufacturers Association declared in 1986 that Memphis was the American city best prepared to cope with a hazardous materials emergency.

Pro-Serve Fire (Brooks Road)

During August 2006, the Memphis Fire Department faced an Extraordinarily complex hazardous materials incident that was complicated by the addition of fire. Three Chemical Manufacturing facilities were involved that produced and stored a wide range of hazardous chemicals.

August 2, 2006 at 13:47 hours, Memphis firefighters were dispatched to a fire next to Valley Products 384 East Brooks Road. First arriving companies radioed there was smoke showing at Pro-Serve, 400 E. Brooks Road. Size-up determined a small fire involving a pallet of sodium chlorate and a forklift, which had been extinguished by the sprinkler system before firefighters arrived. Employees at Pro-Serve had been exposed to smoke and were decontaminated, and given medical evaluations by fire personnel on scene. Clean-up was handled by Pro-Serve. Command became aware of employees at Bucyrus International at 3057 Tranquility had also possibly been exposed to the smoke. Because of limited exposure, Bucyrus employees refused decontamination, medical treatment, and transport. Both incidents were declared stabilized at 15:28 hours.

August 4, 2006 at 16:07 hours, Memphis 911 center received multiple reports of an explosion and heavy smoke in the area of Brooks Road and Third Street. It was initially dispatched as a hazardous materials incident. First in companies reported a fire and heavy smoke at the Pro-Serve building, 400 E. Brooks Road. The first arriving engine company advanced a 2½ inch attack line into the building. Division 1 arrived on scene a short time later and assumed Brooks Road Command. A 2nd alarm was requested at that time. Command decided to cease all offensive operations until more information could be gathered about chemicals inside. A hazardous materials branch was established. Runoff was reported to have entered and contaminated Nonconnah Creek. Hazmat Operations decided to recommend extinguishment of the fire and it had been knocked down at 20:29 approximately 4½ h after companies were dispatched. The incident was declared stabilized at 21:07.

As the scene was downgraded, a fire ruins detail schedule was established. Brooks Road Command was transferred several times. It was discovered that IBC Manufacturing at 416 E. Brooks Road had also been involved in the incident. Due to multiple unknown chemicals stored at that location and the unavailability of IBC personnel, Brooks Road Command gave orders not to enter the structure.

August 5, 2006 at 02:55 hours, Brooks Road Command requested a hazardous materials assignment and Division 1. At 02:56 hours, a 2nd alarm was requested. Brooks Road Command reported fire coming through the roof as well as explosions occurring in what was eventually determined to be the IBC warehouse. Due to the unknown nature of the products, the amount of fire involvement, and environmental concerns from run-off, Brooks Road Command made the decision to let the building free-burn. Efforts were concentrated on isolating the area, evacuating the citizens, and air monitoring. At daylight, aerial surveillance was conducted via helicopter, MSDS sheets were obtained and hazmat teams conducted recon. The intensity of the fire had subsided the decision made to attack the remaining fire that was in the IBC tank farm using limited water and AFFF. At 15:54 hours the fire was reported knocked down. The incident was downgraded and a ruins detail was scheduled including hazmat team personnel.

August 9, 2006 at 16:42 hours, fire companies were dispatched to Pro-Serve 400 E. Brooks Road. Initial arriving companies reported light smoke coming from the tower area. United States Environmental Services (USES) and the Center for toxicology and Environmental Health (CTEH) had been on scene since the earlier incident on 5th conducting clean-up and air monitoring operations. They reported that the material on fire was not water reactive. The company officer made the decision to attack the fire with a 2½ inch hose line.

Hazmat had been dispatched and was conducting air monitoring. Brooks Road Command was set up at Brooks and South Center Road and a 2nd alarm requested at 17:04 hours. Command ordered companies out of the building until more product information was obtained. Hazmat personnel placed unmanned monitors in place to extinguish the remaining fire. The fire was reported knocked down at 20:04 hours and a ruins detail was set up and a fire department presence was maintained until September 1, 2006.

Memphis Fire Department has a rich history, and the hazmat team was one of the early pioneers in hazmat response development. Their extensive and diverse hazmat exposures provide daily challenges for the current hazmat team (*Firehouse Magazine*).

Milwaukee, WI Hazmat Team

Milwaukee is the largest city in the State of Wisconsin and the fifth largest in the Midwest. It is the seat of Eponymous County and lies on the western shore of Lake Michigan.

The cities estimated population was 585,589 in 2020, in a geographical area of 96.84 miles^2. Of those 96.17 miles^2 are on land and 66 miles^2 are on water. Milwaukee is the main cultural and economic center of the Milwaukee metropolitan area which has an estimated population of 2,043,904. The metro area is the second most densely populated area in the Midwest, surpassed only by Chicago.

Fire Department History

In 1837 a volunteer fire department was formed in Milwaukee. As the city grew, so did the fire department eventually boasting 8 hand drawn engines, 2 hook and ladder companies and 2 hose companies. The first steam engine was purchased in 1861. The new engine required an engineer to operate so the career fireman was hired. During the advent of the steam engine and the onset of the civil war, the volunteer organization was replace with a career department. A full-time department began on January 1, 1875.

Today's Modern Department

Chief Mark A. Rohlfing was the first chief hired from outside the Milwaukee Fire Department (MFD) on May 3, 2010. Chief Rohlfing spent most of his 27year career as chief of the Omaha, NE fire department. From Omaha he went to Rapid City for two and one-half years before coming to Milwaukee.

> ***Author's Note:*** *I had the pleasure of meeting Chief Rohlfing while he was in Rapid City when I visited their hazmat team in 2009.*

Under Chief Rohlfing's leadership, 900 uniformed personnel provide service from 31 fire stations. MFD is organized into five battalions and one deputy chief who is the shift commander. Firefighters operate 30 engines, 9 trucks, 2 rescues, 12 ALS Medic Units, and 3 hazmat vehicles. Other apparatus includes a Heavy Urban Rescue Team, Dive Rescue Team, Fire Investigation Unit, Tactical EMS, and Incident Command Unit with a drone. A fireboat operates from April through November and is staffed by Engine 1. Engine companies are staffed with 4 personnel. Truck companies are staffed with 4 except Truck 1 (dive team) with 5 and Truck 16 (hazmat) with 5. Both rescue units are staffed with 5 each.

Hazmat Team History

MFD established a response unit that responded to petroleum spills before the advent of organized hazmat response as we know it today. During 1966 the MFD Repair Shop converted a spare pumper into the Petroleum Dispersal Unit. This unit carried an electric oil skimmer and oil boom for use in the Port of Milwaukee and local rivers. This unit remained in continuous operation until 1989. The functions of the unit were taken over by the hazardous materials team.

Milwaukee's Hazmat Response Team was organized in 1980 with 7 team members operating out of a station wagon assigned to the training academy's battalion chief. Their first hazmat unit was a 1973 Grumman-Cortex (1983–1987) formally used as a Navy recruiting van (Figure 5.77). The next vehicle (1987–1995) was a 1980 Ford-Medicoach formally used as a blood collection vehicle.

Hazmat Team

Milwaukee's Hazmat Team is housed at Station 33, located at 4515 W. Burnham St., in the village of West Milwaukee. Since 1991 Milwaukee has provided fire protection under contract to the Village of West Milwaukee.

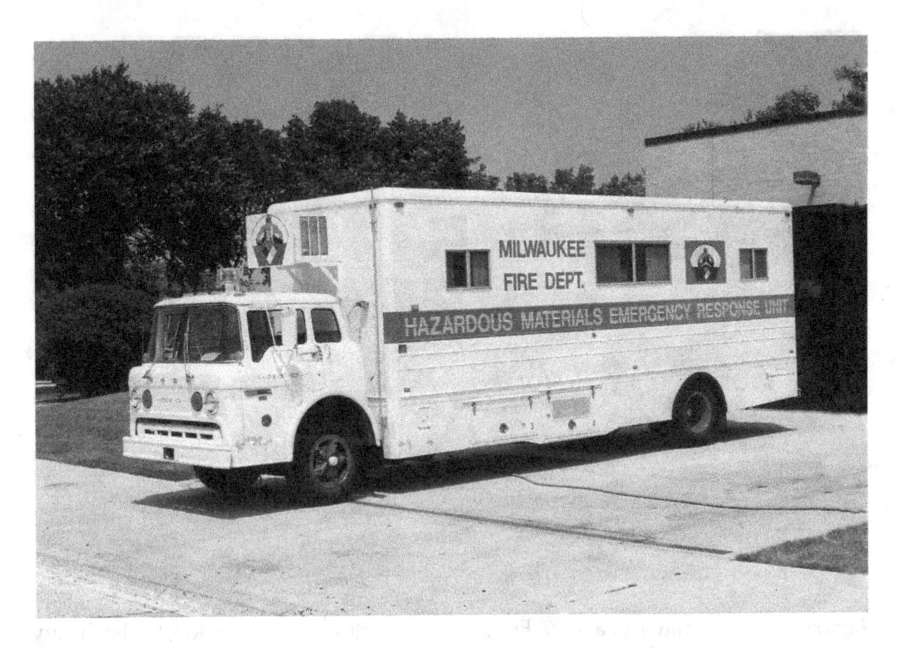

Figure 5.77 Milwaukee's first hazmat unit was a 1973 Grumman-Cortex (1983–1987) formally used as a Navy recruiting van. (Courtesy: Milwaukee Fire Historical Society.)

Apparatus housed at Station 33 includes, Engine 33, Truck 16, Hazmat 1, and Hazmat 2. Hazmat 1 is a 1997 Freightliner tractor with a Marion trailer heavy duty rescue body (scheduled for replacement in 2020) (Figure 5.78). Hazmat 2 is a 2007 Ford F-350 4-door cab with Knapheide utility box (pulls decon trailer). The Canadian Pacific Railroad has loaned a foam trailer to the MFD for use at Bakken crude oil train derailments. It carries 275 gallons of 3% Alcohol Resistant AFFF, 750 gpm portable pump, 10,000 gallon portable water tank, and foam nozzles and fittings. The foam trailer is housed at Station 33.

Engine companies carry a 5 gallon bucket of dry clay absorbent as well as a 5 gallon bucket of industrial detergent. Hazmat 2 responds to spills less than 25 gallons of gasoline, Hazmat 1 will respond to spills greater than 25 gallons. Twenty-five hazmat techs are on duty city wide. Nine are at Station 33, 4 on the engine and 5 on the truck. There are 80 technicians on the MFD. The Hazmat Team is not dedicated, crews from E33 and T16 provide staffing for the hazmat units when needed. If mutual aid is required, it is available from Tier II Teams, City of Racine, City of Madison and Tier III Teams, Washington County, and Sheboygan County.

- Tier I Team has more training and monitoring equipment in radiation and terrorism

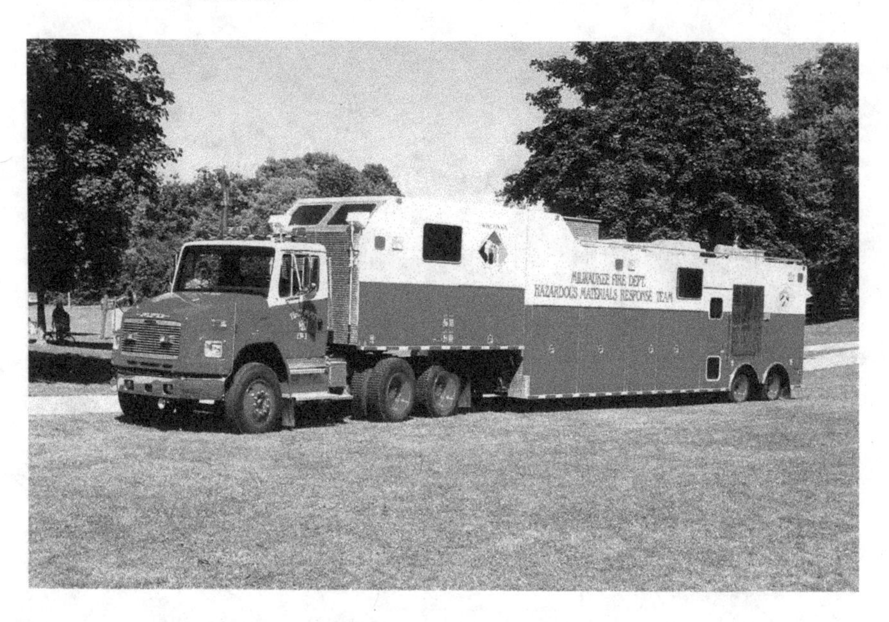

Figure 5.78 Hazmat 1 is a 1997 Freightliner tractor with a Marion trailer heavy duty rescue body (scheduled for replacement in 2020) (Courtesy: Milwaukee Fire Historical Society.)

- Tier II Team (formally called Level A) can make Level A entries.
- Tier III Team have the least capabilities and are generally county teams that will do fuel spills and some identification of unknowns but not much mitigation.

A full hazmat response gets 3 engines, 2 trucks, 1 rescue, 1 medic, 1 safety officer, 2 BC, plus Hazmat 1 & 2. There is also a limited response where only the first due engine, ladder, medic unit, BC, and Hazmat respond.

PPE, Equipment, and Training

Level A Kappler Suits, Level B suits are Tyvek coveralls. Respiratory protection is MSA SCBA with 1 hour bottles and Positive Pressure Air Purifying Respirators (PAPAs). In-suit communications are through portable radios operating on an analog channel.

Monitoring and detection equipment include AreaRae × 4 (one with gamma); MultiRae × 4, (2-with H_2S, 1 with NH_3, 1 with Cl_2); Qrae 4-gas × 2; ToxiRae × 4, (CO_2, SO_2, NH_3, HCN); MiniRae × 2: First Defender RMX S1; TRU Defender FTX; ADM300; Ludlum × 2; Polimaster; Inspector RAD × 4; Nuke Alert × 2; FLIR identiFINDER R100; identiFINDER R100; Honeywell NH_3/Cl_2; ChemPRO 100 i; Extech Temp gun; Raytek Temp gun; TIF 8900 Combustible Gs detectors; TIF XP-1A refrigerant leak detector; Jerome J405 mercury detector; Draeger tubes; DropEx Kit; BADD Kit. Basic CO Detectors are carried on Engines and Trucks. BC carry a better CO Detector.

Train requirements for technicians are the 80 hour IAFF Technician Class. Firefighters are trained to the Operations Level utilizing the IAFF Hazmat Operations Class. Canadian Pacific Railroad has provided, on loan, a training trailer for use in training on valves and hardware associated with railroad tank cars.

Reference Resources

ERDSS-Chemical companion, CAMEO, Wiser for Windows, Marplot, Aloha; Boldfrank's Toxicologic Emergencies, ERG: NIOSH, Guardian Plume, Sigma-Aldrich.

Hazardous Materials Exposures

Transportation routes include I-94, I-43; Canadian National, Canadian Pacific, Union Pacific, and Wisconsin Southern railroads; Port of Milwaukee; Kinnickinnic, Menomonee, and Milwaukee Rivers and tributaries; Mitchell International Airport. Fixed facilities include Miller

Brewery which has large amounts of anhydrous ammonia. Mateion Chemical Company manufacturers various chemicals.

Incidents

Schwab Stamp & Seal Acid Spill

On February 4, 1903, an acid spill occurred at the Schwab Stamp & Seal company. About 2:00 p.m. an acid carboy broke open on the second floor of the facility. Employees scattered after calling the fire department. A leaking carboy of nitric acid leaked creating corrosive and toxic vapors. Cleaning up the mess was only a salvage job for the trucks and Insurance Patrol. Other companies were sent home. Chief of Department Foley and Trucks 1 & 2 located the leaking carboy and carried it outside. Then companies spent time scattering saw dust on the floor, raking it up, and tossing it outside. After venting the building, Captain White of Truck 1 became ill. During the next few hours, man after man complained of choking up. At 9:00 p.m. was at his bedside. Captain White died followed by Ed Hogan.

A little earlier Tom Droney of Chemical 1 had responded to another alarm. While there he collapsed in the snow. Revived he went back to the firehouse where within a short time he also passed. By then Chief Foley was sick. Shortly after White's death he remarked I'll bet $100, that I'll be dead by morning. He was right, he died at 4:15 a.m. The fumes had seared the men's lungs dooming them to slow suffocation.

Marsh Wood Products

On April 20, 1926 a fire occurred at the Marsh Wood Products was the scene of several earlier fires. A small fire got started in the boiler room where a huge bin held tons of sawdust used for fuel. Engines 14, 3, 19, 20, Fireboat 15, Trucks 8-4, and District Chiefs 3 & 4 responded just after lunch. Only a few employees were in the plant. Although no fire was showing when companies arrived, there was a thin haze of smoke and several sprinkler heads were operating. Some companies began to pick up. Engine 14 and Truck 8 dug through the smoldering sawdust looking for sparks. Then a blinding flash occurred. A dozen men were afire from head to toe and ran and stumbled from the building. They came out screaming and writhing in pain. Engine 3's crew who had been outside picking up their line went to help their brothers. A third alarm was sent in. Engine 14 and Truck 8 were wiped out.

The first man to die of his injuries was Stanley Strezeminski of 14's. Lt. Tom Hanlon passed away just after 11 that night. Next was Al Schultz of T-8 who died in the next morning. Dead that night at Deaconess Hospital was Ambrose Skorzewski (E-14). Three days later, his brother Captain John Skorzewski (T-8) succumbed. Last of the six to die was George Liefert of E-14. At least 9 others were hospitalized (*Firehouse Magazine*).

Naval Air Station Corpus Christi Texas: Protecting the Largest Helicopter Repair Facility in the World

History of NASCC

Naval Air Station Corpus Christi (NASCC) was established in 1941, with a primary mission of aviation training with both initial and advanced pilot training. This mission is carried out by Training Air Wing Four, and its four squadrons. NASCC occupies a total of 5,622 acres (9.78 miles^2) with three outlying landing fields and the Corpus Christi Army Depot (CCAD), its largest tenant command. CCAD is located in three geographic locations on the facility with over 6 million feet of space and assigned aviation assets. In addition, there are residential neighborhoods on the base where military members and their families live. Total population is approximately 3,400 with peak time population during the day exceeding 7,000 personnel. The majority working at NASCC are at the CCAD. Also on the base is a U.S. Coast Guard facility with both fixed wing and helicopter aircraft on site.

Background CCAD

CCAD was established in 1961 and is the industry leader for repair and overhaul of helicopter, engines, and components in Army Aviation. It is the largest rotary wing repair facility in the world. CCAD has the largest potential fire and hazardous materials risk on the base. This is due to the complex industrial nature of their operations. Even though they are the highest risk, CCAD's operations are conducted in a safe and highly regulated manner. As a result, there is an extremely low incidence of fire and few significant hazardous materials releases. High severity/low probability risk is managed in several ways: First, aggressive fire prevention and public education focusing on each and every CCAD employee; second, training and excellent equipment. Most incidents are mitigated by the initial response of appropriate resources.

NASCC Fire Department

Naval Air Station Corpus Christi's Fire Department was established following the commissioning the facility March 12, 1941. The first flight training began on May 5, 1941. NASCC Fire Department is comprised of a total of 67 personnel, including two civilians under the leadership of Chief Ruben C. Perez, John T. Morris, CFO now retired, was chief during my visit there (Figure 5.79).

> **Author's Note:** *Not long after I visited NSACC and met Chief Morris, I got word he was retiring. Fortunately, for me, we remained friends on Facebook and while visiting Corpus Christi in the fall of 2018 my wife*

Figure 5.79 Not long after I visited NSACC and met Chief Morris, I got word he was retiring. Fortunately, for me, we remain friends.

and I had dinner at his home and met his wife. John loves to fish and cook. He made a wonderful gourmet dinner. He lives within sight of the Gulf of Mexico. What a wonderful time. Thanks John and enjoy your retirement.

Department tasks are divided into Four divisions: Administration; Operations; Fire Prevention and Pubic Education; and Fire Training, Health, and Safety. Uniformed personnel operate from five stations on the main installation and three at Outlying Landing Fields (OLF), 5.5, 13.6, and 95 miles from the main installation. Fire and EMS personnel operate 2–1,250 gpm. engines, a 100′ platform, combination hazmat/heavy rescue unit, BLS medic unit, and 6 ARFF units. Other units include Special Operations Support Trailer, 1,000 gallon AFFF Trailer, and Light and Air Trailer.

Additional automatic response has been provided to NASCC by the City of Corpus Christi Fire Department (457 times since 2007). Automatic response has also been provided by F&ES to the City of Corpus Christi (355 times since 2007). Beyond actual emergency response, the departments frequently train together, including live fire training (structural and ARFF), technical rescue, and hazardous materials to improve interoperability. NASCC F&ES Fire Department became a Commission on Fire Accreditation International (CFAI) accredited agency on March 17, 2015. Additionally, NASCC F&ES was selected as the 2014 Department of

Defense Fire Department of the Year (Medium Category), and the 2015 Nay Region Southeast Fire Prevention Program of the Year. NAS CC F&ES responded to 1,100 total incidents in 2015.

Hazmat Team History

Department of Defense (DOD) initiated a DOD F&ES certification plan in 1994. Following this change in policy, the Hazardous Materials Team at NASCC was formed.

Hazmat Team

NASCC's Hazardous Materials Team and apparatus are located at Station 1, 10800 D St. Bldg. Corpus Christi, TX. Special Operations Unit 1061 is a 2011 Pierce Velocity HDR configured as a Hazmat/Heavy Rescue truck (Figure 5.80). This is a dual purpose vehicle that carries specialized equipment and supplies for response to hazardous materials incidents and technical rescue incidents.

Support equipment is transported utilizing a Wells Cargo Trailer (Support 1081) 8 feet × 20 feet, which carries mass decon tents, over-pack barrels, bulk absorbents, low pressure air bags, and other related equipment. Additional responding units provide manpower and support of

Figure 5.80 Special Operations Unit 1061 is a 2011 Pierce Velocity HDR configured as a Hazmat/Heavy Rescue truck.

the incident itself. Special Operations 1081 is equipped with a 2 cylinder (4 bank, RSP Space Saver Model 100 A High pressure, 6,000 psi) cascade system. Power and lights to operate the equipment on 1081 is provided by a 25 kW Onan generator and a 9,000 W (210 total lumens Night Scan) Light Tower. Communications capability includes UHF, VHF, and 800 MHz, which allows interoperability with other units in Nueces County.

Mutual aid is available and given by Corpus Christi Fire Department Hazmat who have 54 trained technicians and 18 on duty at any given time. For WMD incidents the 6th Civil Support Team from the Texas National Guard would be requested for an actual event. Engine, Truck, and ARFF vehicles carry a 5 gallon bucket containing dry absorbents, absorbent pillows, and absorbent pads. NASCC HazMat responded to 31 incidents on base and 1 mutual aid to Corpus Christi in 2015.

PPE, Equipment and Training

Hazardous Materials PPE consists of:

- Kappler Zytron 500 full encapsulated vapor protective suits (Level A)
- Lakeland Chemmax liquid splash protective suits (Level B)
- Butyl gloves
- Guardian Neoprene gloves
- Silver Shield gloves
- Nitrile gloves
- Hazmax Boots

Respiratory protection is composed of Scott Safety Air-Pak 75 CBRNE SCBA with 4,500 psi 60 minute cylinders. In-suit communications provided by Motorola XTS 5000 portable radios and Savox Talon throat mic and over ear speaker for use with the Motorola XTS 5000.

Monitoring Equipment

- MultiRae Plus, 5 gas detector with 0_2, LEL, H_2S, CO, and PID sensors
- MultiRae Lite, 6 gas detector with 0_2, LEL, H2S/C0, HCN, and PID sensors
- MiniRae 2000, Photo Ionizing Detector (PID) with 10.6 eV lamp for detecting volatile organic compounds (VOCs), including poisons and combustibles
- MiniRae 3000, PID with 10.6 eV lamp for detecting VOCs, including poisons and combustibles
- Ludlum M 2241-2ERK Radiation Survey Meter
- Canberra ERKADV Radiation Survey Meter

- pH Paper for detecting Acids and Bases
- Fluorine Paper for detecting Fluorine
- Advnt Pro Strip 5 Biological Detection Kit for detecting Anthrax, Ricin Toxin, Botulinum Toxin, Y pestis, and SEB.
- AHURA First Defender, a Raman handheld chemical identifier
- Fisher Tru Defender, a FTIR Handheld Chemical Identifier

All firefighters at NASCC are trained to the Hazmat Technician Level as required by Navy F&ES. Training is based upon NFPA 472: Standard for Competence of Responders to Hazardous Materials Weapons of Mass Destruction Incidents. Certification is provided through IFSAC. Fifty-one Hazmat Technicians are available to the department and a minimum of 9 are on duty at any given time at Fire Station 1. Selected personnel receive specialized training annually to ensure that knowledge levels and skills are developed for an effective response. Remaining technicians are also available as part of a coordinated response.

Reference Resources

Computer resources include, U.S. National Library of Medicine's Wireless Information System for Emergency Responders (WISER) on line database, and U.S. EPA CAMEO. Hard copy reference materials include DOT ERG, Hazmat IQ, CDC NIOSH Pocket Guide, Hawley's Condensed Chemical Dictionary, ASX's Dangerous Properties of Industrial Materials, Texas Tier II Chemical Report, and Safety Data Sheets (formerly MSDS's). CHEMTREC is also used via phone contact.

Hazardous Materials Exposures

Primary hazardous materials at NASCC include, but are not limited to Nitric acid, Sodium Cyanide, Potassium Cyanide, Sodium Hydroxide, and Chlorine Gas. Large quantities of Kerosene (JP-8) are used to support over 400,000 annual flight operations. Transportation Hazmat exposures are generally limited to the previously mentioned chemicals. Deliveries of JP-8 occur on a daily basis to a 600,000 gallon tank farm, and escorted to the fuel farm by Navy law enforcement personnel.

Incidents

In spite of the potential for a hazardous materials incident on base, the controlled environment of the facility results in few actual emergencies. Primarily, this is a result of prevention efforts coupled with engineered containment systems. F&ES has a strong working relationship with base environmental and industrial hygiene agencies as well as law enforcement and is instrumental in limiting the impact of incidents.

The most significant incident outside of the controlled areas, occurred on April 18, 2013 and involved a transportation vehicle carrying 500 gallons of nitric acid. Nitric acid in addition to being very corrosive is also a strong oxidizer. NASCC Hazmat responded to a report of nitric acid leaking from a truck adjacent to the hazardous waste facility. Initially, the tank was leaking slowly, approximately 2–4 gpm from the discharge valve. Leaking nitric acid was being caught in a 60 gallon catch basin by the driver. He was attempting to offload product to available 320 gallon totes. The product discharge valve failed followed by an uncontrollable loss of product. An initial evacuation of 150 feet was placed in effect as the acid overflowed containment and reacted with the asphalt causing a visible plume.

Asphalt is a petroleum product and reacts with oxidizers producing a potentially dangerous situation where the fuel and oxidizer, two sides of the fire tetrahedron, are present. Heat from the reaction could be the ignition source and cause a fire. All of the contents of the tank were released before offensive operations were started. Because of the release of all the product the isolation distance was increased to 500 feet. During the incident a cold front moved through the area and the wind initially out of the SSE at 10 mph changed to the North at 30 mph. Affected effected facilities were sheltered in place. Mitigation was a combination of adding soda bicarbonate to the spill and bringing in a contractor to clean up the spill. This incident stopped flight operations for a period of time and took 3hours and 18 minutes to stabilize.

Nebraska Regional Hazmat Teams

Resources for fire, EMS, and hazardous materials responses in rural areas often are limited by community size, personnel availability, and distances between communities. Hazmat response, being a much more specialized resource in terms of amounts of training required and equipment necessary to handle an incident, is much more limited in many cases in rural areas than fire and EMS resources.

Hazardous materials exposures, on the other hand, are not limited. In fact, hazardous chemicals may be present in much greater quantities in rural areas than in many larger communities with adequate response resources. Even the smallest rural communities may have exposures from agricultural pesticides, anhydrous ammonia, propane, petroleum products, and other hazardous materials. Ethanol manufacturing plants are springing up across the country in rural areas, creating large volumes of flammable ethanol in fixed storage and rail transportation, in particular.

Throughout the state, there are 10 hazardous materials teams from paid, part paid, and volunteer departments that can be activated by the Nebraska Emergency Management Agency (www.nema.ne.gov) (Figure 5.80A). At such times, teams are placed on the state payroll to respond to incidents anywhere in Nebraska, including rural areas.

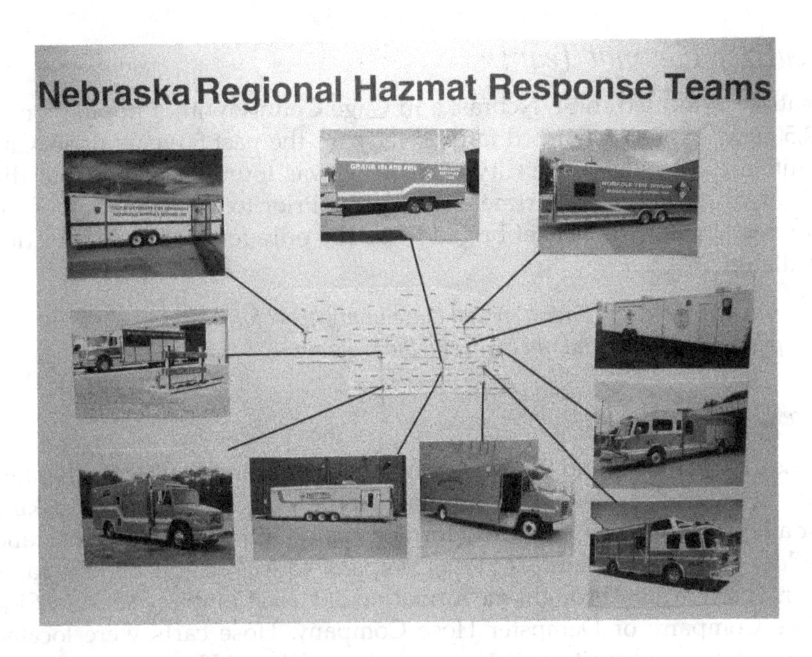

Figure 5.80A Nebraska Regional Hazmat Teams.

Nebraska also has a state Hazardous Incident Team, which comprises the State Patrol, State Fire Marshal, and Department of Environmental Quality.

> ***Author's Note:*** *This team was formed in 1987 by then Governor Kay Orr to provide a diverse response to Hazardous Materials Incidents throughout the State of Nebraska. At the time I was working for the Nebraska State Fire Marshal and was one of the original members of the team. My duty station was changed from Ord to Grand Island so I would be near Interstate 80, the major East-West route through Nebraska.*

The purpose of the team was to provide assistance to local emergency response agencies in the event of a hazardous materials incident, either directly or indirectly. During that time period, there were only teams in Omaha, Lincoln, Grand Island, and Hastings. In the beginning, team members were basically an advisory group and did not get involved in mitigation. Team members were trained in Hazmat Response, the Incident Command System, and other specialized response training. NHIT would respond to any type of incident transportation or fixed facility. In addition to the advisory role the team can be a liaison with state agencies or other resources required to mitigate the incident. The team is not responsible for clean-up but will assist the Incident Commander in making sure the incident is cleaned up in a safe and effective manner.

Beatrice Hazmat Team

Beatrice is located in SE Nebraska in Gage County with a population of 12,500, which has remained fairly steady for the past 50 years or so. City limits are 8.8 miles^2. The City of Beatrice was founded in 1857 but did not organize a Fire Department until 1866. Prior to the formation of the fire department the bucket brigade was the only form of fire protection available.

Author's Note: *I have a special place in my heart for Beatrice as it is the place of my birth and my early childhood home.*

Fire Department History

Hose Companies #1 and the J.B. Kilpatrick were formed in 1866. Firefighters calling themselves the Hooks approached the City Council in 1887 asking for a hook and ladder truck and formed the Rough and Ready Hook and Ladder Company. Volunteer Hose Company #3 was established the same year. The year 1888 brought the formation of Hose Company #5, West Side Hose Company or Dempster Hose Company. Hose carts were located around town and at city hall. During 1904 the Ward 4 Hose Company was formed.

In 1909 the Beatrice Volunteer Fire Department was formed with the construction of a new fire station at 114 North 5th Street. During 1910 a partially paid department was formed, and the Peter Pirsch & Company's combination hose and chemical wagon were purchased. Horses were needed to pull the new apparatus, and Roc and Doc a pair of dapple gray's were purchased. Motorized apparatus was added in 1916 in the form of a white Pirsch. In 1932 a new Seagraves was able to pump 1,000 gpm into a fire to assist in quicker extinguishment. Soon after a modern aerial device with a turn table with a 75 feet ladder height was purchased. Beatrice Fire & Rescue disbanded their volunteers in 2015 and are now a completely career fire department.

Today's Modern Department

Today's Beatrice Fire and Rescue Department is led into action by Chief Brian Daake. Beatrice Fire & Rescue operates from one station at 310 Ella Street, with a new one in the planning process to replace the old 1965 building with much needed additional room. Chief Daake commands 23 personnel with a maximum of 7 on duty each shift. They also employ two part time paramedics. On duty personnel operate 1 Engine, 1–100 feet Truck, Light Rescue, Hazmat Unit and Trailers, and five Medic Units. All firefighters are EMT's or Paramedics. When on duty personnel cannot adequately handle the type or volume of incoming calls, off duty personnel are recalled. Mutual aid is summoned as needed. Beatrice Fire and Rescue also houses and maintains the Beatrice Rural Fire Department Equipment and provides drivers to take them to the scene of an emergency. Medic units handle Beatrice and the bottom half of Gage County. In November of 2018, voters approved an additional half-cent local sales tax to help finance a new fire station for Beatrice Fire and Rescue. The new station will replace the cramped quarters the department has used in the lower level of the historical Beatrice Municipal Auditorium since 1965.

Hazmat Team History

Beatrice Fire & Rescue started thinking about their hazardous materials team in 1992. Following a fumigation accident at Booth Feed Supply in Beatrice in the 1990s that killed one person, the idea of a Hazmat Team in Beatrice started to develop. There were numerous hazardous materials exposures, within and transportation systems though, the community and they wanted to be able to handle any incidents that might occur. Located in town were trucking terminals, two railroads, farmers coop, and a fertilizer manufacturing plant just west of town. Hazmat equipment was purchased in the budget year 1993, which included chemical suits, monitoring instruments, decontamination equipment, and hand tools. Beatrice's first hazmat unit was purchased from Federal Surplus and was an old Air Force Panel Van, which was painted red. The current vehicle was purchased in 2001.

Hazmat Team

Beatrice Fire and Rescue Hazardous Materials Response Team is a part of the State of Nebraska Regional Hazmat Response System and can be called anywhere in the state that they are needed. When hazmat mutual aid is needed, Lincoln, Columbus, and Bellevue are closest. Help is also available from the National Guard Civil Support Team in Lincoln as well. Beatrice Hazmat responded to 39 calls in 2018. Typically, they respond to 30–60 calls depending on the year. Total calls include fuel spills, gas odors, and leaks. Combustible Gas Indicators and absorbent materials are carried on the engine and truck. Beyond that capability the hazmat unit responds to all other hazmat incidents. Once a hazmat call exceeds the ability of on duty personnel to handle, the entire fire department is re-called to duty. There are 15 total hazmat technicians on the department. Their hazmat unit is a 2001 Utilimaster with an on board generator (Figure 5.81). They also have a utility trailer.

Figure 5.81 Beatrice's hazmat unit is a 2001 Utilimaster with an on board generator. (Courtesy Chief Brian Daake)

Equipment and Training

Hazmat team and firefighting respiratory protection is accomplished with Scott air packs and 60 min air bottles. Through the use of their air cascade trailer, they can also utilize supplied air directly to the hazmat PPE.

Currently, 18 of the firefighters are hazmat techs and the goal is to have all personnel trained to the tech level. Each tech received an 80 h training course with in-service training on a regular basis.

Hazardous Materials Exposures

Transportation exposures include the Burlington Northern Santa Fe Spur line that services the city. Major highways include U.S. Highways 77 and 136. There are pipelines north and south of the city for natural gas and an anhydrous ammonia pipeline in the Rural Fire District. Fixed facilities include agricultural related propane, anhydrous ammonia, pesticides, and ammonium nitrate. The power plant utilizes sulfuric acid. Other chemicals include chlorine, cryogenic liquids, and fuel oil.

Incidents

Booth Feed Supply Pesticide Incident

On September 6, 1993, Beatrice Regional Hazardous Materials Team was dispatched to Booth Feed Supply in Plymouth, NE. A building was being fumigated and an employee had entered the area being fumigated and was deceased. Fumigation contractors had activated the fumigant; Fumitoxin Coated Pellets on Saturday September 4, 1993 at approximately 1,600h. The active ingredient in the pesticide was aluminum phosphide, which in contact with water, moisture from the air, acids, and many other liquids form phosphine gas. According to the Centers for Disease Control (CDC), phosphine gas is colorless and a lung damaging agent. Moderate poisoning causes weakness, vomiting, pain, chest pain, diarrhea, and dyspnea (difficulty in breathing). Symptoms of severe poisoning may occur within a few hours to several days. Exposure results in pulmonary edema (fluid in lungs) and may lead to dizziness, cyanosis (blue or purple skin color), unconsciousness, and death (Beatrice Fire Department).

Columbus Hazmat Team

Columbus is located in East Central Nebraska on U.S. Highway 81, 55 miles NW of Lincoln and 71 miles West of Omaha. Columbus is home to Andrew Jackson Higgins, Higgins Industries, and the Higgins Memorial. In 1964, Dwight D. Eisenhower called Andrew Jackson Higgins "the man who won the war for us". Without Higgins's famous landing crafts (LCPs, LCPLs, LCVPs, LCMs) the strategy of World War II would have been much different and winning the war much more difficult. It was the Higgins landing craft that made the D-Day landings at Normandy possible. During 2019 the 75th anniversary of D-Day was remembered in ceremonies throughout the world.

Fire Department History

Columbus, Nebraska is located in the northeast portion of the state in the southeast corner of Platte County. The City of Columbus covers an area of 10.08 miles^2 with a population of around 22,000. The town with a rich American Indian History and was founded in May 1856 by the Columbus Town Company and named after Columbus, Ohio. The Columbus Volunteer Fire Department was established on August 3, 1873 by a group of concerned citizens who wanted a fire department. Forty-seven men signed the Charter Role and established Hose Company #1, which was later expanded to Hose Company #1 & 2. On January 3, 1874 the Pioneer Hook & Ladder Company was organized. On July 20, 1911, Columbus placed their first motorized apparatus into service. "Old Smokey" as it was affectionately called remained in service until 1921.

Today's Modern Department

Columbus has a combination fire department with 15 career and 50 volunteers organized into 4 volunteer companies under the command of Chief Daniel Miller. The volunteer companies carry the historic names of those that started out the fire department. Columbus operates out of two fire stations with three engine companies, one truck 115 feet (a second on order), heavy rescue four medic units, and a hazmat trailer pulled by a semi truck. Columbus also houses and maintains the Columbus Rural Fire District apparatus and provides drivers to respond to rural calls. Additionally, they provide service to the Duncan Rural Fire District. A rural station north of town houses an engine and truck and is operated by Columbus Rural Fire District. Annual alarms average 1,800 per year with approximately 85% of those emergency medical.

Hazmat Team

Charlie Lewis Station #2 is the hazmat station, located at 424 8th Street. This station houses the hazmat tractor trailer (Figure 5.82), Class A & B Foam pumpers and the ethanol plant trailers and totes. There are an average of five hazmat personnel on duty each day. There are also 12 paramedics on the department.

On average, Columbus Hazmat responds to 12 hazmat calls a year with an additional 50–75 fuel spills, gas leaks, and other odor investigations. Engine companies carry absorbent material. If a spill is from any vehicle larger than a passenger vehicle, the Hazmat Team is called in. Mutual aid is available for hazmat from the State Regional Team and surrounding departments and Norfolk.

PPE and Training

PPE is composed of Kappler for Level A & B. Respiratory protection includes MSA SCBA, APRs (Cartridge respirators). In-suit communications is accomplished with Motorola push to talk.

Figure 5.82 Charlie Lewis Station #2 is the hazmat station, located at 424 8th Street houses the hazmat tractor trailer.

Technicians are trained with a 120h class provided by the Nebraska State Fire Marshal Training Division. Some personnel have also gone to the New Mexico Tech Explosives School and the Nevada Nuclear Test Site for Radiation training. Firefighters are trained to the FFI level and receive Hazmat Operations Training and do decon on incident scenes. The goal is to get all career personnel trained to the Technician Level.

Hazardous Materials Exposures

The Union Pacific Railroad runs through the City of Columbus and parts of Platte County. Major highways for the transportation of hazardous materials are U.S. 30 & 81. A major natural gas pipeline is in the Columbus Hazmat first response area, and the Keystone petroleum pipe line is within their regional response area. Columbus is home to the third largest ethanol plant in the United States.

Grand Island Hazmat Team

Despite the name, Grand Island is not surrounded by water. Rather, it is in central Nebraska, about 74 miles from the geographical center of the state and 77 miles from the geographical center of the lower 48 states. Grand Island has a population of 52,000 and covers 27 miles^2. In 2010, the Nebraska State Fair was moved from Lincoln to Fonner Park, a horse-racing track and casino in Grand Island, so the community's population can easily soar above 150,000 during the fair and other events. Grand Island is also the county seat for Hall County.

Fire Department History

Volunteer fire protection in Grand Island was organized in 1874 with the first fire company, followed several months later by the first ladder company. By 1876, membership grew to 25. In 1908, equipment consisted of hand-drawn hose carts, a ladder wagon, and a hand-operated pump that took water from cisterns. In 1909, the first horse-drawn mechanical truck was purchased. The first motorized fire apparatus, a chemical hose truck, was purchased in 1913 for $4,150. Volunteers staffed the Grand Island Fire Department until 1921, when the department became fully paid.

Today's Modern Department

Today's Grand Island Fire Department is led by Chief Cory Schmidt and staffed by 68 uniformed personnel. Grand Island firefighters operate from four stations, using two engines, one rescue, a truck, a heavy rescue, two medic units, and a hazardous materials trailer. The department responded to 5829 calls for service in 2019. Of those, 4750 were EMS, 865 fire related calls 50 were structure fires, 37 outdoor fires, and 127 were hazmat responses, including fuel spills, gas odors, and leaks.

Hazmat Team History

Grand Island formed its Hazardous Materials Response Team in 1985. The team's first response vehicle was a converted United Parcel Service delivery van (Figure 5.83) that was donated to the fire department. The primary response area is Hall County plus approximately 800 miles^2 through mutual aid agreements with area fire departments. Additionally, the team may be activated by the state Emergency Management Agency and respond as needed.

Hazmat Team

Following the 9/11 terrorist attacks, Grand Island's team became a part of the State of Nebraska response team. When called on for a state activation, the team may page additional off-duty technicians to assist in the response. Grand Island's team is not dedicated and personnel from Engine 4 and Medic 4 staff the response trailer when a hazmat incident occurs.

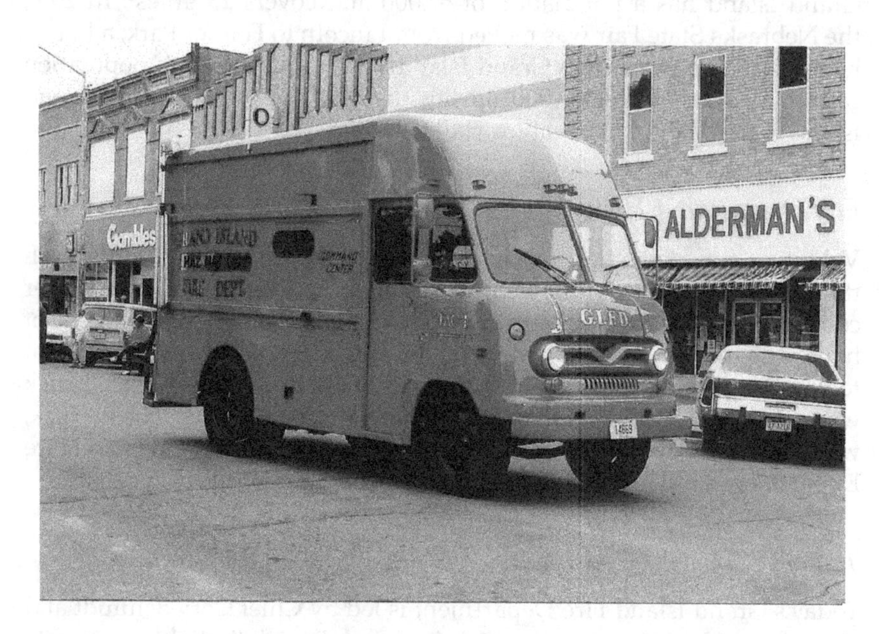

Figure 5.83 Grand Island formed its Hazardous Materials Response Team in 1985. The team's first response vehicle was a converted United Parcel Service delivery van.

Three to five hazmat technicians are on duty each shift, and additional personnel are called back to duty when necessary.

Currently, there are 28 trained hazmat technicians on the fire department. Grand Island is working toward training all uniformed personnel to the technician level. For now, firefighters and medics not trained to the technician level are trained to the operations level. Personnel are also trained to conduct decontamination to free technicians for mitigation duties. In addition to the Engine 4 and Medic 4 personnel, a hazmat response in Grand Island's primary response area brings the shift captain, battalion chief, and the engine and medic unit in the district where the incident occurs. If mutual aid is required, the Hastings Fire Department Hazmat Team is 23 miles to the south and Columbus Fire Department is 60 miles to the northeast. Grand Island also works and trains with the 72nd Civil Support Team of the Nebraska National Guard in Lincoln, 100 miles away.

Grand Island's current hazmat response vehicle is a 24-feet Wells Cargo trailer pulled by a 2003 Ford F350 pickup truck (Figure 5.84). The unit is housed in its own building behind Fire Station 4 on Grand Island's west side. Engine companies carry absorbent materials for cleaning up fuel and oil spills. The hazmat team is called in as needed to assist. All hospitals in Nebraska have decontamination capabilities with coordination through the Nebraska Emergency Management Agency and the University of Nebraska Medical Center.

Figure 5.84 Grand Island's current hazmat response vehicle is a 24-foot Wells Cargo trailer pulled by a 2003 Ford F350 pickup truck.

PPE, Equipment, and Training

Level A and B protection is provided by DuPont Tychem with flash protection for working around flammable materials. Breathing apparatus used is Scott self-contained breathing apparatus (SCBA) with 1-h bottles and positive-pressure air-purifying respirators (PAPRs) for WMD responses. The team also carries 3M full-face respirators with organic vapor, acid gas, ammonia methylamine, and multi gas cartridges. In-suit communications is provided by Kappler headsets that are used with normal radio communications equipment.

Each team member has a personal protective equipment (PPE) bag with gloves, boots, and coveralls to be used during an incident. Equipment is loaded on the response trailer based on the personnel responding and the anticipated type of hazardous materials involved. Personnel are developing a tag system to place on certain types of equipment to indicate what equipment is taken on a specific type of response.

In the building where the hazmat trailer is housed, monitoring and detection equipment is calibrated and stored along with other equipment until needed for a response. Monitoring and detection equipment carried by the Grand Island hazmat team includes Chemical Agent Test Kits, photo ionization detector, TMX-412 multi-gas detectors, APD 2000 chemical warfare agent detector, radiological emergency response kits, Biowarfare Agent Detection Devices (BADD) kit, Hazmat ID, and Haz Kat. The Nebraska Department of Health provides sampling kits to the state's hazmat teams. These kits provide equipment and supplies to gather samples of weapons of mass destruction (WMD) to be analyzed at the state labs.

Technician-level training is done through the Nebraska Emergency Management Agency's 80-h course. Every 2 years, team members take a 40-h refresher. Additional specialized training is obtained through the Center for Domestic Preparedness in Anniston, AL, for chemical and biological terrorist agents and in Pueblo, CO, for railcar incident training.

Reference Resources

Reference resources are now only computer and internet based, they no longer use hard copy materials. Grand Island Hazmat Team members use incident forms for various positions in the Hazmat Incident Command structure. Using the prepared forms simplifies recording information, tracking personnel, gathering needed information, developing required site safety plans, and managing the incident scene. Forms are used for the incident commander, safety officer, entry group supervisor, and hazmat operations branch officer. The team has developed a Mutual Aid Link-Up Questionnaire to document information about an incident before and during responses to scenes that in some cases may take long periods. Gathering predetermined types of information before arriving on scene saves time and lets the members prepare for what they may be dealing with once on scene. The form contains three major headings, Dispatch Information, On Scene Information, and Post-Incident Information that help members write after-action reports and return equipment and supplies to service.

Hazardous Materials Exposures

Major hazmat transportation exposures are Interstate 80, U.S. Highway 281, Nebraska Highway 2, and the Union Pacific and Burlington Northern Santa Fe railroads. No major pipelines are present in the primary response territory. Chemicals transported through Grand Island are many and varied in nature. They include anhydrous ammonia, propane, chlorine, and pesticides. Fixed-facility exposures include the Case-New Holland manufacturing plant, Chief Industries, Swift Meat Packing facility, and McCain Foods processing center.

Incidents

Grand Island's hazmat team has responded to several major incidents in recent years, including a chlorine emergency at the YMCA, a chemical fire at a foundry in Hastings, an ethanol spill involving a railcar in Wood River, chlorine at a water treatment plant and a gasoline spill at an ethanol plant. The team also is called on periodically to assist law enforcement in the evaluation and cleanup of illegal methamphetamine labs (*Firehouse Magazine*).

Hastings Hazmat Team

Hastings in located in south central Nebraska and is the county seat of Adams County. It covers an area of 14.8 miles^2 with a population of approximately 25,000. Hastings is home to Hastings College and it is known as the town where Kool-Aid was invented by Edwin Perkins in 1927. During World War II, it was also home to the largest Naval Ammunition Depot in the United States.

Today's Modern Department

Hastings Fire & Rescue operates from two stations with 27 career and 32 part time firefighters under the leadership of Chief Brad Starling. They operate with two engines, one Quint, 1–100 feet platform truck, rescue, hazmat trailer, and medic units. During 2019, Hastings Fire & Rescue responded to 2,920 fire incidents, 3,031 EMS incidents, and 59 hazmat incidents.

Hastings Fire & Rescue houses and maintain the Hastings Rural Fire Protection District apparatus covering an additional 185 miles^2 outside the city limits in Adams and Clay Counties. They also provide advance life support medical response in Adams, Clay, Webster, and Furnas Counties. Personnel include 4 administrative staff, 27 fulltime on shift firefighters, and officers, of which there are 12 FF/Paramedics and 9 FF/EMTs. Hastings Fire Department is authorized by the City Council to have up to 32 part time firefighters. Currently, there are six part-timers on staff.

Hazmat Team History

In 1940 the "City Service Truck" was placed into service carrying sawdust and equipment to handle any gasoline or oil spill. Under Chief J.C. Mitera the Hastings Fire & Rescue began developing their hazardous materials team in the 1980s. Initially, they were trained and equipped for the Operations Level. Training mostly centered on chemicals within their response area. The first hazardous materials unit was Squad 7, which had been purchased in 1977 as a rescue truck. Squad 7 was replaced in 2002 by Rescue 1. Additionally, they have three trailers, SERT 1 a 28 feet Hazmat Response Trailer (Figure 5.85), SERT 2 a 16 feet spill trailer, and SERT 3 an 18 feet mass decon trailer. These trailers are towed by Cobra 2, a 2001 Ford F-350 with a utility body. Cobra 3 and FU 3 can be used to tow trailers when needed. There are three levels of hazmat response:

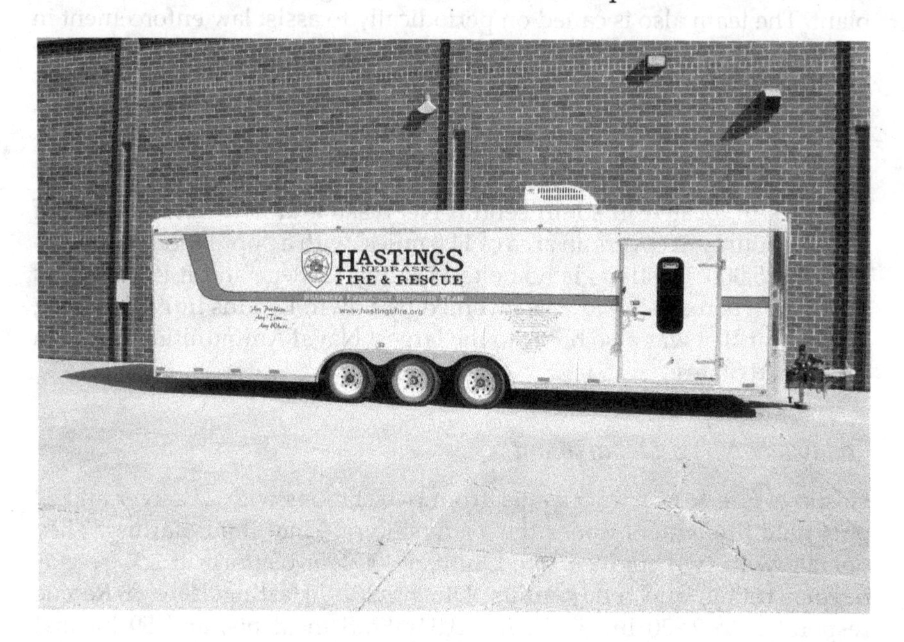

Figure 5.85 SERT 1 a 28 foot Hazmat Response Trailer is their main hazmat unit.

- 1–11 Hazmat-All off duty personnel.
- 2–11 Hazmat-Mutual Aid From Grand Island
- 3–11 Hazmat-Mutual Aid From GIFD, Red Willow Western Rural, and or the Nebraska 72nd CST.

In April 2002, Hastings Hazmat Team became 1 of 10 regional hazmat response teams in the State of Nebraska. All Hastings Firefighters are required to become Hazmat Technicians. They receive an initial 80h certification class and are required to take a 40h refresher class every 4years. An additional 40h of continuing education/training is required annually.

Hazardous Materials Exposures

Hastings has a number of transportation systems and fixed facilities where hazardous materials can be found. Hastings is located at the intersection of U.S. Highways 6 and 81. They are 12 miles south of Interstate 80. A petroleum pipeline and tank farm sets a few miles north of Hastings on U.S. Highway 81. Also located on the east side of Hastings are four ethanol plants and a coal fired power plant. Agricultural chemicals such as anhydrous ammonia, pesticides, and propane are present in Hastings as well. Union Pacific and Burlington Northern Santa Fe run through the City of Hastings with a variety of hazardous materials.

Incidents

Natural Gas Explosion and Fire

On February 10, 1979, a natural gas explosion and fire destroyed three downtown Hastings buildings. Over 100 buildings experienced broken glass. Over 100 firefighters from Hastings and six other fire departments battled the three alarm fire. The city's downtown was evacuated and people sheltered at the city auditorium. It was determined that the gas leak occurred because of cracking of aging 8 inch cast iron pipes caused by extreme cold. Leaking gas entered the buildings by way of old sanitary sewer lines, gas service lines, and a ditch line. The gas was likely ignited by a boiler in the basement of the barber shop. Gas leaks continued for several days and more evacuations were conducted for areas of downtown Hastings. There were no fatalities or serious injuries.

Naval Ammunition Depot Explosions

The largest explosion during the depots operation occurred at 9:15 a.m. on September 15, 1944, when the south transfer depot of the railroad line blew up, leaving a crater 550 feet long, 220 feet wide, and 50 feet deep.

Reportedly, nine servicemen were killed and fifty-three injured. Those killed were Coast Guard S1/C Bert E. Hugen, and Navy S1/C Leslie Williams, S1/C Freeman Lorenzo Tull, S1/C Willie Williams, S2/C Daniel Casey, S2/C Frank William David, S2/C Samuel Burns, S2/C Clarence Randolph, and S2/C Ulysses Cole, Jr. There is still speculation that the number of dead and injured was higher. The blast was felt as far away as Kansas and Iowa. There was damage in all the towns around. A portion of the roof at the Harvard school caved in, injuring ten children.

The earthen barricades in front of the storage igloos loaded with explosives held, preventing an even greater loss of life and property. Newspaper accounts of the explosion are limited due to the wartime security issues involved (Hastings Fire Department).

Lincoln Hazmat Team

Author's Note: *Lincoln was my home from early childhood. One of my best friends in grade school's dad was a Lincoln Fire Fighter and driver at Station 2 formerly on "O" Street downtown. I became interested in Lincoln Fire Department through my friend and was hooked when there was a kitchen fire next door late one afternoon. I heard the fire engine sirens and looked out to see my friends day stopping at the hydrant at the top of the hill and laid hose down the street to the house next door. Wasn't a major fire, but I knew I wanted to be a Lincoln Firefighter when I grew up.*

I did become a firefighter, but my parents moved away from Lincoln when I was a junior in high school to Illinois. We lived in Dundee and it was there as a volunteer that my fire department career started. After retirement I moved back to Nebraska and through a series of events became friends with Tod Allen Chief of the Crete Volunteer Fire Department. Tod was also a fire apparatus driver on Truck 1 at Station 1 in downtown Lincoln. He invited me to come and ride with them on "B" Shift, which I did and continue to do. So, in a sense, I finally did become a Lincoln Firefighter. Thanks to all of the firefighters and officers at Lincoln Fire and Rescue for making me feel at home and a part of them. This has been my dream come true.

Lincoln is the Capital of the State of Nebraska and home to the University of Nebraska at Lincoln the flagship school of the University of Nebraska System. The Capital City is located in SE Nebraska and is the county seat for Lancaster County. It covers an area of approximately 95 miles^2 with a population of over 293,905 in 2020.

Fire Department History

The first of Lincoln's Fire Companies originally banded together February 9th, 1872. They numbered between 40–50 men and boasted one "Steamer"

(Chapin #1), two hose carts, and paraphernalia to outfit a bucket brigade. On June 4, 1877, the membership rose to 69 men, continuing to increase until finally it became necessary to divide the force into two divisions, termed respectively, No. 1 and No. 2, with specially appointed Captains.

On August 28, 1875, the City purchases the first two horses to pull the steamer.

During 1879, the volunteer firefighters responded to 17 alarms, saving a total of $90,750 worth of property, suffering an incredibly small loss of only $2,250 on the whole. A citizen's petition was finally presented to the City Council in December, 1885, calling for the establishment of a full-time paid Fire Department. The Council quickly voted to establish one full-time, paid company to co-exist with the two volunteer companies.

The first paid Fire Company was located at Engine House #2, on the corner of 10th & 'Q' Streets. The volunteer fire companies would last just one more year before fading forever from the scene. The Lincoln Fire Department began its existence as one hose company with the equipment consisting of only one hose cart and two horses. The steamers and other apparatus remaining with the volunteer companies.

Friday, January 22, 1886, marks the date of the very first paid Fire Department alarm.

In January, 1887, the size of the Lincoln Fire Department was increased to three companies, adding 35 men to the roster, many of whom, not too surprisingly, had been experienced members of the volunteer companies.

The Lincoln Fire Department of 1920, under the leadership of Chief Neil T. Sommer, consisted of 52 men assigned to the four different fire stations throughout the city: Station No. 1, Headquarters, at 10th and 'Q' Street; Station No. 2 at 2300 'O' Street; Station No. 3 at 1223 'F' Street; and Station No. 4 at 844 North 27th Street. One of the most significant events that occurred was when the last horses were sold in 1919. The replacement of a company's horses by a gasoline engine was generally a sad occasion.

Today's Modern Department

Under the leadership of Interim Fire Chief Michael Despain, the Lincoln Fire Department has 279 uniformed personnel who operate 16 engine companies, 4 truck companies, 8 medic units, Air Cascade 14, and 1 hazardous materials company from 16 stations. Paramedics are assigned to each engine company. In addition to the hazardous materials unit, the Lincoln Fire Department also maintains a decontamination trailer and several decontamination equipment trailers for use at local hospitals for decontamination or used in areas where more mobility is required. No more tears baby shampoo is used in the shower system in the trailer because it is a very mild soap. The decontamination trailer is kept at the Training Center.

Lincoln Fire & Rescue responded 26,166 calls for service in 2018. EMS calls accounted for 20,148 of those responses or 77% of all department runs. Fire related calls were 2,878 or 11% and 1,047 hazardous materials incidents in 2018. In addition to hazmat and terrorism response the large decontamination trailer is used for rehab at fire incidents where needed. Decontamination equipment trailers would be transported and operated by mutual aid companies from rural departments surrounding Lincoln.

Lincoln fire provides training for hospital personnel and rural firefighters for conducting decontamination using the trailers. Lincoln is also home to Urban Search and Rescue Nebraska Task Force 1 (USAR-TF1). Hazardous materials technicians on the USAR Team do not normally respond to hazmat calls in the city, but are available as resources if needed. Each engine company carries 2–3 bags of absorbent materials for cleaning up small fuel spills and a patching kit for gasoline leaks at auto accidents. Larger spills would require the hazmat company response with 200 additional pounds of absorbent on the unit and an additional 500 lb at the station.

Hazmat Team

Lincoln's hazardous materials unit is located at Station 14, 1435 NW 1st Street. The unit is a 1994/2011 Pierce with a refurbished back and new cab in 2011 with a command center in the crew cab (Figure 5.86). The New HM14 was purchased from University of California Davis in 2020 and had 7,000 miles. The passenger side has all of the dress out equipment which is

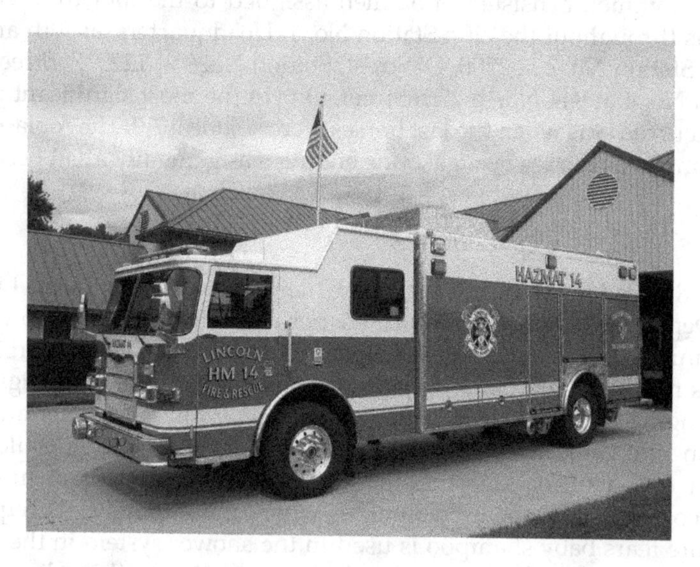

Figure 5.86 Lincoln's hazardous materials unit is located at Station 14, 1435 NW 1st Street. The unit is a 2011 Pierce with a command center in the crew cab.

stored in an inside area at the rear of the Command Center. This area has its own exterior door for entry into the area. A binder has been compiled with information regarding team members PPE sizes. Compatibility glove and PPE charts are kept with the dress out kit. Each vehicle compartment has a listing of equipment posted on the inside of the compartment door. Lincoln Fire Departments Logistics Section put in many hours making the New HM14 "Lincoln Ready".

Additional apparatus at Station 14 includes Engine 14 and Air Cascade 14. Engine 14 is a 2017 Smeal, made in Snyder, NE, with a 1,250 gpm pump. Station 14 is approximately 9 years old and features private bedrooms for all on duty firefighters. It is also home to a police sub-station, community room, exercise area, and the fire department training division offices. Lincoln apparatus has a two color emergency light system utilizing red on one side and blue on the other. In Nebraska, all emergency vehicles including police, fire, and EMS use the red/blue system.

The hazardous materials unit is not staffed full time. When a hazardous materials call is received the fire fighters from Engine 14 staff the hazardous materials unit and the engine is placed out of service. All firefighters at Station 14 are hazardous materials technicians. In addition, firefighters at Station 2 located at 33rd and Holdridge Street and Station 9 at Cotner & Vine Streets are also hazmat technicians. Station 14 generally has 5 personnel on duty and Stations 2 & 9 have 4 each for a total of 13 hazmat personnel on duty each shift.

PPE, Equipment, and Training

Personnel Protective Equipment used by Lincoln's Hazmat Team for Level A is the Kappler Tychem Responder and for Level B Kappler Tychem encapsulated and non-encapsulated suits. Tyvec sleeves have been provided for use in protecting turnouts when working at fuel spills. They are disposed of following use. Respiratory Protection is provided with MSA SCBA and 1 h bottles (carried on Air 14), MSA negative pressure cartridge respirators and MSA PAPR's. In-suit communication is provided by GE Throat mike. (Skull mikes are on order) Radios in use are Macom P7100 and are on the 800 MHz band. Hand signals are posted inside the door of the dress out compartment of the hazmat unit. A new radio system is planned for 2019.

Truck companies carry MSA Passports and CO monitors. The Health Department has a Mass Spectrometer for chemical analysis that they can bring to an incident scene. Monitoring Instruments & Identification Equipment carried on Lincoln's Hazmat Unit includes MSA 4-gas, PIDs, pH paper, and RAE System. Equipment for monitoring for Terrorist Agents includes RAD [radiation detection backpack, APD-2000 (2)] detects chemical warfare agents, pepper spray, and mace; Guardian BTA (Bio Agents); Draeger, CMS; Ludlum, RAD Alert; M 256 A Kits; M 8–9 Papers; and the CDV-777.

Hazardous materials technicians in Lincoln go through a 40 h technician level training program. Some technicians have also received WMD technician level training through the Center for Domestic Preparedness (CDP) which is located at the former Fort McClellan Army Base in Anniston, Alabama, the railroad training school at Pueblo, Colorado, the Department of Energy site outside Las Vegas, Nevada, and the EPA training programs. Eight hour quarterly drills are conducted to keep skills sharp.

Research Resources

Research resources on board the hazmat unit include NIOSH Pocket Guide, DOT ERG, Emergency Handling of Hazardous Materials in Surface Transportation, EPA Meth Lab Book, SAX's Manual, Merck Index, Condensed Chemical Dictionary, Farm Chemical Handbook, Chris Manuals, and the CAMEO computer program. They have also prepared a binder with a listing of all local contacts and contact information. Diesel spills and natural gas leaks make up a large number of hazardous materials responses in Lincoln. Interstate 80, a major East West transportation route runs through Lincoln as well as U.S. Highways 6, 34, and 77. The Burlington Santa Fe Railroad also runs through Lincoln and has a rail yard on the West end of the city.

Hazardous Materials Exposures

Lincoln has several major trucking company terminals. Several grain elevators are located in Lincoln with farm chemicals, anhydrous ammonia, and propane plants associated with the agricultural industry in the State of Nebraska. Several bulk petroleum tanks are located in the Lincoln response area. Plating companies use cyanide, acids, and caustic materials. Chlorine is stored and used at water treatment plants and public swimming pools. Goodyear has a belt and hose plant in East Lincoln with various hazardous materials.

Lincoln has a decontamination trailer, and several smaller trailers with a decontamination tent, water heaters, and other equipment for mass decontamination. The only thing needed to support the tent decontamination is a generator and water supply. Tents are equipped with lighting, a heater, and sump pump as well as a rack for rolling non-ambulatory patients through on backboards. Police sometimes use the tents for crime scene investigations during inclement weather. The inspection division has a robot available for explosives, terrorist agent, and hazardous materials incidents. Check with the Lincoln Fire Department for information on other WMD equipment.

Incidents

Picric Acid Incident

On June 24, 2011, a container of Picric Acid, a very unstable explosive material when dried out, was removed from a Lincoln residence. The owner, who was a Physics professor and had collected some chemicals from some of his classes. He had passed away and when his family was going through his things they found a container which had a hand written label "solid picric acid dry handle with care and was dated 1973". A robot was used to remove the container from the residence and placed in a bomb containment vessel. It was later detonated in a bunker in the Lincoln Air Park on the west side of town. Precautions were taken along the travel route by police blocking intersections and escorting the material to a place of safety. The incident was concluded without any issues.

Rail Car Hopper Gondola Scrap Metal Fire

On February 13, 2012, a fire occurred in a hopper car in the BNSF West "O" Street rail yard. Some of the scrap metal included combustible aluminum and magnesium. It was suspected that smoke from the fire may contain phosphine from the burning metals. Downwind monitoring indicated its presence. Levels were however, below OSHA PEL limits so no evacuation was conducted. Because of the potential for explosion when water is applied to some burning metals, it was decided to utilize Class D or Dry Chemical Agents. However, large amounts of those materials were not readily available. Because of the high sides on the rail car, they could not use an end loader to cover the fire with sand. So they decided to turn the car over and dump the burning metal on the ground and cover it with sand and cover the sand with flooding amounts of water. The fire was controlled and extinguished without further incident (*Firehouse Magazine*).

Norfolk Hazmat Team

Norfolk, Nebraska is located in northeast part of state in Madison County. Norfolk has a population 25,027 in 2020. and an area of 110 miles^2. An additional 4,228 people live in rural areas around the city.

Fire Department History

Norfolk Fire was established in 1884 as a volunteer department. Over the years, it evolved into a career department with its first paid chief hired in 1950.

Today's Modern Department

Today the department is a combination of career and paid on call firefighters (volunteers). Nine career personnel max are on duty each shift. Minimum staffing is 7 on any given shift. Norfolk Fire & Rescue provides service to the City of Norfolk, Norfolk Rural Fire District, and EMS to rural areas as well. Under the leadership of Chief Scott Cordes, they operate two Engines, two Trucks, two Rescues, and four medic units. If call type or volume is beyond the ability of on duty staff, volunteers are called in or career personnel are recalled. They operate out of two staffed stations and one unstaffed satellite station. Norfolk Fire & Rescue houses and maintains the Norfolk Rural Fire Protection District apparatus and responds to rural calls with a driver for the apparatus. The department responds to approximately 2,150 calls for service each year with fires accounting for 350 of those calls.

Hazmat Team

Ten personnel are dispatched on a regional request by Nebraska Emergency Management. Hazardous materials equipment and apparatus are housed at Station 1, 701 Koenigstein Avenue. Apparatus includes Engine 2 a 2013 Ford F-550, a 2014 Hazmat One Trailer (Figure 5.87),

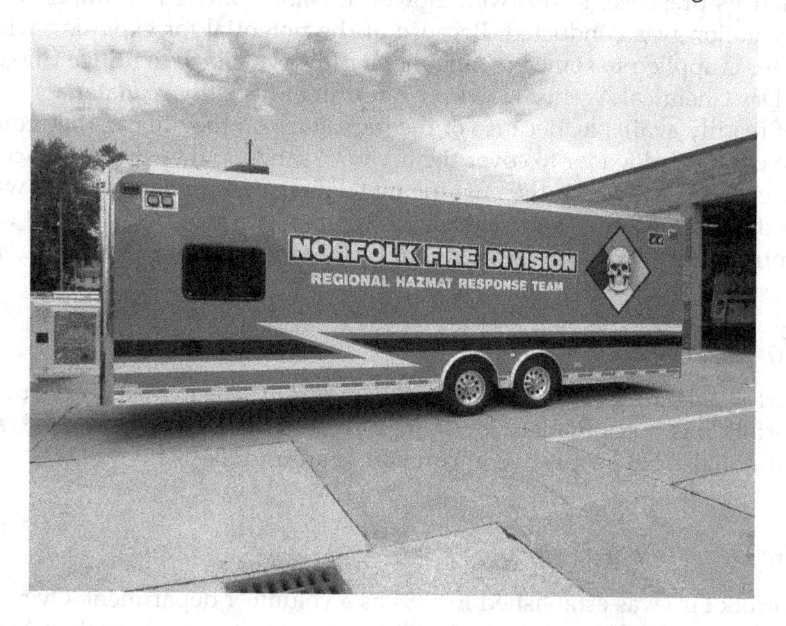

Figure 5.87 Hazardous materials equipment and apparatus are housed at Station 1, 701 Koenigstein Avenue. Apparatus includes Engine 2 a 2013 Ford F-550, a 2014 Hazmat One Trailer.

Utility 1 a 2016 Ford F350 along with decon and foam trailers. On average, Norfolk's Regional Hazmat Team responds to 18 calls a year with 3 outside the city and 15 in the city and rural fire district. Seven to nine personnel are on shift if all personnel are working.

When mutual aid is necessary for hazmat responses, area fire departments have personnel trained to the Operations Level who can assist with decon and other functions. Additional assistance is available from the Nebraska State Fire Marshal, Nebraska State Patrol, and other units of the State Regional Response Teams. Lincoln's National Guard Civil Support Team is also an available resource.

PPE, Equipment, and Training

PPE is Kappler Zytron 300, 500, and Kappler Frontline 500. Respiratory protection used is MSA Firehawk 4,500 psi with 60 min bottles, APRS MSA G1s. In-suit communication accomplished with Motorola Portable Radios. Monitoring equipment includes: RAE Systems Wireless, attach to ProRae Guardian Software, and Raelink; Mesh System; AreaRaes; Multi-Rae Pro w/gamma; Multi-Rae Pros in HCN; Q-Rae 3; Multi -Rae six sensor meters (with HCN); Multi-Rae Plus; Rae Colorimetric tubes; Ludlum 2241-1 Kit; CD V-77 Kits; Entrylink Search Camera; Sper Scientific Basic alt meter 840087; Megger & Ex Tech Eart Testers (Ground Resistance); M-8 & M-9 Chemical Agent Detector Paper; Ph paper and sticks; Hazmat Smart Strip Badges; and Haz-Chem Kit. Soon to be added are Area Rae Pros and Multi-Rae Lite.

All career personnel, including new personnel, are trained to the Technician Level as a part of their initial training. Technician training entails an 80 h technician class, 40 h refresher every 5 years, and continuing education throughout the year. Decon is performed by Operations Level Personnel.

Reference Resources

Reference materials include computer and hard copies. Engine 4 carries floor dry for fuel spills. Hazmat is called based upon the motor carrier and situation.

Hazardous Materials Exposures

Norfolk is serviced by the NE Central Railroad. Major highways include U.S. Highways 81 & 275. Fixed facilities include Helena Chemical, Nucor Steel, Cardinal Industries, NE Nebraska Ethanol lant, Milk Industries, Valero Pipeline, and Zoubek Oil.

Hazmat Incidents

Protient Propane Fire

On December 10, 2009 at 07:03 p.m., Norfolk Fire and Rescue was challenged with a once in a career type of incident when an alarm came in for a possible explosion at 704 Omaha Avenue in Norfolk. Initial information indicated a large building fire with flames and heavy smoke showing. However, upon arrival, responders encountered a 30,000 gallon fixed propane tank on fire (Figure 5.88). First order of business was to evacuate 7–8,000 people in harm's way. Norfolk Police and Firefighting staff began evacuating a 1/2 mile radius of homes and businesses. Mutual aid was summoned to cover city fire stations and the Nebraska State Patrol (NSP) provided a helicopter for aerial surveillance of the incident scene. Command staff reported that this was the game changer in the successful handling the incident, which was concluded without any injuries to the public or emergency responders. The incident was brought under control by 12:00 hours (Norfolk Fire Department).

North Platte Hazmat Team

Fire Department History

North Platte's first organized fire protection began with a volunteer fire brigade organized shortly after the arrival of the Union Pacific Railroad in November of 1866. Initial membership included 12 volunteers who were all employees of the railroad. They responded primarily to prairie fires.

Figure 5.88 Initial information indicated a large building fire with flames and heavy smoke showing. However, upon arrival responders encountered a 30,000 gallon fixed propane tank on fire.

To become a member you had to own your own leather bucket. North Platte was incorporated as a city in 1874, and the first fire brigade was reorganized into the North Platte Volunteer Fire Department in 1887. This department holds the distinction of being the oldest volunteer organization still in operation without any interruption from its inception, not only in North Platte, but also Lincoln County. In the early days the volunteers were called to service by the sounding of the Union Pacific's shop whistle. North Platte's first career firefighter was Amiel Traub appointed in 1918.

Today's Modern Fire Department: Protecting Largest Rail Yard in the World

Today the North Platte Fire Department is a combination organization with 40 career members and 36 volunteers. All 39 career firefighters are trained to the Hazardous Materials Technician Level. Volunteers train three times per month and are called to duty to provide additional personnel to supplement the career force for working structure fires and wild land fires. Volunteers also handle much of the fire prevention and public education functions of the department. They are organized into three shifts with a Sergeant and second Lieutenant for each shift. Career personnel work 24h shifts with a captain, lieutenant, and three firefighters at Station 1 and a Captain and three firefighters at Stations 2–3. A Fire Marshal, Battalion Chief and Assistant Chief round out the Chief's command staff. North Platte Fire Department is organized into three companies, which originated with the inception of the volunteer brigade in 1887. Buffalo Bill Hook and Ladder started by a $100.00 donation from William F. "Buffalo Bill" Cody.

Other companies include the Cody Engine and Hinman Hose. Each company is comprised of 12 members; 3 members of each company serve as line and support officers. Richard Paul Pedersen Jr. a department member for 40 years and the chief for 20 years now retired. Dennis Thompson was promoted to take his place. North Platte Fire Department covers an area of 369 miles^2 and responded to 3,823 alarms for service in 2019; 536 fire; 3,287 EMS; and 52 hazmat. Response equipment includes three engine companies, one truck company, one heavy rescue/hazmat, hazmat trailer, three medic units, tanker, light unit, and grass rigs.

Hazmat Team History

Hazardous materials response began in North Platte in 1995–1996 with an operations level capability. During 2002 a memorandum of understanding was established with the Nebraska Emergency Management Agency (NEMA) to transition from an operations level team to a technician level team. NEMA provided funding for the North Platte team and they became part of a regional system of teams in the State of Nebraska.

When a regional team is activated by NEMA, those personnel are paid by the state as well as back fill personnel to cover those sent on the hazmat response.

Hazmat Team

North Platte's hazardous materials unit and equipment are housed at Station 3, which is located at 3501 W. second on the west side of North Platte near the Bailey Rail Yard. The hazmat team is not dedicated so personnel at Station 3 will man the hazmat unit when a call comes in. Their response vehicle is a 2005 Freightliner, which doubles as a heavy rescue vehicle as well (Figure 5.89). Additional equipment is carried in a 20 feet cargo trailer. Also housed at Station 3 are a Quint, two medic units, and a grass fire rig. Hazmat responds to all hazardous materials incidents in the city including fuel spills. Mutual aid is available from other regional hazmat teams in the state.

PPE and Training

Level A entry suits are DuPont Tychem and Level B are also made by DuPont. Respiratory protection is provided by MSA with 60 min bottles for hazmat operations and N95 masks, Particulate Filtering Face Piece Respirators. In-suit communication is accomplished with Motorola Throat Mike 1250s.

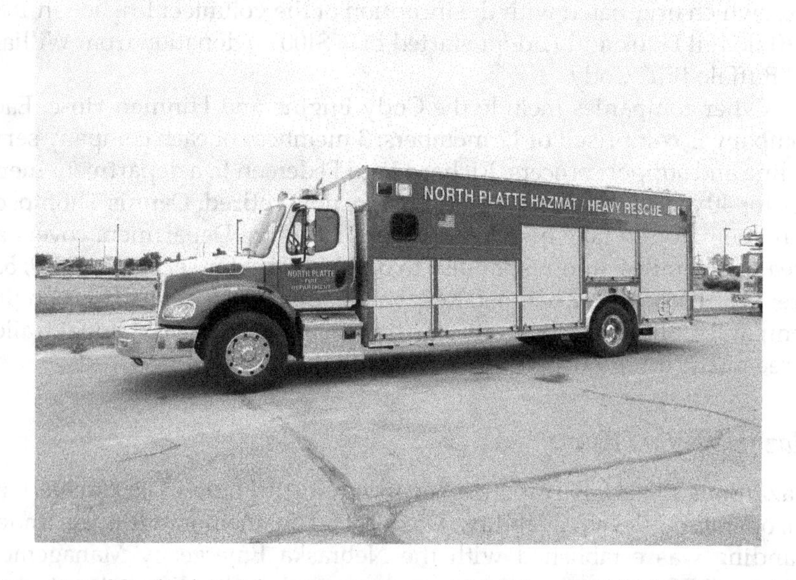

Figure 5.89 North Platte's hazmat response vehicle is a 2005 Freightliner, which doubles as a heavy rescue vehicle.

All career firefighters on the North Platte Fire Department have been trained to the technician level. Volunteers are trained to the operations level. Each career firefighter goes through an 80h technician course and additional training is taken at Anniston, Alabama for weapons of mass destruction, Pueblo, Colorado's tank car school, and Las Vegas, Nevada for radiation training.

Hazardous Materials Exposures

North Platte is located on Interstate 80 and major highways US 83 and US 30. Union Pacific Railroad has one main line and four additional lines that run through the heart of the city. Their team may also be dispatched to the Burlington Northern Santa Fe (BNSF) railroad line outside their district through the regional response system. While much of the goods transported through Bailey Rail Yard are non-hazardous commodities, a wide variety of hazardous materials are shipped through the yard on a regular basis. Bailey Rail Yard is the major fixed hazardous materials exposure with rail cars of hazardous materials passing through the yard and 4 million gallons of diesel fuel storage capacity.

Each year the railroad uses approximately 254 million gallons of diesel in the rail yard. Fixed foam fire protection has been provided by Union Pacific for the diesel fuel storage tanks. Other fixed hazardous materials exposure in North Platte include a Wal-mart Distribution Center, a tank farm with five pipelines 2 miles south of town, a Liquefied Natural Gas (LNG) facility at a truck stop, anhydrous ammonia, propane, and agricultural chemicals.

Incidents

Bailey Rail Yard Fire

During October of 2012, North Platte Fire Department and Hazmat Personnel responded to an oil storage tank fire at the Bailey Rail Yard shortly after 1:00 in the afternoon (Figure 5.90). The 15,000 gallon tank caught fire from a suspected short in a heater located just above the liquid level, which ignited the vapors in the tank. Pressure built up in the tank causing a tank rupture sending the top of the tank 150 feet away. There were no injuries but some minor damage occurred to a nearby building. The Union Pacific Foam Trailer was placed in operation to fight the fire with foam.

Bailey Rail Yard

Bailey Rail Yard, located on the west side of North Platte, Nebraska is the largest railroad classification yard in the world. It was named in honor

Figure 5.90 During October of 2012 North Platte Fire Department and Hazmat Personnel responded to an oil storage tank fire at the Bailey Rail Yard shortly after 1:00 in the afternoon.

of former Union Pacific Railroad President Ed H. Bailey. Bailey Rail Yard covers an area of 2,850 acres with a total length of 8 miles, well beyond the borders of North Platte. If the University of Nebraska Cornhuskers were to play football at the Bailey Rail Yard, they would have enough room for 3,097 football fields! Placed end to end, railroad track within Bailey Rail Yard would reach 315 miles and would cover the distance between North Platte east to Omaha, Nebraska on the Iowa border along the Missouri River.

Omaha is the location of Union Pacific's world headquarters. Each 24 h 10,000 railroad cars are handled in the Bailey Rail Yard. North Platte is home to 2,600 Union Pacific workers including those who work in the yard and shops and those who operate over-the-road trains. Union Pacific routes range from Chicago, Memphis, New Orleans, and Milwaukee on the east to Seattle, Oakland, Los Angeles, and Long Beach on the west coast. Bailey Rail Yard does not have a fire brigade or hazmat capability in house so the North Platte Fire Department handles all fires and hazardous materials emergency responses. Union Pacific contracts with a cleanup company to deal with non-emergency hazardous materials spills and cleanup of spills handled by the fire department.

The Union Pacific Railroad which was a part of the Transcontinental Railroad is the largest railroad in North America. Union Pacific is celebrating its 150th Anniversary. During World War II the North Platte Canteen operated by volunteers and funded totally by private funds served over 6 million members of the armed forces during 10 min troop train stops in North Platte. During that short amount of time, service personnel were given food items, magazines, and newspapers to take with them. North Platte was also home to "Buffalo Bill" Cody who organized his famous Wild West Show from his ranch north of the Bailey Rail Yard. His show was transported via rail and wagon.

Union Pacific Railroad and the North Platte Fire Department work closely together to handle hazardous materials incidents as well as fire emergencies that may occur at the Bailey Rail Yard. Union Pacific's 2,700 personnel at Bailey have been trained to the hazardous materials Awareness Level. According to the North Platte Fire Department Union Pacific has a great safety program for spill prevention. Union Pacific and the fire department have an excellent working relationship. The fire department is only responsible for the stabilization of a hazmat incident in the rail yard. Pre-designated entry points have been identified at Bailey depending upon the location of an incident for fire department and Union Pacific personnel. The Hazardous Materials Manager for Union Pacific at the Bailey Rail Yard assists the North Platte FD whenever needed. He provides technical support, especially for tank car issues, as well as some training for fire department personnel who respond to Bailey. Recently the Union Pacific purchased a foam trailer for use by the fire department that will be housed in a special building at the Bailey Rail Yard. It carries 275 gallons of AFFF Alcohol Type foam concentrate for ethanol and other polar solvent chemical fires and spills. Foam can be delivered from a master stream nozzle mounted on top of the trailer and two hand lines. Approximately, 300 feet of both 2½ and 1¼ inch hand lines are carried on the trailer. Water can be supplied from a hydrant, fire truck or pumped from a 10,000 gallon portable tank or a static water source. If additional foam is needed, Union Pacific has trailers in Salt Lake City, Utah and Des Moines, Iowa that can be mobilized and mutual aid from the Burlington Northern Santa Fe railroad is available if they are close by. There is another 350 gallons of alcohol type foam at the North Platte airport.

Union Pacific Hazardous Materials Manager for Bailey Rail Yards, to the Continental Divide in Colorado and Wyoming, Western Kansas and 3/4 of Nebraska provided training to the North Platte Fire Department on use of the foam trailer. The trailer is available to the fire department for use off site of Bailey Rail Yards if railroad property is affected or if life and safety are in jeopardy elsewhere. Foam would have to be replaced by the responsible party for the latter. Union Pacific Railroad provides approximately 2,500 emergency responders with training annually in the areas served by the railroad. During 2012, Union Pacific completed its 50th class at the Association of American Railroads Security and Emergency Response Training Center in Pueblo, Colorado. They also participate in industry-wide whistle-stop training tours and TRANSCAER (Transportation Community Awareness and Emergency Response) (*Firehouse Magazine*).

Omaha Hazmat Team

Located on the banks of the Missouri River at the Eastern end of Nebraska, Omaha is the largest city in the state. Omaha sets on the West side of the river across from Council Bluffs, Iowa. The Omaha Fire-Rescue Department provides fire, rescue, hazardous materials, and EMS services to 506,022 people within 193.18 miles^2. Including the suburban areas adjacent to the city, the Omaha-Council Bluffs Metropolitan area is home to over 1.2 million people. In addition, through mutual and automatic aid fire, rescue and hazmat service extends into Douglas County and other surrounding areas.

Fire Department History

Omaha's organized fire service began on May 2, 1860 with the formation of the Pioneer Hook & Ladder Company #1. On July 10, 1866 due to the increased growth of the city a second fire company was formed, Fire King #1 which operated a hand engine purchased by the city from Davenport, Iowa. Engine companies 2 & 3 were soon organized to assist the existing companies. Continued rapid growth resulted in the consolidation of fire companies in the city into the Omaha Fire Department on April 23, 1875.

The newly created fire department was volunteer except for the offices of chief engineer (who received a salary of $1,500 per annum) engineers, drivers and stokers. In addition to the Omaha fire companies the Union Pacific Railroad formed the "Durant Engine and Hose Company No. 1," in January 1868 for fire protection of railroad properties. This fire company was not part of the Omaha Fire Department but was subject to its call upon special notice from the chief. The organization was purely voluntary, however, no one but employees of the Union Pacific Railroad were eligible for membership. Upon several occasions the company was called upon by the Omaha Fire Department for assistance. The first paid firefighters in the Omaha Fire Department started work in 1867.

Today's Modern Department

Modern Day Omaha fire companies are housed in 24 stations located throughout the city. Omaha Fire has approximately 650 uniformed personnel led by Fire Chief Daniel Olsen Omaha firefighters operate 24 engine companies, 9 truck companies, 16 medic units, 1 rescue unit, 2 hazmat units, 6 battalion chief units, and 1 paramedic shift supervisor unit. Additional equipment includes six brush units, one command vehicle; one technical rescue truck, two mass casualty trailers; and one Medicart. EMS and rescue calls make up about 3/4 of the 61,353 annual total alarms. Fire calls account for approximately 5% and other types of alarms account for the other 22%. Omaha Fire does not provide fire protection to Eppley Airport, which has its own fire department operated by the Omaha Airport Authority.

Hazmat Team History

The Omaha Fire Department began organized hazmat response in the city in 1985. Initially four firefighters were assigned to an engine company with dual responsibility for fire suppression/EMS and hazardous materials response. During the mid 1990s the hazardous materials team became dedicated with four team members assigned to the hazardous materials unit per shift. Their first hazmat response unit was a delivery truck type vehicle. The hazmat team was assigned to a station along with an engine company. Engine company personnel were assigned to a back-up hazmat role in addition to normal engine company responsibilities.

Hazmat Team

In 1995 the Omaha Fire Department created a Special Operations Program which included hazmat, Rope Rescue, Trench Rescue, and Rapid Intervention Response. A standard type rescue hazmat vehicle was purchased in 1998 to increase the effectiveness of the team. Additional equipment was purchased to carry out the missions of the Special Operations Team. In 1999 a WMD trailer and associated equipment were placed into service. The trailer carries extra PPE, miscellaneous hazmat support equipment, decontamination tents, and decontamination equipment. A second hazmat vehicle was purchased in 2005 giving Omaha Fire the capability to place two dedicated hazmat teams into service. Both hazmat units are housed with an engine company that is also staffed with four hazmat technicians. The assigned engine companies serve as back-up for Hazmat and Special Operations tasks.

Omaha's Hazmat Team responds to an average of 1,200 calls per year including fuel spills, gas odors, and leaks. Engine companies carry an oil pick up absorbent product that is a pumicite volcanic ash derivative. All engine companies also carry a surface washing agent (Petro Green) in a (2) gallon sprayer. A 2/1 ratio of water to Petro Green is used. Engine companies will call in the hazmat team if the spill is larger than what they can handle with on board supplies. Any spill other than a hydrocarbon fuel spill will result in the hazmat team responding regardless of size of the spill. Rescue 1 (Hazmat 1) is housed at Station 60, located at 2929 S. 129th Avenue. Rescue 1 is a 2005 Emergency One with a Cummins Diesel 500 HP engine (Figure 5.91). Also housed at Station 60 is Engine 60, a 2010 Crimson with a 2,000 gpm pump and a Technical Rescue Trailer. Rescue 33 is a 2013 Rosenbauer Rescue Body housed at Station 33 (Figure 5.92), 3232 South 42nd Street. Engine 5 2016 Rosenbauer 2,000 gpm housed at Station 5, 2209 Florence Blvd.

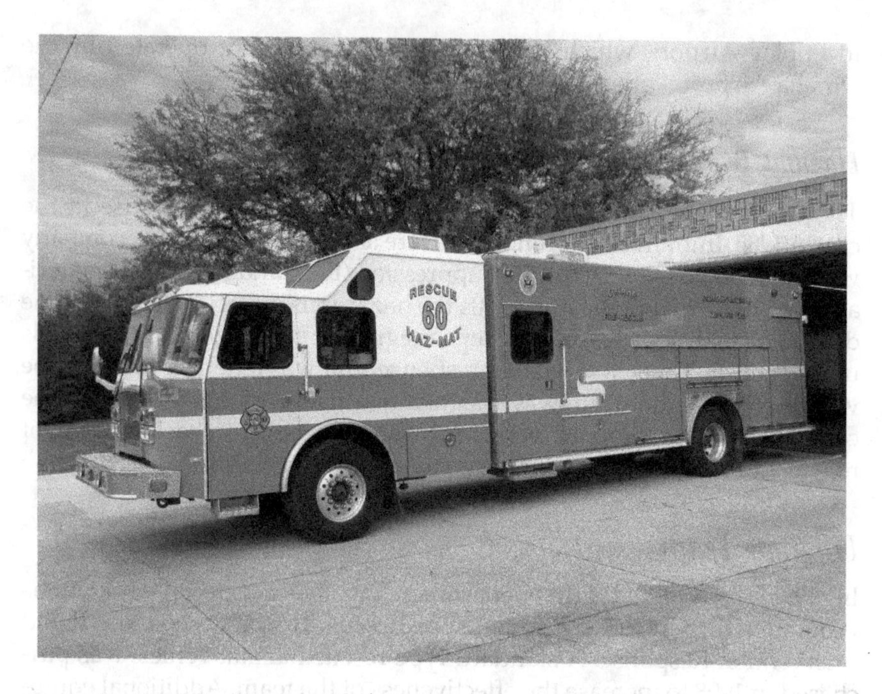

Figure 5.91 Rescue 1 is a 2005 Emergency One with a Cummins Diesel 500 HP engine.

Figure 5.92 Rescue 33 is a 2013 Rosenbauer Rescue Body housed at Station 33.

Most of the equipment carried on the hazmat vehicles is typical for hazmat and WMD response situations. On a daily basis, each hazmat unit is staffed with four technicians. Each engine company located at Stations 5 and 60 is also staffed with four technicians. So on any given day the ideal staffing would be 16 technicians between the hazmat units and engine companies. There are 48 total hazmat technicians assigned to Stations 5 and 60 on all three shifts.

There are approximately 25 additional Hazmat Technicians on duty on each shift scattered about on other apparatus around the city that can be called into service if needed. Mutual aid for hazmat incidents is available from the Nebraska USAR Task Force 1, Omaha Police Department, City of Omaha Police Department, Douglas County Health, Union Pacific and National Guard Civil Support Team. Omaha's Hazmat Team has mutual aid agreements with surrounding counties/towns, and they are part of the State MOU Hazmat response plan.

PPE, Equipment and Training

PPE worn by the Omaha Hazmat Team for Level A One Suit-Saint Gobain Kappler Zytron and for Level B Non-encapsulated Kappler. Respiratory Protection is provided with Scott SCBA 30 min and 1 h bottles, 5 min escape air packs and Supplied Air Respirators. In suit-communications is accomplished with Motorola APX 6000. Research Resources include Laptop, IPAD, ERG and TIER II Reports.

Monitoring instruments & identification equipment used by the Omaha Hazmat Team includes Draeger CMS Kit, Mini Rae PID Meter, Multi Rae 5 Gas Meter, Industrial Scientific T82 CO Monitor, Industrial Scientific M40 4 Gas Monitor, Haz Cat Kit, APD 2000, Smith's HazMat ID, Smith's Gas ID, Ludlum 2241 radiological detector, and the ThermoEberline FH406-L Radiological Monitor.

Hazmat personnel are required to be trained to the technician level IFSAC/Proboard. Hazmat Technicians must also go through 40 h of Rope Rescue/high angle, 40 h Confined Space training and 24 h of Trench Rescue Training. Omaha Fire Instructors are used for Rope Rescue/Confined Space initial and refresher training. They also use a contractor from Colorado (Innovative Access Inc.) for training as well. Hazmat personnel assigned to Stations 5 and 60 (16 per shift, 48 total) receive 24 h of annual HAZWOPER training. All Omaha firefighters are trained to the Hazmat Operations Level IFSAC/Proboard.

Hazardous Materials Exposures

Transportation exposures in the Omaha area include Interstates 80, 480, and 680. The Missouri River is a major barge traffic route on the East side of the city. Omaha is also a major rail center with potential exposures from chlorine, ethanol, various liquid chemicals, and pressurized and non-pressurized chemicals. There are approximately 220 facilities in the Omaha area that must file Tier II reports each year. Some of the major chemicals listed are anhydrous ammonia and sulfuric acid. Various pipelines carrying hazardous materials can also be found throughout the response area.

Incidents

Omaha Fire Department Hazmat/SP OPS units have been assigned to do air monitoring for numerous high profile events some which include NCAA Men's College World Series, NCAA Men's Basketball first/second Rounds, Sweet 16 and Elite 8 Tournaments, NCAA Women's Volleyball Final Four, USA Olympic Swim Trials, US Senior Golf Open, and Presidential/Vice Presidential Visits. Major incidents last 10 years include Hospital Leak, Food Processing Leak, International Nutrition Collapse, Rail Car Explosion, and Hair Product Plant Leak (*Firehouse Magazine*).

Scottsbluff Hazmat Team

Scottsbluff was laid out in 1899 across the North Platte River from its namesake a bluff that was named after Hiram Scott a fur trader with the Rocky Mountain Fur Company. He was found dead in the vicinity on the return trip from a fur expedition. The namesake bluff is now a U.S. National Park called Scotts Bluff National Monument. Scottsbluff is located in western Nebraska in the great plains region of the United States and the Panhandle region of Nebraska. Scottsbluff is the largest city in the Nebraska panhandle population is 15,039 and the city covers an area of 6.27 miles^2. Gering a smaller community founded in 1887 is located next to Scottsbluff and the two towns have grown together to form the seventh largest urban area in Nebraska.

Fire Department History

Scottsbluff Fire Department was formed in 1900 as a volunteer department. In 1916 the first career firefighter was hired. This individual was tasked with maintaining and driving first motorized fire truck. The combination one career and volunteer fire company met the community's needs until 1926. Two additional drivers were hired in order to form three shifts. During the 1940s the department evolved into a 24 h career department with several officers. In 1953 the city and rural departments joined forces and worked side-by-side. By the 1960s, there were seven firefighters per shift, staffing two stations. During an economic down turn, Station 2 was no longer maintained and six personnel staffed Station 1. In 1995 the partnership with Scottsbluff Rural Fire District ended. During 2018, all volunteer services ended due to lack of participation.

Today's Modern Department

Scottsbluff Fire Department is led by Chief Thomas Schingle with 17 uniformed personnel, 5 per shift with 1 chief, 1 assistant chief, and 1 fire prevention officer. Firefighter work a 48/96 shift schedule. Scottsbluff FD operates one engine company (one in reserve) and one rescue BLS. Manning is with a minimum of 2 and the truck company is not staffed. Other equipment includes Dive Rescue, 32 feet hazmat trailer, fire investigation unit, and one brush unit. During 2018, firefighters responded to 2,026 calls for service. EMS calls accounted for 1,316, fires 64, hazardous condition (no fire) 303, false alarms 108, and miscellaneous 372.

Hazmat Team History

Scottsbluff placed a hazardous waste disposal Unit into service in 1988. Around 1996 personnel began obtaining technician-level training. The National WMD survey in 1999 prompted the team to push for more technicians. The State of Nebraska began developing State Emergency Response Teams (ERTS) around the same time. Through a memorandum of understanding, Scottsbluff Fire became one of the ten SERTs for hazardous materials response in the State of Nebraska.

Hazmat Team

Scottsbluff's hazmat trailer is located at 1801 Avenue B in the main fire station. The trailer is a 32 feet Pace Cargo Sport (Figure 5.93). Scottsbluff averages three hazmat calls per year. Absorbent materials are carried on the engine. Four up to five hazmat technicians are on duty each shift. Thirteen of the department's firefighters are trained to the technician level. When mutual aid is required it would come from North Platte or Red Willow Western Rural Hazmat Teams.

PPE, Equipment, and Training

Various Level A & B chemical suits are carried that are the same as other teams in the state response system. In-suit communications is accomplished with Motorola HT1250 radio with Motorola mic fixed to mask.

Respiratory protection utilizes Draeger SCBA with 1 h bottles, which is the only type of protection they carry. Monitoring and detection equipment includes Smith's detector LCD 3.2 E, Ludlum 2242-2 radiological detector, Ahura First Defender, Hazmat ID, Draeger tubes, and HazCat.

Hazmat team members receive 24 h of refresher training every 2 years.

Figure 5.93 Scottsbluff's hazmat trailer is located at 1801 Avenue B, in the main fire station. The trailer is a 32 feet Pace Cargo Sport. (Courtesy: Scottsbluff Fire Department.)

Reference Resources

Reference materials both electronic and hard copy include CAMEO, Emergency Care for Hazmat Exposure, Radiological Emergency Response Handbook, and others.

Hazardous Materials Exposures

Transportation highway routes include I-80, Nebraska 92, U.S. 26, and 365. Union Pacific and Burlington Northern Santa Fe railroads service Scottsbluff. Fixed facilities include compressed and liquefied gas storage, commercial cleaning storage, chemical storage, bulk fuel storage, chemical storage and transport, and welding supplies.

Incidents

On November 2, 2000 a Burlington Northern Santa Fe train derailment released benzene/dicyclopentadiene near the Simplot Soilbuilders facility in Scottsbluff. There were no fires, injuries, or exposures to the hazardous materials (Scottsbluff Fire Department).

McCook Red Willow Western Rural Fire Protection District Hazmat Team

Author's Note: *Red Willow Western Rural Fire Protection District was one of the most interesting visits I have made on my travels across the United States visiting fire departments. This department is all volunteer, but you would not know it unless someone told you. The volunteers I witnessed during my visit were dedicated beyond the normal, very well equipped in terms of apparatus, tools and ingenuity. This department is a family operation both in terms of blood and brother hood and sister hood. Chief Bill Elliott his daughter Billie and her husband Jeffrey Cole are the heart and soul of this department. While I only visited with three of the volunteers, I have no doubt that the rest of them are cut out of the same mold.*

I did not get to meet Chief Elliott, but I could feel him with us at every step of the visit. They have prepared very well to deal with the hazards they have to face in terms of hazardous materials response and fire response. They are located in a very rural part of Nebraska however they have been known to travel far and wide to help other departments in time of need. They have some of the best wildfire equipment that I have ever seen. They have made good use of surplus equipment available from the government and their entire operation from what I witnessed is FIRST CLASS. I would strongly encourage anyone who travels anywhere near McCook, NE to stop and visit. You won't be sorry you did.

Red Willow County is located in the SW corner of Nebraska on the Kansas Border. The county seat is McCook and the RWWRFPD Station is also located on the Southwest side of McCook. McCook has a population of 7,898 and covers an area of 5.39 miles^2. Red Willow County outside of McCook has a population of 3,157 and covers an area of 718 miles^2. McCook has its own fire department that covers the city and works with RWWRFPD as needed.

Fire Department History

RWWRFPD was started in 1977, following a fire which destroyed a rental home north of Perry. It is a completely volunteer fire department. Don Olsen who owned the rental property and five other men started the RWWRFPD. During 1978 the department moved into the first rural station with equipment supplied by the Nebraska Department of Forestry. The fire protection district covers an area of 235 miles^2, 156 in Red Willow and 78 in Furnace Counties. McCook Fire Chief Don Einspahr acted as chief for both departments training volunteers and keeping the newly formed district up to date.

In 1983, Chief Einspahr offered to train a volunteer to replace him as chief. At that time the rural department relocated equipment and a truck to the city fire department and assist Einspahr with city calls until an agreement was made between the two administrations. Gene Wegner, one of the first volunteers took over as fire chief in 1985. On July 17, 1991 the current fire station was purchased at the Golden Rod Trading Post on Federal Avenue. In 1993, Bill Elliott was named chief and remains the chief to this day. Bill Elliott was responsible for the formation of the hazmat team and continues to keep the department on the leading edge of firefighting and hazmat response in Western Nebraska.

Today's Modern Department

In 1993, Bill Elliott was named chief and remains the chief to this day. Bill Elliott was responsible for the formation of the hazmat team and continues to keep the department up to date on training and equipment. Elliott and Co Chief Mike Allen are the longest members on the department with Chief Elliott serving 41 years and Chief Allen 44 years. Presently, there are 25 volunteers and 2 junior members on the department. RWWRFPD responds to an average of 50 fire calls a year and of those approximately 10 are hazardous materials related. RWWRFPD personnel operate one engine, one truck, one rescue, three tenders 2,500 gallons each, four brush units, two Mules and Argo, all for brush. They do not respond to EMS calls. Hastings and the State Patrol are available for Mutual Aid.

Hazardous Materials History

During 1990, RWWRFPD formed the first hazardous materials team in the area. Specialized equipment and a truck were purchased for the new hazmat team.

Hazmat Team

RWWRFPDs Hazmat Unit is a 1998 Freight Liner Chevy (Figure 5.94). An equipment trailer is towed Chevy 3500 with a crew cab and compartment body for carrying equipment. There are 12 hazardous materials technicians on the team. The City of McCook does their own responses to fuel spills and odor complaints. Their pumper carries floor dry for spills in the district. Ten of the department members are Hazardous Materials Technicians and other volunteers are trained to Awareness & Operations. Mutual aid if needed is available from Hastings or North Platte.

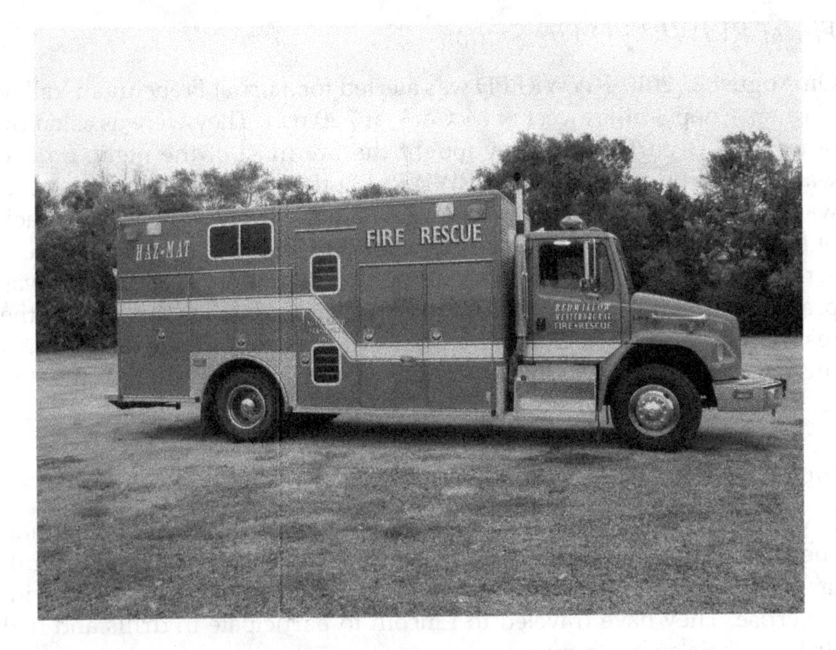

Figure 5.94 RWWRFPDs Hazmat Unit is a 1998 Freight Liner Chevy.

PPE, Equipment and Training

They use Zitron and Kappler PPE for Level A&B. MSA SCBA are used with 1 h bottles. The in-suit communication is accomplished with hand signals and portable radios and ear mikes. Monitoring equipment includes Multi Rae, 5 gas, 4 gas, Draeger Chips, and Smith ID. Technicians 80 h courses following a 40 h Operations course and 8 h Awareness. Firefighters receive Awareness and Operations training. Some of the personnel have gone to the rail car school in Colorado.

Reference Resources

Reference materials are a combination of computer and hard copy.

Hazardous Materials Exposures

Major highways in the district for transportation of hazardous materials include U.S. Highways 6, 34, and 83. Burlington Northern Santa Fe railroad goes through the initial response areas. An above ground natural gas pipeline also exists in the fire district. Fixed facilities include two ethanol plants; Agricultural facilities for propane, anhydrous ammonia, and pesticides; and Meth Labs.

Fire at RWWRFPD Fire Station

On August 27, 2010, RWWRFPD was alerted for a fire at Frenchman Valley Farmers Coop 3 miles west of McCook at 7:20 p.m. They were assisted by other local departments. They fought the fire most of the night. Smoke was spotted by a passer-by at RWWRFPD the next morning and a call was paged out at 06:00 hours. Apparently the fire burned itself out for lack of fuel and there was very little fire when McCook City Fire Department arrived on scene. It started in a pumper in the station. Fire damage was pretty much limited to the apparatus with smoke damage throughout the rest of the station. The State Fire Marshall investigated and determined the cause to be electrical in the pumper that was in the station. Estimated loss to the pumper and station was $150,000.

Incidents

RWWRFPDs Hazmat Team has responded to many incidents of common chemicals found in almost in any community. Because they are familiar with these materials handling them has been quicker, safer, and with expertise. They have traveled to Lincoln to participate in drills and had drills in their own community.

Propane Tank Leak McCook

On August 17, 2017, 10:30 a.m., Red Willow Hazmat Team responded to McCook with 13 team members to assist the McCook Fire Department with a propane leak. The 350 gallon tank was in a residential area. During filling of the tank, it was discovered that a seal around a gauge had failed and the leak began. Neighbors were evacuated from the area. Vapor from the leaking tank was suppressed by spraying and misting water over the tank. Electricity was shut off in the area to control ignition sources and prevent the propane from igniting. It was determined the tank could not be moved from the area and there was no way of shutting off the leak. Patching was attempted but unsuccessful. Mitigation could only be accomplished by allowing the tank to vent its 200 gallons of product into the atmosphere. By 3:30 p.m. the tank was nearly empty and air monitoring of the area showed insignificant levels of vapor.

One home did show levels within the explosive limits of propane and was ventilated before the evacuation was lifted. Residents were allowed back into the neighborhood and the power was turned back on. Fire Chief Harpham of McCook Fire Department praised RWWRFPD for their quick response and assistance. "They are one of only ten teams in the State of Nebraska. We are fortunate to have them in McCook and their expertise in the area of hazardous materials went a long way in mitigating this incident. We can't thank them enough."

Hydrochloric Acid Spill Trenton

On March 29, 2018 while driving on a the River Road 5 miles east of Trenton, NE a tanker hauling hydrochloric acid flipped over while making a turn on a on a gravel road at the corner of Roads 714 and 369. Small amounts of diesel fuel and motor oil leaked from the truck. Trenton Fire & Rescue was called to the scene around 7:00 a.m. Upon recognizing, this accident involved hazardous materials; the RWWRFPD Hazmat Team was summoned for assistance. McCook Fire Department and Community Hospital were put on alert in the event the situation worsened.

Examination of the tanker revealed no hydrochloric acid was leaking. The decision was made to upright the tanker with four wreckers. It took 11 h to mitigate the incident safely and there were no injuries or complications.

Anhydrous Ammonia Leak McCook

An ammonia leak occurred at French Man Valley Coop (Figure 5.95). McCook City Fire Department responded and called for RWWRFPDs Hazmat Team. Personnel arrived and surveyed the situation. Technicians made an entry in Level A PPE and shut off a valve. Someone stealing anhydrous left the valve open. Anhydrous ammonia is used in the manufacture of methamphetamines (RWWRFPD).

Figure 5.95 An ammonia leak occurred at French Man Valley Coop in McCook. (Courtesy: RWWRFPD.)

New Orleans Hazmat Team

Prolog: *News media and others have said you couldn't appreciate the devastation caused by Hurricane Katrina in New Orleans from the news video footage and photographs presented in the media. What an understatement! I had the opportunity to visit New Orleans on December 23, 2005 to tour the city and talk with hazmat team members. While we toured the devastation caused by Hurricane Katrina my camera spent more time in my lap than taking photos. I did take photos, but nothing I saw that day could be representative of scope of the disaster in a photograph. It is one thing to see media coverage of Hurricane Katrina, but to be there and experience the aftermath first hand was an experience I will never forget. I had made several visits to New Orleans over the years and love the city so much. It is like no other, at least in the United States. Then to visit her after Hurricane Katrina was heart breaking.*

Fire Department History

Founding

On Good Friday in 1788, March 21, at approximately 1 p.m., a great fire occurred in New Orleans. A candle on an alter in the home of Don Jose Nunez, the Colony Treasurer accidentally came in contact with lace curtains. Gale force winds blowing in the area fanned the flame, lighting Nunez' home on Chartres Street totally ablaze. In those days, bells on the Church of Saint Louis called citizens to assembly during emergencies. Catholic tradition says, however, that bells never toll on Good Friday. As a result, no alert was sounded. Within a matter of five hours the fire burned itself out. A tally of damages showed that four-fifths of the Colony lay in heap of ashes. What had taken 70 years to accomplish, was gone in a mere five hours.

In April 1829, a group of men joined to together to fight fires, the first serious step to organize a first class Volunteer Fire Department in New Orleans. These volunteers went under the name of the Firemen's Charitable Association (FCA), a title it kept for 62 years until the first paid fire-fighting force, the New Orleans Fire Department, came into existence on December 15, 1891.

Career Department

Chief Engineer Thomas O'Connor, leader of the volunteers, remained to become the first Chief of the New Orleans Fire Department, bridging the transition, and continuing to lead it for the next 20 years. Henri Buckman was selected as the foreman and first chartered member of the company. Buckman was also present when the New Orleans Fire Department went into service in 1891. Because of his 62 years of devoted, efficient service to the FCA, Henri Buckman was officially declared the father of the Fire Service in New Orleans.

The first fire fought by the paid New Orleans Fire Department occurred on February 17, 1892 at 9 p.m. at the corner of Canal and Bourbon Streets. Kinked hose lines delayed the attack on the fire which extended to the next building. As at all major fires, a large crowd gathered, among them many volunteer firemen, who pitched in to help out their comrades-in-arms. Leaping over the restraining lines, they straightened the kinked hoses, enabling the fire at A.W. Schwartz' General Store to be extinguished.

Today's Modern Department

New Orleans Fire Department is led by Fire Commissioner Timothy McConnell and operates out of 31 fire stations. Firefighters operate 22 engines, 5 trucks, 8 squirts, 1 Quint, 2 rescues, 2 squads, and 1 hazmat unit.

Hazmat Team History

The New Orleans Fire Department established a Hazardous Materials officer in the mid 1970s for training and coordination of response personnel. As more emphasis was placed on hazardous materials incidents, the New Orleans Fire Department established a dedicated Hazardous Materials Team and custom-designed an apparatus in 1989. Since the inception of the team the focus has been on specialized training and equipment to maintain an effective response to varies types of hazardous materials incidents.

Several acts of terrorism occurred in the United States during the 1990s; these events highlighted the need for domestic preparedness by first responders. The Hazardous Materials Team began a transformation in the 1990s from one of chemical response to a more diverse role. The incidents now include Chemical, Ordinance, Biological and Radiological (COBRA), and Weapons of Mass Destruction (WMD).

Hazmat Team

The Hazmat Unit has three personnel assigned to it per platoon, One Captain, One Operator, and One Firefighter. Should a Hazmat incident occur and more personnel be needed, the Rescue and/or Flying Squads may back up the Hazmat team.

Under ordinary circumstances, a visit to the New Orleans Hazmat Team would begin at Station 7 located at 1441 Saint Peter Street (Figure 5.96) in the Treme-Lafitte Neighborhood. There Hazmat 1 and Hazmat 2 are headquartered along with Rescue 1, Engine 7, a mass decontamination trailer, and other hazmat support equipment. However, Station 7 became one of the casualties of Hurricane Katrina with 4 feet of water inside that

Figure 5.96 Under ordinary circumstances, a visit to the New Orleans Hazmat Team would begin at Station 7 located at 1441 Saint Peter Street.

Figure 5.97 Following the hurricane Station 7 was uninhabitable due to mold, bacteria, and structural damage. (Courtesy: New Orleans Fire Department.)

remained there for 10 days. Station 7 was uninhabitable due to mold, bacteria, and structural damage (Figure 5.97).

The building housing part of Station 7 was built around 1900 as a bar with living quarters above. Upstairs rooms were later changed to accommodate a brothel. The building fell into disrepair and a fire occurred in the attic at some point. Charred wood framing members still remain today. Eventually the city purchased the building and converted it into a firehouse, adding a three bay apparatus floor, office area, and day room. As of the publication date of this book, Station 7 was still undergoing repairs and a projected re-occupancy date was unknown.

Hurricane Katrina

My day in New Orleans began by following directions from Chief Woodridge to the site where Hazmat 1 was located. I proceeded to an area along the Mississippi River behind the Hilton Hotel on the river side of the flood gates next to the Cruise Ship Sensation. There I found Hazmat 1 parked next to a security fence in front of several bright yellow tents which turned out to be the temporary headquarters for the hazmat team (Figure 5.98). New Orleans Hazmat 1 is a 2000 custom built American LaFrance. They also respond with a Chevy Suburban purchased with federal grant funds for assessment and control (Figure 5.99). Both units have satellite communications capability that had ironically been installed just

Figure 5.98 Security fence in front of several bright yellow tents were the temporary headquarters for the hazmat team.

Figure 5.99 New Orleans Hazmat 1 is a 2000 custom built American LaFrance. They also respond with a Chevy Suburban purchased with federal grant funds for assessment and control.

1 week before Katrina hit. New Orleans Hazmat normally responds for Hazmat and WMD calls to the parishes of Orleans, St. Bernard, Jefferson, and Plaqumines.

Captain Don Birou was the officer in charge when I arrived. We sat down at a table in his tent quarters and I began to hear the most amazing stories of victim rescue firefighter survival and hazmat response following Hurricane Katrina. Having spent over 28 years in the emergency services I never dreamed I would hear about fire department or hazmat personnel having to operate under such desperate conditions within the United States. Some of the conditions I was about to hear sounded like something right out of Hollywood. Firefighters in New Orleans who remained there to deal with the aftermath of Hurricane Katrina are heroes in the truest since of the word. Following the hurricane they survived, operated, and rescued thousands of victims under impossible conditions. This would be a day burned into my mind for as long as I live.

Awaiting the arrival of Hurricane Katrina, New Orleans' hazmat personnel were staged in the city's convention center to ride out the storm. Once the storm subsided, personnel emerged to survey the aftermath of the Category 3 storm. Surprisingly, there wasn't the catastrophic damage many had expected. It looked as though New Orleans had dodged the bullet once again. Power was out but back-up systems designed for

24–48 hours of operation in the emergency communications center and other facilities in most cases were in full operation.

Reports began to filter in on Monday August 29th that water was flowing over some levees and some were beginning to fail. The city rapidly began to fill up with water in low lying areas (much of New Orleans is below sea level) as Lake Pontchartrain swollen by the storm surge of the hurricane began to flow into New Orleans. The real disaster was just beginning. Ultimately, almost 80% of the city would be under water including the 911 and communications centers and many of the city's fire and police stations. Nearly 70% of New Orleans fire stations were under water or damaged. Some had been looted. At the time of my visit, 12 of the 33 fire stations in the city were operational along with 2/3's of the fire apparatus. Following the storm nearly 75% of firefighter families were living outside of New Orleans and most lost their homes and belongings.

Prior to Hurricane Katrina the New Orleans Fire Department had no boats. Their only fire boat had been decommissioned a few months earlier and would have been useless any way because of its size and location (on the Mississippi River). Firefighters were asked and many brought their own boats to work with them the day of the hurricane. Thousands of people were rescued by fire fighters using their own personal boats. The city was isolated by the floods. There was no way for mutual aid to reach the city by road, everything was flooded. Since the hurricane, they have acquired 18 boats (15 foot' flat bottom metal boats) through a donation from the Leary Foundation.

On August 31, 2005, backup batteries and generators for the department's communications network went dead. There was no way to replenish the fuel or batteries needed to keep the communications network on line. Firefighters who were on duty the day the levees broke were stranded in the city and on their own once radio and telephone communications systems went down. They were without food and water and in many cases without quarters. All cellular and regular phone service was disabled by the hurricane. Firefighters were forced to use written messages delivered by runner to exchange information. Satellite phones were the only means of communications and only two of them were available.

Captain Don Birou was able to get through to his sister at Phillips/Conoco Emergency Operation Center in Houston. He expressed the dire communications situation in New Orleans and asked if there was any help that could be provided. Captain Birou's sister made some contacts and within 2 days there were 10 hand held satellite phones at FedEx in Baton Rouge. According to Captain Birou, once received these "10 handheld phones became the most important items on the fire department". After about 5 days the direct connection on their Nextel phones began working.

Nine hazmat specialists on duty the day of the hurricane ultimately worked 24/7 until December 18th except for a mandatory rehab September 6–12! Personnel from the New York City Fire Department were assigned to New Orleans Hazmat during the rehab period. As New Orleans slipped into anarchy and police protection was limited or non-existent the Superintendent of the Fire Department told firefighters who had personal weapons to carry them and they became their own police force for protecting themselves and other personnel.

Each new day began with hazmat team members foraging for boats, fuel, flashlights, tire repair kits, food, and water to accomplish the rescue mission before them and for survival themselves. During the first 2½ weeks, their time was spent on rescue operations. One of the rescue operations at the Louisiana State University Dental School turned out to be 100 plus firefighters, police officers, and New Orleans Health Department Paramedics who were using the structure as a safe haven from hurricane Katrina. Flood waters from the broken levees surrounded the Dental School trapping the personnel for 4 days.

When not conducting rescue operations the hazmat team members spent time evaluating a list of 93 Tier II Facilities with Extremely Hazardous Substances (EHS). They also searched through 430 facilities without EHS substances looking for railcars and other hazmat containers with leaks or damage. They also searched for "orphaned" containers carried away from facilities by the flood waters. On a New Orleans City map a grid system was drawn to identify areas of the city that needed to be searched. As of December 7, 2005, over 10,817 containers of hazardous materials had been located and evaluated.

Examples of containers are listed below:

- Glass Bottles Lab chemicals, solvents, corrosives, oxidizers, flammable liquids, dangerous when wet phosphoric materials, toxic, radioactive, and unknowns.
- 30–70 lb Cylinders LPG, Propane
- 150 lb Cylinders Oxygen, acetylene, helium, argon, and nitrogen
- Drums, pails, Hydrocarbon fuels, oils, solvents, corrosives, oxidizers, flammable liquids, dangerous when wet phosphoric materials, toxic, radioactive, and unknowns.
- 200–5,000 gallon Hydrocarbon fuels, oils, & unknown, tanks, or boats-land locked

All containers found were checked and marked. Five major Level A entries were conducted for atmospheric testing, evaluation and discovery of hazards in the following locations:

- New Orleans Medical Center of Louisiana (Charity Hospital)
- University Hospital

- LSU Dental School
- Dennis Sheen Transfer Company (Following fire and explosions)
- 2321 Timoleon Street (unknown release at debris removal site)

The Louisiana National Guard 62nd Civil Support Team and the EPA on scene coordinator and START teams eventually reached the city as waters started to recede and were an excellent resource during hazardous materials operations. They provided personnel and hazmat equipment and helped the firefighters survive the first couple of weeks. Civil Support Teams and a small EPA group were assigned under the command of New Orleans Hazmat for operations within the city. Hazmat personnel from Illinois Departments of Carpentersville, Chicago, Chicago Heights, and Decatur assisted New Orleans Hazmat in addition to Gonzales, Louisiana. Approximately 2,000 firefighters from around the country supplemented the entire New Orleans Fire Department in the weeks following Katrina.

There were over 3,000 railcars in Orleans Parish during Hurricane Katrina (Figure 5.100). Of those, 640 contained hazardous materials and 105 of those were derailed by the flood waters. Two releases occurred from railcars, one at the CSX rail yard releasing ethylene oxide and another at Air Products Company releasing hydrogen chloride. The CG Railway Company is a Mexican operation where railcars are shipped from Mexico

Figure 5.100 There were over 3,000 railcars in Orleans Parish during Hurricane Katrina. (Courtesy: New Orleans Fire Department.)

by specially designed barges to off load in New Orleans. Four of the tracks in the CG Rail Yard were under water with five derailed cars. Consists provided by CG Railway Company showed there were 14 railcars containing hazardous materials at their New Orleans facility.

Materials in the railcars included methylamine, anhydrous ammonia, potassium hydroxide solution, and chlorine. Strong winds and storm surge also caused several cars to derail in the CSX rail yard. One of the derailed cars contained fuming sulfuric acid. Other nearby cars contained methyl acrylate monomer, heptanes, and combustible liquids. Several barges were also washed over the levees by the storm surge. One such barge was about 2/3 full of benzene. The barge traveled approximately 100 feet from the levee across an industrial yard, a four lane highway, through power poles and through a swamp. No damage occurred to the barge and there were no leaks. The barge still rested where it landed when I visited awaiting removal of the benzene by the owners.

Hazmat team members responded to one explosion with fire at 3200 Chartres Street with phosphorus, arsenic, calcium, and tin metal involved. They also responded to the Mandeville Street Wharf fire that destroyed a large portion of the massive warehouse. It is estimated that over 1,000 natural gas leaks occurred, a total that was likely much higher. Other missions included Hazmat Response for Presidential visits on October 10 and 11. Ammonia releases from two separate companies. Two leaks occurred at New Orleans Cold Storage and one at Browns Velvet Dairy.

As time went on and rescues were concluded and the waters receded, hazmat team members focused on identifying hazardous materials in buildings and homes. As more people returned to the city, call volume increased for hazardous materials discovered in debris piles. Refrigerators and other appliances with Freon had to be identified so that contractors could remove the Freon before disposal. Fumigation contractor procedures were assessed by hazmat personnel and permits issued for fumigation operations to insure they were conducted safely.

EMS personnel from the New Orleans Department of Health, about 60% of Fire personnel including the hazmat team and other City Hall employees were living on the cruise ship Sensation (Figure 5.101), including the Superintendent and Assistant Superintendent of the fire department. Police personnel were living aboard the other cruise ship at the same dock. Most of the hazmat team members spent Christmas 2005 away from their families. Life aboard the cruise ships included room, meals, and a nightly movie. There were no phones or television. Following the removal of the cruise ship firefighters were provided with FEMA trailers to be used until their stations are repaired or replaced. Personnel are took

Figure 5.101 EMS personnel from the New Orleans Department of Health, about 60% of Fire personnel including the hazmat team and other City Hall employees were living on the cruise ship Sensation.

things one day at a time and hoped that it wouldn't be too long before things begin to return to normal in New Orleans.

Norfolk Virginia Hazmat Team

Norfolk is an independent city in the Commonwealth of Virginia with an estimated population of 242,234 in 2020 and an area of 66 miles^2. The city is located at the core of the Hampton Roads Metropolitan area, named for the large natural harbor of the same located at the mouth of the Chesapeake Bay. Norfolk has a long history as a strategic military and transportation point. The largest Navy base in the world, Naval Station Norfolk, is located in Norfolk along with one of NATO's two Strategic Command headquarters. The city also has the corporate headquarters of Norfolk Southern Railway, one of America's principal Class 1 railroads, and Maersk Line, Limited, which manages the world's largest fleet of US-flag vessels.

Fire Department History

> **Author's Note:** *I have included more history here than I normally would, but I thought this was really interesting.*

The first reference to fire matters in the Towne of Norfolk is contained in a 1730 act of assembly which prohibited wooden chimneys. On September 15, 1736, Norfolk became a borough by act of a royal charter. In 1740 the worshipful court ordered "... Doth (that) orders on ye (the) breaking out of a fire the drum shall be beat". The exact location of the drum is presumed to be centrally located within the borough. On September 14, 1751, the court ordered the remittance of 100 lb sterling to Ennis & Hope, London "for one fire engine complete" and for buckets and other utensils usable with said fire engine. The fire engine was built by Newsban & Rag of White Chapel, London, England, the engine's dimensions were 5 feet long, 18 inches wide, and 12 inches deep. Other comparable cities already had similar safeguards in place. Philadelphia purchased a "hand operated" engine dated as early as 1719 and New York City in 1731. Both were purchased from London, England. When filled with buckets of water and manned by 4–6 men the engine was capable of throwing a 1/2 inch stream at a distance of 30 feet. The pumper had to be lifted to be turned around.

On January 1, 1776, at 3:15 p.m. English gunboats under the command of Lord John Murray Earl of Dunmore attacked Norfolk from the Elizabeth River. Under cannon fire, English troops came ashore. Over a 3 day period the troops went house to house burning two-thirds of the Borough (nearly 900 homes). Warehouses from which riflemen fired upon the landing party were the original target. But the flames could not be controlled. Women and children ran for safety grabbing any possession they could. All privately owned engines along with the City Mace and civic records were moved to the Kemps Landing (Kempsville) in Princess Anne County for safekeeping. In 1783 the court ordered the town sergeant to return the engines and dig numerous public wells for added fire protection. That same year appointment of caretakers for the wells was made. About this time there were numerous fire engines in possession of private individuals without an organized force.

Union Hose Co. – 1797

In 1797, Fire Co. 1 was organized by Dr. Robert Archer. There is no record but it is believed this was the origin of the Union Hose Company. Eventually the crew was commanded by Foreman Martin Ryan (our second paid Fire Chief). The station house was located on Main Street next to the Court House, near Nebraska (Atlantic). Around 1846 the firehouse was relocated to Fenchurch Street, near Bermuda.

Phoenix Fire Co. – 1824

In 1824, James M. Steed, James S. Garrison, Paul Repiton, Jacob Vickery, and others organized the Phoenix Fire Company. The Phoenix was equipped with two suction engines, built by Smith of NYC. The fire engine was a very small piano style machine; working with side levers, using 2 inch leather hose could throw a 3/4 inch stream of water. The fire company members performed continuous duty until merging with the Hope and Aide Fire Companies in 1846. The engine house was built downtown on the lower end of Talbot, where the Aid Fire Company's engine house was later built following the merger.

Work song of the Phoenix

When the Franklin was in the mud and the Union was in the mire looking at the Phoenix boys a putting out the fire.

Franklin Fire Co. – 1827

The origin of the Franklin Fire Company may go back as far as 1803. Mr. Olyphant, a cashier of the United States Bank and prominent figure under Joseph A. Barren, John Myers, James S. Garrison, and others, performed continuous duty until merging into the Hope and Aide Fire Companies in 1846. The Company obtained a privately owned engine that was kept in a stable on Freemason Street. During that year the cornerstone for their engine house was laid on Main Street near Gray (Atlantic Street) then again to Bank and William Streets. The three companies remained the entire complement of the fire fighting forces of the town for 20 years.

At all fire previous to 1849, the fire engines were worked by African American men, who relieved the monotony of their work with song extemporized often by themselves. They had one general song, the chorus of which was:

Head in his hat, foot in de pan,
Old Capt. Soot, de dry dock man.

The work song was a compliment to Navy Yard Commodore Smoot.

Hope Fire Co. – 1846

Both the Phoenix and Franklin Fire Companies disbanded after a large fire at Main and Old Market Square and formed the Hope Fire Company. N. Calvart King was placed in command. A Hope member's son wrote that the Hope was located in the Town of Berkley. The Company contracted with John Rogers & Sons of Baltimore, MD, for the building of a suction engine at a cost of $700. The Company had an auxiliary engine of its own called the "Civil". The Hope Company was taken over by a group from the Aide Company by the name of the "Stingers". The organization was a success and it was soon necessary to relocate the membership and newly acquired apparatus. The new firehouse was relocated to Cove Street, near Church. In 1867 the Company obtained an Amoskeag suction engine.

Aid Fire Co. – 1846

The Aid Fire Company was an auxiliary company responsible to supply water to the Union Fire Company. The Aide placed an order with P&A How of Norfolk to build an engine. The fire engine failed to draft and was sold to Hampton. The firehouse was located on Granby St., near Freemason. The Aid Company encouraged gatherings at the firehouse and soon required larger quarters and developed into a full-fledged fire company. They relocated downtown to the north end of Granby, near Talbot Street. They became an outstanding fire force in the city: however, contention and discord soon arose and it was not very long before the discontents of the Aid Company formed a separate company – the "Stingers".

Relief Fire Co. – 1846

The Relief Fire Company was an auxiliary company responsible to supply water (like the Aide Fire Co.) to the Union Fire Company. The Relief Fire Company was located on Holt Street, near Church.

First Chief Engineer of the Volunteers – 1846

Finlay F. Ferguson was made Chief Engineer (Fire Chief) of this "volunteer" fire organization. After a big fire on Main Street, the insufficiency of the department was evident, and needed reorganization. The new organization was a failure; however the fault was thought to be in the lack of equipment. Chief Ferguson visited the Harmony Engine Co. in Philadelphia. There he purchased their double-ended Agnew engine with gallery for the sum of $400. Upon the return trip to Norfolk the engine was found not to have suction capabilities. It was a good machine, however, and cheap. It was housed with the Union Fire Company.

United Fire Co. – 1850

The United Fire Company was located on Water Street. Foremen Thomas Kevill commanded this new fire force with men of the disbanded Union Company. The United was the second white fire company, organized in Norfolk. Following a disastrous fire on Water Street in 1948 the need for fire protection in that locality was necessary. The Company was known as "real good efficient workers". When the Civil War broke in 1861, all but four members (because of their family size) volunteered for duty. The United volunteered as a heavy artillery company for the Confederate army. In 1866 the City procured an Amoskeag fire engine and it was placed in-service with the United.

Work song of the United

Down with the red and up with the green, for that is the color of our machine

The Civil War

On March 4, 1861, Abraham Lincoln was elected as the 16th President of the United States. When Civil War broke out on April 12, 1861, at Fort Sumter, soldiers were needed at the battle front. Norfolk answered the "call to arms" and furnished a number of military companies and a large number of men to the Confederate Army. In fact, it is estimated nearly two-thirds of the soldiers from Norfolk were firemen. Volunteer Firemen Carter Williams, Wright, Lakies, Icenbice, Carter, Cussick, and John Bonfanti where among the names whose noble and heroic acts of bravery on the battle field will always be remembered.

In an attempt to reduce the North's great naval advantage, Confederate engineers converted a scuttled Union frigate, the U.S.S. Merrimac, into an iron-sided vessel rechristened the C.S.S. Virginia. On March 9, 1862, in the first naval engagement between ironclad ships, the Monitor fought the Virginia to a draw, but not before the Virginia had sunk two wooden Union warships off the Norfolk coast line.

Federal Confederate soldiers took charge in 1862. They made efforts to organize the volunteers. Steam power was introduced with the purchase of a first-size Amoskeag steam engine that was assigned to The United Fire Company. A hose connection from a boiler in the firehouse to the boiler on the pumper kept steam pressure up to 25 lb. As the horses pulled away the hose was disconnected and as the engine cleared the station the engineer sloshed kerosene on the excelsior and kindling which was under the boiler and struck a match to it. As the steamer swung around into the street smoke was pouring from the boiler and by the time it reached the fire, steam pressure would be high enough to cause the safety valve to pop.

The Civil War ended in 1865. General Lee's troops were surrounded and on April 7, Grant called upon Lee to surrender. The two commanders met on April 9, and agreed on the terms of surrender. On April 13, Lincoln attended a play at Ford's Theatre and was shot. John Wilkes Booth, who hoped to throw the country into political chaos, had shot the President. The soldiers returned to home to find a few changes. The volunteer department rosters changed annually following election of officers:

Feuding Fire Companies – Hope and United, September 16, 1871

The Virginian newspaper reported on September 16, 1871: About 9 o'clock on Saturday evening an outrage occurred in this city which, if the perpetrators are not brought to justice a foul blot will be cast upon the fair fame of our community, and which, occurring at this peculiar time, cannot be too strongly reprobated. For some time past a feeling of ill-will has been nursed and kept warm by several unfortunate collisions. On Saturday night it culminated in a wanton assault made by a number of the members of one company upon the other, and in the affray several parties were wounded by pistol shots.

Some 20–40 members of the United Hose Company in revenge for an alleged assault committed upon one of their members the previous night, proceeded to the Hope engine house. After ringing the bell the United took the hose reel out of the engine house. Only one member of the Hope was at the engine house at that time, and as he ran out he was fired upon twice. When he made his way back to the station house the alarm was sounded. The crowd marched down to Main Street, where a party of the members of the Hope stood collected nearly in front of the Atlantic saloon. A quarrel ensued, when there commenced an indiscriminate fire of pistols.

But, few of the Hope were armed, but some 20 shots were fired on both sides in the melee, while sticks and stones were freely used. The firing continued for some minutes, and several parties were injured. The attacking party then drew off and scattered. The wounded men were taken in charge and attended to. James Dernay, member of the United, was fatality shot in the back. The ball went through his body and lodged under the skin over the abdomen. Firefighter Dernay was taken to his house on Water Street to receive medical attention. He died 2 days later.

"Nothing short of a paid fire department will meet the exigencies of our position. The feud is too bitter to allow us to believe that these two companies can continue their organized existence without further bloodshed and danger to the public. We can conceive of no apology for this daring violation of the laws, and the community stands aghast at a realization of the fact that so large a number of our firemen have been found ready to participate in a deadly encounter in the chief thoroughfare of our city."

Career Fire Department Is Formed

As a result, the Norfolk City council took action and the volunteer companies were abolished as of December 15, 1871. Thomas Kevill, Foremen of The United Fire Company, was nominated Chief Engineer (Fire Chief) by the newly formed Board of Fire Commissioners. The fire protection territory to cover was not large and the population had not grown. Quick movement to any part of the city was easily reachable. Centralizing the apparatus and associated equipment made sense. Contract was awarded for the erection of a Fire Department Headquarters upon the site occupied by the old Blues and Juniors Hall at Williams and Avon Streets. Upon the completion of the building in 1872 the apparatus of the United Engine Company, Hope Engine Company and Union Truck Company, operated by nine pain men, were transferred to their new quarters.

Today's Modern Department

Norfolk's fire department is led by Acting Chief John DiBacco who is in command of 510 uniformed personnel responding out of 14 stations. Norfolk firefighters are organized into 3 battalions and operate 14 engines, 7 trucks, 2 platforms and 5 aerials, 2 rescue units, and 12 advanced life support (ALS) medic units. Two ready reserve units are available; Medic 6 is also a Bariatric Unit. Trucks and rescue vehicles are also ALS capable. Miscellaneous equipment includes 2 fire boats, staffed by ladder companies when needed, 2 zodiacs, 2 high water trucks, and a bus to provide rescue for persons in low-lying areas. Terrorism response capabilities include a Zumro Mass Casualty decontamination system, which is staffed by stations 7 & 12, Metropolitan Medical Response System (MMRS), Mass decontamination system, Metropolitan Medical Strike Team, and a combination Police/Fire Bomb Squad Team. Norfolk is built on a shell bed and is susceptible to both urban flash flooding and coastal storms. Hazmat Unit is housed and operated by Station 7 personnel.

Hazmat Team History

In the early inception of the squad companies, a certain level of hazardous materials response equipment was carried. In 1985 the squads carried one Butyl Level "A" encapsulated suit on each squad. Each truck carried one suit so the fit was quite interesting. Only the smaller guys could wear it. They were rarely used and we did not have training on how to use them.

Lieutenant Edward L. Senter Jr. of Squad 1 "B" Shift sent a letter, dated February 1, 1989, to then Chief of Department Thomas E. Gardner with recommendations for the formation of a hazardous materials team in Norfolk.

The Norfolk Hazmat Team was officially formed in the late 1980s. This operation was conducted out of a step van (Figure 5.102). Often if we did not have certain gear, we would go back to the station to get it. In 1991 the step van at Squad 1 was replaced with a Salisbury Heavy Rescue and the step van became a support unit. The role of the hazmat team also branched into technical rescue and the teams have been cross trained in technical rescue since that time. That cross training helped during confined space events on board ship as we understood detection and monitoring could make better assessments of the environments.

Figure 5.102 The Norfolk Hazmat Team was officially formed in the late 1980s. The team operated out of a step van. (Courtesy: Norfolk, VA Fire Department.)

On September 29, 2003, the Squads were renamed as Rescue Companies 1 & 2. In 2004 the teams were restructured to include stations 7 and 14 as support stations. The step van was sold and replaced with an international cab hazmat unit. The Rescue Companies were reduced from 4 personnel to 3 personnel with the thought that they would receive additional trained manpower from station 7 or 14. All personnel at station 7 were trained to hazmat technician and all of the personnel at station 14 were trained to the levels of rescue outlined in National Fire Protection Association (NFPA) 2006 and 1670. Rescue company personnel remained cross trained as Virginia Department of Emergency Management (VDEM) hazmat specialists and rescue specialist to include collapse rescue. Stations 7 & 14 received pickup trucks and 10,000 lb support trailers. One trailer now carries the mass decontamination system (Sta-7) and the other trailer carries collapse rescue equipment (Sta-14). Rescue company officers assigned to the squads were Lieutenants. The officers changed as the program grew and eventually became captains. The program now uses Lieutenants again on the rescue companies.

Hazmat Team

Station 7 is the hazmat station and is located at 1211 West 43rd Street home of Rescue Company 1. Hazmat 1 is a 2003 Freight liner with a custom body by E-One (Figure 5.103). Hazmat Station 14 is the 2nd station and it

Figure 5.103 Hazmat 1 is a 2003 Freight liner with a custom body by E-One. (Courtesy: Norfolk, VA Fire Department.)

is located at 1460 Norview Ave and is home to Rescue Company 2. Within the Norfolk Hazmat Team are 19 Hazmat Specialists and 27 technicians. Norfolk's Hazmat Team is a dedicated with a minimum of 8 personnel on duty each shift. There are a total of 54 technicians available city wide. Hazmat 1 has averaged approximately 9 calls per year over the past 10 years. Factoring in hazmat-related calls handled by other companies the total hazmat related calls including gas odors, CO calls, hydrocarbon spills, and others the total averages 435 per year. Oil dry is carried by engine companies for small hydrocarbon spills. The company officer decides if the hazmat team is also needed.

Norfolk's Hazmat Team is a part of the Southside Regional Hazardous Materials Team, which is a state sponsored team. Mutual aid is available from Virginia Beach, Chesapeake, and Portsmouth. Assistance is also available from the Navy Mid-Atlantic Region Hazmat Teams and an additional team from Newport News.

PPE, Equipment and Training

Both Rescue Companies utilize MSA SCBAs along with MSA SABA systems with Air System breather boxes. The latter is used for confined space entry and rescue. Hazmat PPE for Level A, B, and C is Lakeland garments. Additionally they carry Trellchem Level A flash garments as part of the regional hazardous materials team. The MT94 multi-threat garment is used for chemical, biological, radiological, nuclear and explosive (CBRNE) response and is effective in clandestine lab clean up as well.

Rae system products are used for standard gases and photo ionization detection (MultiRae). Also carried are the Hazmat ID for IR spectroscopy and ammonia sensors due to refrigeration systems container anhydrous ammonia in their jurisdiction. Radiation monitors include personal radiation detection (PRDs), OSLDs (optically stimulated luminescence dosimeter for exposure control to personnel). Thermo Back Pack (RID) is used for Gamma and Neutron detections as well as Identifinder for isotope identification. They also have both commercial and military reactors within their response area so they also carry Ludlum 2240-3 instruments to measure Alpha, Beta, and Gamma radiation. Bio Assay systems are used on biological agents. Various colorimetric systems are used for identification of chemicals and their respective families. M8, M9, pH, KI, HF papers are also available. However, these require that responders are able to dip the paper in a sample of the material. APD 2000 is used for chemical warfare agents as well as 256 A Kits.

In the late 1980s the Virginia Department of Fire Programs began teaching some awareness and operations courses. The Virginia Department of Emergency Management started a hazardous materials technician program and formed state teams around the state after a Pentaborane accident killed a worker and injured firefighters in Central Virginia.

Reference Resources

Research of information on chemicals, PPE, chemical characteristics, and other information is available in the form of hard copy materials and electronic data from the internet, smart phones, and IPods. PEAC software is the primary resource along with APPS and a cache of hard reference material (Books).

Hazardous Materials Exposures

Transportation exposures include Marine terminals with heavy intermodal traffic, including box containers, Type 1, 2, 5, 7, tanks. Interstates 64 and 264 provide hazardous materials movement between the Port area and out of the state. Hazardous materials are not permitted in the Chesapeake Bay Bridge and Tunnel. U.S. highways 17 and 56 also provide routes to and from the west. Waterways provide transportation of containers, LPG and LNG carriers utilizing ships, barges, and tugs. There are petroleum and chemical carriers as well. A 24 inch natural gas line runs though heavily populated areas of Norfolk.

Fixed facilities in Norfolk have a wide variety of hazardous materials. The Norfolk International Terminals with intermodal containers and shipping is the 2nd largest port on the East Coast. Refrigeration warehouses associated with the fishing industry and shipping utilize anhydrous ammonia for cooling. Shipyards utilize solvents and hazardous operations aboard ships. Vessels may be bulk petroleum and LPG, for example. Large communications switching centers have very large battery rooms in both high rise and single story facilities. Allied Terminals have large tanks of sodium hydroxide on the Elizabeth River in the Berkley section of the city. Norfolk Southern Coal Piers is the largest coal facility in the northern hemisphere.

Finally, there are numerous military installations throughout the Norfolk area. These include the Navy Atlantic Fleet, both aircraft terminals and marine vessels (4 aircraft carriers, submarines, cruisers, and others some of which have nuclear propulsion systems. Also located in Norfolk is the Little Creek Amphibious Base which houses the east coast Navy seal teams and amphibious assault vessels.

Incidents

Generally, Norfolk Hazmat Team has responded to anhydrous ammonia leaks with evacuations, ship board explosions with mass casualties, million gallon sodium hydroxide tank leaking and failing, Three IED's found on tanks containing perchloroethylene, ship fires involving hazardous materials, tank truck fire, many confined space incidents in shipyards and mutual aid to Newport News on a confined space incident involving the Aircraft Carrier Harry Truman.

Exxon Tank Truck Fire

An MC306 Gasoline tanker truck caught fire on a residential street in Norfolk, VA, in 1984 (Figure 5.104). The truck was carrying 8,500 gallons of gasoline. The truck driver, William Scott, 58, from Virginia Beach, VA, received first and second degree burns on his face, legs, and arms. Two firefighters were suffered minor burns while fighting the inferno which destroyed the Exxon tanker and 21 cars. The blaze created a burning river that flowed through storm drains into the Hague, an inlet off the Elizabeth River. Officials evacuated residents of the 168 unit Pembroke Towers and the 64 unit Hague Park apartment building by the accident site on Colley Avenue in the Ghent neighborhood. Several elderly residents were rolled outside in wheelchairs. Five of those evacuated were treated for high blood pressure and breathing problems. Another 9 people were evacuated from the Ronald McDonald House lodging.

Firefighters extinguished the fire using foam within an hour of the 9 a.m. explosion and fire and residents were allowed to their homes. A man backing a garbage truck out of a driveway at the Ronald McDonald House collided with the Exxon tanker. The tail of the garbage truck

Figure 5.104 An MC306 Gasoline tanker truck caught fire on a residential street in Norfolk, VA in 1984. (Courtesy: Norfolk, VA Fire Department.)

caught the tail of the tanker and the driver said he saw fuel flowing down the street. The truck exploded 5–10 seconds later with a "good sized boom". It is not known what ignited the gasoline (Norfolk, VA Fire Department).

Northwest Arkansas Regional Hazmat Team: Reverts to Everybody for Themselves

Prolog: *During the mid-1990s, Brent Boydston was a student of mine at the National Fire Academy from Bentonville Arkansas Fire Department. It was the beginning of a two-decade-long friendship. Brent was a firefighter/paramedic when I met him, now some 20 plus years later he is Chief of the Bentonville Fire Department, which came as no surprise to me. When I first went to Bentonville it was just a small town that happened to be the home of Sam Walton and his Wal-Mart Stores. Brent took Sam Walton on his last ride as he told me. When I first visit Brent and his family, they had one, maybe two stations. The main station was a couple of bays. Today they have what I call the palace. It is the most amazing fire station I have ever seen in all my travels across the United States and Canada. Bentonville is a small metropolis these days, Northwest Arkansas has grown into mini-metropolitan area with retail giant Wal-Mart everywhere you look, not only in Bentonville but around the world. Thanks to Brent I have been able to watch it grow on my visits to Bentonville, thanks for the memories.*

Northwest Arkansas is one of the fastest growing and most dynamic regions in the United States. It is home to Wal-Mart, J.B. Hunt Trucking Co., Tyson Foods, and the University of Arkansas. With the beautiful Beaver Lake and the majestic Ozark Mountains, the Northwest Arkansas region, inspired by its surroundings, is an up-and-coming leader in business, education, sports, tourism, and arts and culture.

Regional Hazmat Team History

In the mid-1980s, the Northwest Arkansas Metropolitan Fire Chiefs Association contemplated how to comply with the mandates of newly passed legislation by Congress titled the Emergency Planning and Community Right-to-Know Act (EPCRA), also known as SARA Title III. County judges in Arkansas are responsible for hazardous materials response planning in their jurisdictions and are the chief administrative officers in their counties.

The idea was put forth by the Metro Chiefs and county judges for Benton and Washington counties (the largest and most populated in Northwest Arkansas) to form a regional hazardous materials response team. They felt local departments did not have the financial resources or personnel to have individual teams. The plan was taken to each legislative body in the towns and two counties for approval. Initially, there was some resistance to the plan, but in the end all jurisdictions agreed to the regional team and an intergovernmental agreement was signed by all.

At the time this occurred, this was a pioneering concept that was yet to be utilized in many parts of the country. During the initial planning, it was determined that $200,000 would be required to create a team and $150,000 a year to maintain it. Funding for the team was derived from a 38 cents per-capita assessment for each jurisdiction per year to support operational costs. Communities within the response area would pay up to $21,266, depending on the population. Bentonville paid $14,470 in 2010, Bella Vista $10,051, Siloam Springs $5,715, and Rogers $21,266. Washington County's attorney and treasurer handled legal issues and the collected funds for the team.

Six fire departments within the two counties assigned personnel to respond as part of the team. Bella Vista, Bentonville, and Siloam Springs provided four personnel each and a team leader. Fayetteville and Rogers provided five personnel each and a team leader. Springdale provided seven team members and a team leader. Training was conducted in house by a full-time team training officer and a team chief was appointed to coordinate daily operations. The Springdale Fire Department was centrally located in the response area and had available space and agreed to host the hazmat response unit.

Regional Team

The Northwest Arkansas Hazardous Materials Response Team went into operation in 1989 (Figure 5.105) and operated successfully for more than two decades. It should also be noted that the municipal areas within the regional team's protection area border one another, much like the suburbs of larger cities around the country. Response times for the closest personnel and local support equipment was less than 30 min and the Hazardous Materials Unit housed in Springdale might take 30–90 min, depending on where in the two-county area it was responding and traffic conditions.

End of an Era, Hazmat Returns to Local Jurisdictions

In the late 2000s, several chief officers retired and new chiefs were promoted or newly hired in several Northwest Arkansas fire departments. The longstanding concept of a regional hazmat team was no longer supported by all members of the Metro Chiefs Association. Following extensive discussions about the existing regional hazmat team, the chiefs voted to disband the team and return the responsibility for hazmat response to the local level.

Under the new hazmat response plan, each county would be responsible for its own hazmat response. Larger departments would form local

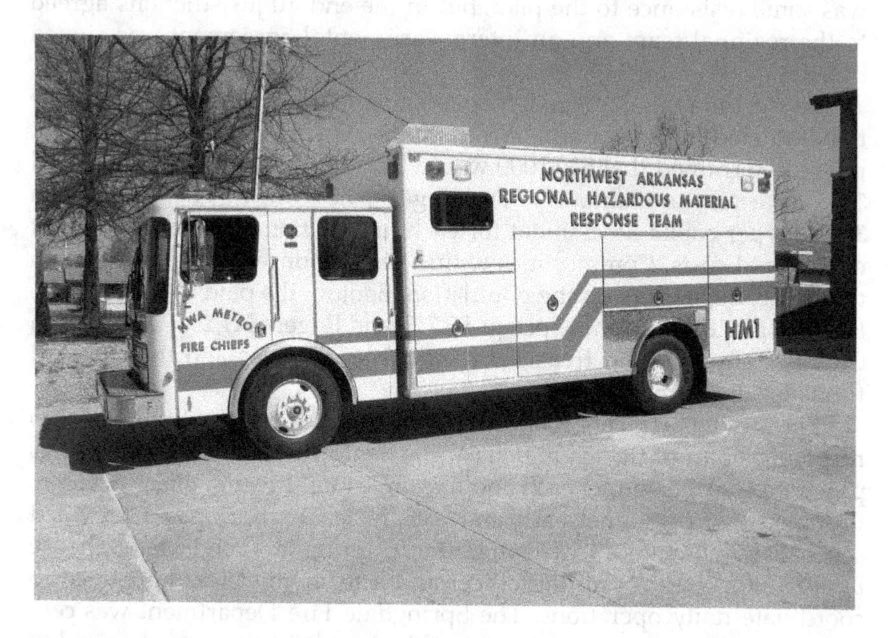

Figure 5.105 The Northwest Arkansas Hazardous Materials Response Team went into operation in 1989.

teams and provide protection for smaller departments and unincorporated areas in each county. Within Washington County, the Fayetteville and Springdale fire departments formed hazmat teams. Portions of Springdale are also located in Benton County. In Benton County, the Bentonville, Bella Vista, Rogers, and Siloam Springs fire departments formed hazmat response teams.

Thus, one regional hazmat team for the two-county area has now evolved into six individual hazmat teams. Personnel who were shared among the departments now staff their own teams, creating the need for additional personnel to be trained and equipped. One response unit and associated equipment for the regional team has become six individual response units and six sets of associated equipment.

Alliance Formed

Since the regional team was disbanded, two departments in Benton County have formed an alliance for the purposes of training, coordination of equipment purchase, and response to hazmat incidents Bentonville and Bella Vista. Located north of Bentonville, Bella Vista is just south of the Missouri state line. Bentonville is south of Bella Vista, west of Rogers, and east of Siloam Springs. Bentonville and Bella Vista provide hazmat response for their jurisdictions, Northwest Arkansas Regional Airport and the cities of Cave Springs, Centerton, and Highfill.

Bentonville and Bella Vista Team History

The startup of the newly created team has cost Bentonville $26,676 in salaries, $25,000 for equipment first year ($15,000 of that was from sale of the regional team's equipment), $16,000 for training, and $55,000 for monitoring instruments (from Homeland Security Grant Program funding). Bentonville placed its hazmat team in operation on January 1, 2013. The Bentonville team responded to 80 hazmat incidents during its first year of operation. Total hazmat response calls have included fuel spills, gas odors, and leaks. Engine companies respond to fuel spills and can handle up to 45 gallons.

Engine and ladder companies each carry a four-gas meter, thermal imager, and absorbent materials to clean up small spills. One person from the hazmat team is on site of any fuel spill. Larger spills require the response of the full Bentonville hazmat team. Usually, six hazmat technicians are on duty each shift. A total of 18 technician-level personnel located throughout the department can be recalled to duty if necessary. Bentonville's hazmat team is not dedicated, so Station 1 personnel man the hazmat unit for hazardous materials calls.

Bentonville Components

Bentonville's population increased almost 80% between 2000 and 2010, with an estimated 2020 estimated population of 55,000, an increase of 12% since 2014. The population increases each weekday when more than 33,000 Wal-Mart employees arrive for work from outside Bentonville.

Fire Department History

The Bentonville Fire Department was founded in 1887 and was a largely volunteer organization until the first driver/operator was hired in 1922. Ambulance service was added to the department's responsibilities in 1977. The department joined the regional hazmat team in 1989 and formed a 12-person technical rescue team in 1995.

Today's Modern Department

The Bentonville Fire Department is under the command of Chief Brent Boydston, who started as a part-time employee in 1984 and was hired full time in 1989. Boydston rose through the ranks and was appointed as chief in April 2012. Bentonville is an ISO Class 2 fire department with 84 uniformed personnel and currently 10 paid-on-call personnel (it has authorization for 12). Bentonville's paid-on-call personnel are often the source of full-time hires on other departments.

The Bentonville Fire Department has expanded from a single station to seven stations. A new Fire Station 1 opened in June 2008 (Figure 5.106) and features a six-bay apparatus room with a vehicle exhaust extraction system, wash bay, 15 bunkrooms, a large day room, a stainless-steel kitchen with a separate dining room, a 1,800 square foot training room, a lobby museum, administrative offices, and a five-level training tower, and is fully fire sprinkled. This station was the 2008 Station Styles Gold Medal winner in the Career Stations category, presented by Fire Chief Magazine. Bentonville responds to over 7,000 calls each year. It operates seven engine companies, three ladders (quints), one 137 foot straight stick and six squads (paramedic ambulances).

Station 1 houses Truck 1, Squads 1 and 10, Rescue 1, Engine 1, Engine 219, Support 1, Battalion 1, and a motorized rescue cart. Fire Station 2, which opened in 1989 and was renovated in 2012, houses Squad 2 and Quint 2. Fire Station 3, opened in 1996, houses Squad 3 and Engine 3 and was designed with a maintenance pit for fire and emergency medical service (EMS) apparatus. Fire Station 4 opened in 2001 as the city's first two-story station. It houses Engine 4, Squad 4, and Brush 4. Fire Station 5 opened in 2005 and houses Engine 5, Squad 5, and Brush 5. This station

Figure 5.106 Bentonville opened a new Fire Station 1 in June 2008. The design won an award in the *Fire Chief* Magazine station competition.

is fully fire sprinkled. Fire Station 6 was built in 2014 and opened in 2015. It houses Engine 6 and Squad 6, Station 7 houses Engine 7.

Hazmat Team

Support Unit 1 is the primary hazardous materials response vehicle for the Bentonville Fire Department and is housed at Station 1 (Figure 5.107). Support Unit 1 is a 2000 E-ONE Cyclone II 21 foot walk-in vehicle. Most of the hazmat equipment is carried in the walk-in area. In-line air and bottle-filling air is provided by an onboard air supply system located on Support Unit 1. The cascade system is capable of refilling 80 bottles. Six hazmat personnel are on duty each day with 26 total technicians on the department. The total number of HM calls include fuel spills, gas odors, and leaks. Engine, squad, and truck companies carry Quick Spill Kits. Spills greater than 20 gallons require a hazmat team response. Bentonville responded to 138 hazmat calls in 2019. Mutual aid if needed is available from Bella Vista, Rogers, Springdale, Fayetteville, and Siloam Springs Fire Departments.

PPE, Equipment and Training

Chemical protective clothing includes Kappler Provent 7,000 coveralls for splash, Kappler Z1000Xp for Level B, and Kappler Z500 NFPA for Level A. Respiratory protection is provided by Scott self-contained breathing

Figure 5.107 Support Unit 1 is the primary hazardous materials response vehicle for the Bentonville Fire Department and is housed at Station 1.

apparatus (SCBA) with 45 min bottles and particulate filters, powered air-purifying respirator (PAPR), and in-line supplied air. In-suit communications are provided by Motorola XPR 6550 digital with PTT, throat microphone, and ear bud attachments.

Monitoring instruments include four-gas Q8Rae II, four-gas MultiRAEs with PID, MiniRAE 3000 PID, ChemRAE, Ahura Defender Ramon, Guardian biological reader, and Smith's Detection Multi-Mode Threat Detector for TIC, CWA narcotics and explosives. Bentonville is in the process of obtaining a grant for a mobile mass spectrometer.

Hazmat personnel training in the Bentonville Fire Department includes the seven ICS courses from FEMA/Homeland Security.

- Chemistry 40 Hours.
- Basic Air Monitoring.
- Advance Air Monitoring.
- Arkansas SERC Technician (Level III) Trained.
- Public Safety WMD response-Sampling Techniques and Guidelines.
- Emergency Response to Domestic Biological Incidents PER 220 (LSU).
- Introduction to EOD/IED (FBI).
- WMD RAD/NUC Course for Hazmat Technicians.

All firefighters are trained to the Operations Level.

Reference Resources

Reference materials used by the Bentonville hazmat team include the Emergency Response Guidebook, National Institute for Occupational Safety and Health (NIOSH), Surviving the Hazardous Materials Incident, WISER Hazmat Pocket Guide, and Hazmat Segregation and Pipeline Emergencies software.

Hazardous Materials Exposures

Primary hazardous materials exposures in Bentonville and Benton County are 10 million square feet of operational space used by Wal-Mart on a daily basis and potential roadway incidents. Chemical, biological, radiological, nuclear and explosive (CBRNE) events are the primary concern with the Wal-Mart complex. Suspicious packages at the Wal-Mart mailing facility are one of the major risks in terms of CBRNE events. Interstate 49 runs through Benton County from Bella Vista at the Missouri border and through Bentonville, Rogers, and Springdale. In addition to any hazardous materials transported on Interstate 49, the TransMontaigne, Razorback pipeline, and terminal on average fill and send on their tankers carrying a combined total of 750,000 gallons of motor fuel per day. This facility also receives 60,000 gallons a day of ethanol used as a motor fuel additive.

On February 2012, a Commodity Flow Study was conducted on Interstate 540 (now called Interstate 49). The study was conducted over 2 days, the first from 11:30 a.m. to 7:30 p.m. and the second day from 9 a.m. to 3 p.m. Results of the study showed various petroleum products, propane, ethyl lactate, oxygen, ammonium nitrate, anhydrous ammonia, methacrylic acid, and other corrosives, nitrogen and CO_2, to name a few. Bentonville has a propane plant for distribution of propane to residential and business occupancies. There are no railroad facilities in Bentonville's primary response area, although two railroads serve Northwest Arkansas. Kansas City Southern mainly transports coal and the Arkansas/Missouri railroad transports hazardous materials through portions of Benton County.

Bella Vista Fire Department

The Bella Vistas Fire department was established in 1969. Their first response area is primarily residential and business occupancies. Under the command of Chief Stephen Sims, 38 uniformed personnel respond out of three fire stations. Station 1 is in the town center with seven firefighters on duty. Station 2, located on the east side, is staffed with three firefighters. Station 3 is in the Highlands area of town and has two firefighters on duty.

A fourth station is in the planning stages. Sims joined the department in 1995 and was appointed as chief in 2006. Bella Vista operates two engine companies, one ladder, four squads, one heavy rescue, a water rescue, and hazmat trailer pulled by a pickup truck.

Bella Vista's population is 28,500 and the fire department covers an area of 63 miles^2 for fire and EMS coverage and 125 miles^2 for hazmat response. It also protects the communities of Gravette, Maysville, Pea Ridge, and Sulphur Springs. There are 38 career firefighters on the department and 20 of those are paramedics and 17 are emergency medical technicians (EMTs). The department also has six paid-on-call firefighters. Each shift, 12 career firefighters are on duty.

Hazmat Team

The Bella Vista Fire Department has nine trained hazmat technicians, with three on duty each shift. Firefighters and EMS personnel are trained to the hazmat operations level. Kappler Level A suits are used by Bella Vista hazmat personnel with SCBA and 1 h bottles. Monitoring equipment includes radiation detectors, photo ionization detectors, and a four-gas monitor. The department relies on Bentonville for additional monitoring capability. An air supply system is carried on the rescue truck. Bella Vista carries a Chlorine A Kit and Midland Emergency Response Kit along with other typical hazmat response equipment. The primary hazmat exposure for fixed facilities is propane. There is no natural gas service in Bella Vista, so all requirements for fuel are met by the use of propane. Highway transportation is the other main hazmat exposure (*Firehouse Magazine*).

Oklahoma City Hazmat Team

Oklahoma City is the Capitol of Oklahoma and located in central Oklahoma. It is one of the largest geographical cities in the United States containing 621 miles^2 and a population of approximately 655,407 in 2020. The Oklahoma City metropolitan population is approximately 1.41 million.

Fire Department History

Oklahoma City Fire Department (OKCFD) organized in 1889 as a small volunteer fire company, led by Chief Andy Binns. They operated one horse-drawn wagon housed in a small frame building to protect 10,000 residents. During 1894 the first paid firefighters were given 50 cents per hour for fighting fires. OKCFD became the official name in 1899. The first motorized apparatus was purchased in 1910.

Today's Modern Department

Today's modern OKCFD, under the leadership of Chief Richard A. Kelley, has approximately 877 personnel operating from 37 stations strategically located throughout the city. Annual alarm totals average around 77,000. Apparatus includes 37 engines, 13 rescue ladders, 2 heavy rescue vehicles, hazardous materials unit, chemical, biological, radiological, nuclear and explosive (CBRNE) unit, decontamination trailers, and mobile air unit. Special teams include hazardous materials, Urban Search and Rescue (USAR), and dive and swift water rescue. USAR team members train their own rescue dogs, while there I met Jeff Hanlon, his dog's name is Willy, who was in the training program at the time (Figure 5.108). OKCFD responds to medical and rescue calls, but they do not transport patients. Transportation is provided by a private ambulance service that is subsidized by the city.

Hazardous Materials Team History

The hazmat team was started by Deputy Chief Jim Henning in 1984. Johnny Davis, David Bowman, and Gary Marrs were selected as the

Figure 5.108 Special teams include hazardous materials, Urban Search and Rescue (USAR). OKCFD does their own rescue dog training in house.

station officers. They were allowed to hand-select their crews. All the crews then went to the National Fire Academy in Emmitsburg to receive their initial training. Station 4 became the hazmat station. It was located downtown at SW 4th and Broadway. There was an engine and the hazmat rig assigned to the station. There were four people assigned to the station on all three shifts. The hazmat was not initially manned full time. When a call came in, they would take the engine and hazmat. The hazmat was an old rescue squad that was refurbished by the crews at Station 4. They were allowed to design and build the inside of the rig. They had an MSA combustible gas indicator, CDV 777 radiation monitors and dosimeters, pH paper, some Level A suits, and some flash suits. They responded on all hazmat calls across the city.

Hazmat Team

Today Station 5 is now the hazmat station. It is located at NW 22nd and Broadway. They have staffing 5 with a minimum staffing of 4 (Figure 5.109) and an OK Homeland Security Regional Response CBRNE unit (not manned full time). A minimum of 7 hazmat technicians are on duty at the station each shift. There are over 100 hazmat techs throughout the city. All other firefighters are trained to the operations level. Tow trucks in the city carry absorbent materials for spills up to 10 gallons of liquids and 5 lb of solids. For spills of greater quantity the hazmat team responds.

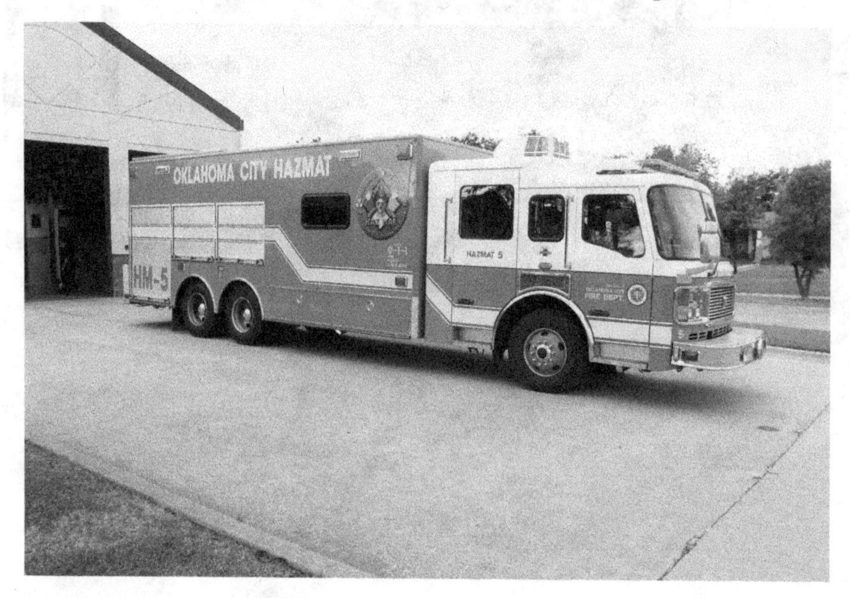

Figure 5.109 Today Station 5 is now the hazmat station. It is located at NW 22nd and Broadway.

A typical hazmat call includes the companies originally called Hazmat 5 and Engine 5. If decon trailer is required, Rescue 8 responds to provide personnel for the trailer. Mutual aid is available from The 63rd Civil Support Team located in OKC and a foam tanker from Tinker Air Force Base. Approximately 50 hazmat technicians are on duty at other stations and if needed, off-duty personnel can be called in. OKCFD Hazmat units carry typical hazardous materials equipment and supplies.

PPE, Equipment and Training

Chemical suits are Kappler Zitron 300 for level A and Tyvek for level B. In-suit communications utilize throat mikes and hand signals. Scott air packs with 1 hour bottles are used for most respiratory protection. Air purifying respirators are also available if needed. Hazmat team members receive technician-level training and hazmat chemistry.

Hazardous Materials Exposures

Transportation hazardous materials exposures include Interstates 35, 40, 44, and 235. State routes 3, 14, and 66. Burlington Northern Santa Fe and Union Pacific railroads cross through the city. Pipelines carrying crude oil and jet fuel are located within the city limits. Fixed facility exposures include Oil Well leaks, Research Park Medical Research, State Department of Agriculture, State Health Department, Tank farms citywide, water treatment facilities, 3-M company, U.S. Foods, Cold Storage OK, ethanol blending facility, Cotton Mill COOP, Miniky and Zerox.

Incidents

B&M Oil Company. The early morning fire involved a building full of 55 gallon drums of various hazardous materials including ethanol, motor oil, and hydraulic oil, among others. Ethanol drums explode creating fireballs.

Another incident occurred involving a MC-306 carrying gasoline that rolled over. Crews from OKCFD applied foam on the tanker. Additional incidents included Bachman Chemical, fish kill in a river, oil well tank fires, white powder incidents, and paraformaldehyde fire where personnel were overcome and transported to the hospital.

Bombing at the Alfred P. Murrah Federal Building

April 19, 2019, was the 24rd anniversary of the bombing of the Mariah Federal Building in Oklahoma City (OKC) which killed 168 and injured over 600. At the time the loss of life was the largest in an act of terrorism in the United States. This was an act of domestic terrorism carried out by Timothy McVeigh and *Terry Nichols* in retaliation for the sieges that occurred at Ruby Ridge, Idaho. Federal agents in a standoff with Randy Weaver resulted in the death of his 14 year old son and wife. Waco, Texas another standoff at the Branch Dividian compound April 19, 1993 resulted in 76 deaths including women, children, and leader David Koresh. April 19, 1775 is also Patriots Day, the anniversary of the rebellion against the British authority at Lexington, MA.

Occurring at 9:02 a.m. the blast from a 4,800 lb mixture of ammonium nitrate, nitro methane, and fuel oil destroyed 1/3 of the nine-story building. The blast was heard 16 miles away and registered 3.2 on the Richter Scale 16.1 miles away. Destruction and property damage occurred to 324 buildings in a 16 block radius around the Murrah building and shattered glass in another 258 buildings. Eighty-six cars in the area were burned or destroyed by the blast pressure. The bomb was placed in a Rider rental truck and parked in front of the federal building in downtown OKC. Timothy McVeigh and Terry Nichols planned, gathered materials, built the bomb, and brought the truck to OKC. They had previously surveyed the government complex and placed a getaway car a few blocks from the federal building. They thought of themselves as revolutionaries and planned to fire the first shot in a new American Revolution.

Response to the Bombing

> **Author's Note:** *While teaching a 1 week Hazmat Incident Command Course for the National Fire Academy for the 63rd Civil Support Team, there was an officer from the OKCFD in the class. Turns out he was one of the first chief officers on the scene of the bombing. He was kind enough to give me a walking tour of the bombing site that had been turned into the National Memorial. He provided an overview of his and OKCFD's actions during the response to the incident. That personal touch made my tour of the site much more meaningful and informational. One of the things I learned was the Federal Building was not the intended target of McVeigh. When McVeigh and Nichols scouted the site and they chose the Federal Courthouse across from the Federal Building. However, when they arrived to place the truck bomb, street work was being conducted in front of the courthouse and the truck could not park there. The federal building became a target of opportunity.*

Calls to the 911 center began at 9:03 a.m. when over 1,800 calls for help were received. Many responders were already in route when they heard the initial explosion. Within 23 min the State Emergency Operations

Center had been activated. Within the "Golden Hour", 50 people had been rescued from the rubble and sent to hospitals throughout the area. During the response phase, there was so much help coming from citizens and emergency personnel, no security zone could be created. A local television station put out a broadcast without being told to, that all available doctors and nurses respond to the scene. This created more problems with accountability. Everyone was focused on saving lives, but because of the circumstances managing the scene became difficult at best. One nurse that responded to the call for help was killed while searching the site for victims when she was hit on the head by debris. During the search someone thought they found another bomb and the site was evacuated along with a 4 block area around the building. This gave incident commanders a chance to gain control and secure the site. ID tags were issued to those who had a legal responsibility to be operating at the site. The last victim was rescued from the building debris by 7:00 p.m.

Eleven FEMA Urban Search and Rescue Teams were dispatched to the site. Communications was another problem during the incident. Groups of responders were not able to talk to other groups. As a result, many did not know what others were doing. People evacuating the scene clogged roads and delayed emergency vehicle response to the scene. At the time of the bombing, Incident Command was new to the OKC department and they were in a learning process. IC involved a culture change and not all officers were using it. Chief Marrs got on the radio during the bombing response and told the officers to set up command and start using ICS. After that ICS was used by the department all the time.

OKCFD did not have an Urban Search and Rescue (USAR) Team at the time of the bombing. In the years following several members received USAR training. FEMA was adding a team and OKC hoped they would be the one. However, Missouri was selected instead. Eventually, the OK Department of Homeland Security funded a state USAR Team. Currently the team is composed of an OKC metro team and a Tulsa metro team. When combined they are a Type 1 USAR Team. Oklahoma Task Force 1 (OKTF1) also has a K-9 team with 8 dogs in OKC and 10 in Tulsa. Firefighter Jeff Hanlon trains OKC USAR search dogs; a dog named Willy is currently in training. Both teams have a swift water rescue component and a Helicopter Search and Rescue Team. Oklahoma Air National Guard flies the helicopters and OK-TF1 rides the hoist.

Investigation

McVeigh was captured just 90 minutes after the explosion by an Oklahoma Highway Patrol officer Charlie Hanger about 62 miles North of OKC. McVeigh's car was missing the rear license plate and was stopped by Officer Hanger. McVeigh was arrested for carrying a concealed weapon. Officer Hanger had no idea McVeigh was the OKC bomber. McVeigh was taken to the Noble County Jail and all of his clothing was collected and put in paper bags. As the FBI began their investigation evidence quickly lead to McVeigh. Evidence collected at the bombing scene lead to a vehicle identification of a Ryder Truck rented in Kansas. On April 20, the FBI released a sketch of McVeigh obtained from the Ryder dealer's description. A motel owner in Junction City, Kansas, recognized the person in the sketch as a guest who registered at the motel as Timothy McVeigh.

The FBI ran arrest records check on McVeigh and found he was in custody at the Noble County Jail in Perry, Ok. Additional investigation led to army buddy Terry Nichols and he was arrested as well. Nichols was tired, convicted, and sentenced to life in prison. McVeigh was found guilty and sentenced to death. He was executed June 11, 2001. FBI Agent Barry Black said of the investigation "It went as it was supposed to".

Oklahoma City Bombing National Memorial

On May 23, 1995, the remains of the Murrah Federal Building were demolished. Planning for a fitting memorial erected in its place carried on for 2 years. Today on the bombing site, there is a memorial and museum honoring the significance of that tragic day, in 1995. According to Kari

Figure 5.110 Two of the most outstanding features of the memorial are the "Gates of Time".

Watkins executive director of the Oklahoma City National Memorial & Museum, "The memorial was really built to remember those who were killed and those who survived and those who were changed forever". Two of the most outstanding features of the memorial are the "Gates of Time" (Figure 5.110). One gate is located at each end of a reflecting pond. The concrete gates are covered with a "Naval" and yellow bronze. On one of the gates is the time 9:01, representing the last moment of peace and on the other gate 9:03 representing the first moments of recovery. For me, the most moving part of the memorial was the 168 bronze chairs, each engraved with the name of one of the victims.

Large chairs represented the adults and small chairs the children. Also on the memorial site is a 90 year old American elm, "The Survivor Tree, tilted by the force of the explosion, but still standing and living in spite of the explosion (Figure 5.112). I have been to the memorial twice (*Firehouse Magazine*).

> ***Author's Note:*** *The second visit I brought my 5 year old granddaughter Abby. My wife and I explained what the memorial was all about. Following our explanation of the chairs, Abby immediately ran to a chair and hugged it. I believe she understood the purpose of the memorial (Figure 5.111).*

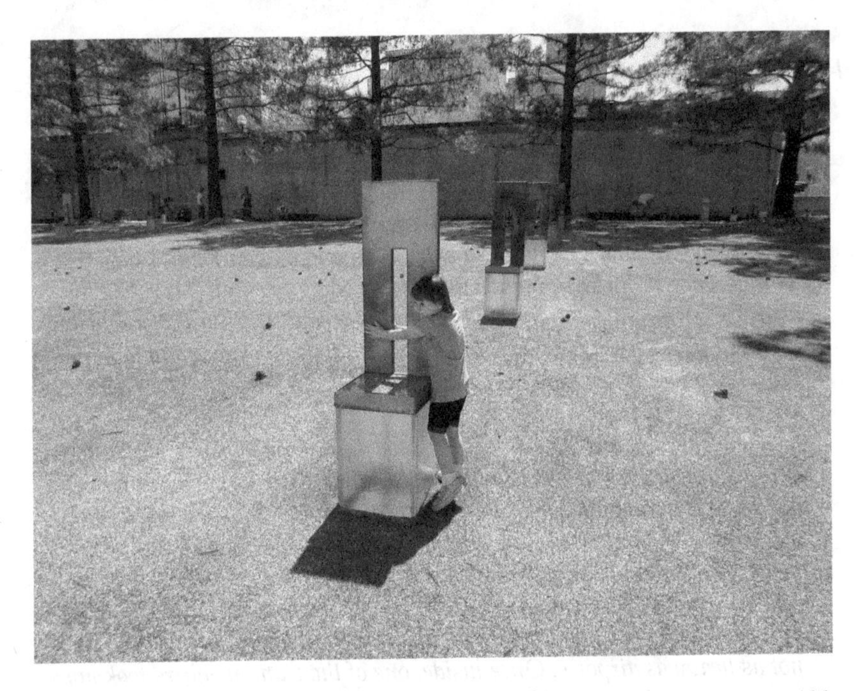

Figure 5.111 Following our explanation of the chairs and their meaning, Abby immediately ran to a chair and hugged it.

Figure 5.112 The Survivor Tree, tilted by the force of the explosion, but still standing and living in spite of the explosion.

Pentagon Force Protection Team

Prolog: *During a visit to the Charles County, MD, hazmat team, I found out that many of the team members worked day jobs in various career departments in the Washington D.C. Metro area. One such team member was also on the Pentagon Force Protection Team. During the visit, he asked me if I would do a story on the Pentagon team. I thought why not, not really knowing what that was going to entail. He called the team leader on the spot and told him and he said, let's do it. This visit involved a drive to Washington D.C., which was not one of my favorite things to do in those days. Actually the trip was to Arlington, VA, where the Pentagon is located.*

Once there I met the team at their headquarters, which is not on the Pentagon facility. We spend quite a bit of time there talking and taking photos. Then came the VIP tour of the Pentagon. This occurred on a weekend, so there were not the normal thousands of workers there, which made the tour much more enjoyable. I could not drive to the Pentagon, so I rode with the team chief. All of the response apparatus was also brought along for photographs. Went through security, that oddly was not as heavy as airports. Once inside, one of the team members took my camera and said he would give it back in the two places where I could take photos, the 911 memorial room and the Medal of Honor Room.

From this point on, I cannot talk about what I saw or was told. Exception, I did see the Secretary of Defense's Office. Once back outside, I was able to photograph the 911 memorial on the west side of the Pentagon and we took photographs of the apparatus and personnel (Figure 5.113). This visit had to be one of the highlights not only of my career but my life.

The Pentagon, located in Arlington County, VA, is the headquarters of the U.S. Department of Defense and the world's largest office building in terms of square footage 6.5 million square feet, of which 3.7 million is used for offices. The building has five floors above ground and two below ground. Approximately 23,000 military personnel are employed at the Pentagon, along with 3,000 non-defense support personnel.

While the building is massive looking from the outside, once inside you do not get the feeling you are in such a large building. The building is almost a city within itself with many areas that resemble a shopping mall. There is an athletic club, 20 fast-food outlets, a full-service restaurant, and many retail shops and professional services. The Pentagon also has its own Metro subway station and bus terminal. The interior of the building contains a pentagon-shaped, five-acre courtyard. During the Cold War, the Russians thought there was a missile silo in the courtyard and, for a time, a bull's-eye was placed at the center of the courtyard for them to view from their satellites. This was eventually replaced by a snack bar.

Figure 5.113 Outside the Pentagon on the West side is a memorial to those who died in the September 11, 2001 attack on the facility.

Ground was broken for the construction of the Pentagon ironically on September 11, 1941, exactly 60 years before the hijacked American Airlines Flight 77 was intentionally crashed into the west side of the building by terrorists. (The building was dedicated on January 15, 1943.) The 9/11 crash killed 189 people, including 5 terrorists, 59 passengers and 125 people at work in the Pentagon. Following the 9/11 attack, the present-day Pentagon Force Protection Agency (PFPA) was established and charged with safeguarding the occupants, visitors, and infrastructure of the Pentagon Reservation. The agency is made up of law enforcement personnel (U.S. Pentagon Police); criminal investigative and protective services agents; threat-management agents; chemical, biological, radiological, nuclear and explosives (CBRNE) technicians; and anti-terrorism/force protection and physical security personnel. This column focuses on the CBRNE Response Division.

Team History

Before the CBRNE division was created, it was preceded by a Bio Agent Testing (BAT) team. Today's CBRNE division is composed of full-time personnel with military and fire/hazardous materials service backgrounds. For now, the CBRNE division is quartered off-site in a former warehouse in Arlington. A new home for the division and its equipment and vehicles is under construction in the maintenance area of the Pentagon Reservation. This facility will enable the division to house all of its response vehicles, which now are kept in a fenced area.

Presently, there are 32 dedicated response personnel and 45 total in the CBRNE division. They respond to the Pentagon and its facilities and protect a daytime population of up to 35,000 people. On an annual basis, the members respond to several hundred calls.

Force Protection Team

CBRNE division vehicles are designated for functional assignments during a response. Each vehicle carries initial-response equipment and personal gear for assigned technicians. The Rescue/Rapid Intervention Team Emergency Response Vehicle (RIT ERV) also carries emergency medical service (EMS) equipment. Three 2009 Ford F-550/Fouts Brothers ERVs are designated as follows: one ERV for entry operations, one ERV for decontamination, and one ERV for Rescue/RIT. The largest vehicle in the CBRNE fleet is a 2003 International 7400/KME hazmat squad that carries large items, including bulk supplies, additional technical decon equipment, additional personal protective equipment (PPE), mitigation supplies, and rehab equipment and provides a command/research area.

The 2002 GMC Mobile Analytical Laboratory is used to perform on-site analysis of samples using bench top gas chromatography/mass spectrometry (GC/MS) and biological sample analysis (Figure 5.114). This vehicle was originally outfitted for the National Guard Civil Support Teams and carries the same types of equipment. The remaining response vehicles are Chevrolet Suburban of various years. One Suburban contains a command module for the hazmat branch officer or incident commander to perform incident command and accountability functions (Figure 5.115). Part of the CBRNE division is the Explosive Ordnance Disposal (EOD) unit, which is the only civilian bomb squad in the U.S. Army.

Robots

One of the many assets maintained by the CBRNE division is a Remotec MK3 Mod 0 robot provided by the Hazardous Devices Division. The robot is currently being outfitted with monitoring equipment for chemical warfare agents, toxic industrial chemicals/toxic industrial materials (TICs/TIMs), and radiological materials. The robot will be used to survey an incident prior to team members making an entry. This will allow the gathering of basic atmospheric readings, enhance incident size-up, and assist in the selection of PPE. The robot will also be used to provide video and audio from the "hot zone" to the command post. The robot can also be

Figure 5.114 Author Robert Burke during visit to Pentagon.

Figure 5.115 Pentagon Force Protection Team Fleet and team members.

used to haul equipment into the "hot zone," reducing the workload of responders wearing PPE.

PPE, Equipment and Training

Chemical protective suits used by the CBRNE division are Trellchem by Trellborg for Level A and Tyvec F for Level B. They also carry military Saratoga and Lanx chemical protective over garments. The Saratoga suit employs spherical carbon technology to provide effective body protection from all known chemical/biological warfare agents. LANX Fabric Systems use a technology of polymerically encapsulated activated carbon, a new and unique technology that provides extremely uniform carbon distribution and chemical protection. Respiratory protection used for chemical suits are Scott self-contained breathing apparatus (SCBA) and M40 APR masks and powered air purifying respirators (PAPRs). In-suit communications are provided by Scott Envoy Radiocom system. Monitoring equipment used by the CBRNE division includes:

- Chemical – MultiRAE, Ppb RAE 3000, Area RAE, ChemPRO 100, LCD 3.3, Miniwarn, Colorimetric Tubes, pH paper, potassium iodide starch paper, M-8 Paper, M-9 Tape, HazCat Kit
- Analytical – Hapsite, Griffin 450 GC/MS, HazMat ID
- Biological – Bio-Check 20/20, Handheld Assays, Razor
- Radiological – Identifinder, Raider, Ludlum, AN/UDR-14, Polymaster GN Pager

Initial-entry instruments are carried in black suitcases for easy access during an emergency. Those instruments include Ludlum and accessories; Miniwarn and pump attachment; Chempro 100; digital camera; MultiRAE; Identifinder; plus additional batteries, chargers, and consumables.

Experienced personnel are hired from the outside and hazmat certification is conducted in house. Pentagon police officers are trained by the CBRNE division to the Hazardous Materials Operations Level and in the use of Level C chemical and respiratory protection.

Reference Resources

The CBRNE division uses a wide variety of hazmat research materials that are common to all response teams. Electronic resources include the HazMaster G3, COBRA, and WISER software.

Much of the infrastructure inside the Pentagon for communications, monitoring, and security is secret. Photography is not allowed except for the interior 911 Memorial and the Medal of Honor room. Fixed monitors for chemical radiological and biological agents are very advanced and extremely sensitive. Portable instrumentation can identify any unknown substance.

Hazardous Materials Exposures

Major hazmat exposures include the swimming pool, heating and cooling systems, and the fuel farm on the east side of the Pentagon Reservation. Terrorist or criminal acts may present a wide range of threats involving chemical, biological, radiological, nuclear, and high-yield explosive materials. All commercial vehicles entering the Pentagon Reservation are screened for CBRNE materials. Additionally, 100% of the mail sent to the Pentagon is screened. Outside is a memorial to those who died in the 911 attacks.

Most common chemicals found in any city would also likely be found in the Pentagon as part of daily operations at the facility, transportation hub, and parking lot. The majority of responses at the Pentagon involve suspicious letters and packages or those containing white powder. Protests are common at the Pentagon, and during such events the CBRNE team establishes a decontamination capability as a precaution. Any mitigation or clean-up that may be required as a result of a hazmat or CBRNE event is conducted by the Pentagon Safety Office. Pentagon police will determine whether a response by the CBRNE division is required. If environmental issues arise, the CBRNE division works with Environmental Protection Agency (EPA) regional teams and is assisted by the EPA for environmental crimes. Mutual aid is provided by the Arlington County Fire Department (*Firehouse Magazine*).

Philadelphia, Pennsylvania Hazmat Team

***Prolog:** My first visit to Philadelphia occurred on the weekend between a two week class at the National Fire Academy in 1981. I took 4 two week classes and two weekend classes that year. During that visit I saw all of the usual landmarks in the city and stumbled across the Firemen's Hall Museum at 147 N. 2nd Street. Since that time I have numerous visits to Philadelphia after becoming friends with 3 former National Fire Academy students: Chief Bill Doty, Hazmat Battalion Chief, Chief Michael Roeshman, Hazmat Administrative Unit, and Tom Micozzie, Hazmat Coordinator Delaware County, PA. Between the three of them I have learned more about Philadelphia and Delaware County then I imagined. Spent many days riding with them and was introduced to several Philly treats, including the Philly Cheese Steak Sub and Rita's Italian Ice, Gelati.*

One thing I learned and found very interesting, the state has a burn injury reimbursement program. When any firefighter in the state is burned in a fire, they are given money. Philadelphia has a tradition that when a firefighter gets their burn money, they buy and cook a celebration dinner for the crew at the station. One day when I was riding with Chief Doty one of the Battalion 1 firefighters got his burn money and we were invited to eat with them. What a spread, steak, shrimp, and all the trimmings. Thanks to my friends for all the memories, Philly has a big place in my heart.

Philadelphia, sometimes known just as Philly, nicknamed "City of Brotherly Love" is the largest city in the Commonwealth of Pennsylvania. It is located in far Southeast Pennsylvania bordering both States of New Jersey and Delaware. With an estimated population of 1,591,800 in 2020. Philadelphia is the 8th largest city in the United States and has a geographical area: land area of 134.8 and 8.53 miles^2 of water. Philadelphia is bordered on the Southwest and Southeast by the Schuylkill and Delaware Rivers, respectively. Metropolitan Philadelphia has an estimated population of over 6 million makes it the 8th most populous area in the U.S. Philly played an instrumental role in the American Revolution as a meeting place for the founding fathers of the United States. Delegates signed the Declaration of Independence in 1776 at the Second Continental Congress, and the Constitution at the Philadelphia Convention of 1787.

Fire Department History

Following an extensive fire in Philadelphia, December 7, 1736, Benjamin Franklin founded the Union Fire Company, also known as "Bucket Brigade" (Figure 5.116). It was the first formally organized all volunteer fire company in the colonies. The first full-fledged volunteer firefighter

Figure 5.116 Following an extensive fire in Philadelphia, December 7, 1736, Benjamin Franklin founded the Union Fire Company, also known as "Bucket Brigade".

in America was Isaac Paschall. Career fire service in Philadelphia is considered to have descended from the Union Fire Company. "Ordinances of 1840, 1855, and 1856 established a City Fire Department which was a voluntary association of independent fire companies which, in return for subsidies, accepted the direction of City Councils.

An ordinance of 29 December 1870 established Philadelphia's first fully paid and municipally-controlled Fire Department, administered by seven Commissioners chosen by Councils. The Department went into service on 15 March 1871. The Commissioners were abolished and the department placed under the control of the Department of Public Safety as the Bureau of Fire in 1887 in compliance with the 1885 Bullitt Bill and enabling ordinance of 1886. The Fire Marshal, first appointed on 1864, was a member of the Bureau of Police until 1937 when his office was removed from it and placed directly under the Director of the Department of Public Safety. In 1950, it was transferred to the Bureau of Fire.

The City Charter of 1951 abolished the Department of Public Safety and established the present Fire Department. At that time its inspectorial duties were transferred to the Department of Licenses and Inspections.

On 14 February 1972, the Office of Emergency Preparedness, which had been organized in January 1952 as the Philadelphia Civil Defense Council, with the Mayor as Director, merged with the Fire Department and the Office was placed under the direct jurisdiction of the Fire Commissioner."

Today's Modern Department

Under the leadership of Commissioner Adam K. Thiel, who became the departments 20th Fire Commissioner in May 2016 the city operates out of 63 fire stations. Philadelphia's Fire Department has 3,000 plus uniformed personnel who operate 56 engine companies, 27 truck companies, 3 heavy rescue, 55 medic units, 2 fireboats, the hazardous materials unit, and 6 foam units.

Hazardous Materials Team

Philadelphia's Fire Department hazardous materials unit responded to 1,009 hazardous materials calls in 2018. Of those, 998 were Level 1, 10, Level II, and 1 Level III. Small spills of hydrocarbon fuels and gas odor and leak calls are handled by companies that carry absorbent and dispersal materials along with monitoring equipment. The company officer determines at what point the Hazmat Team is called in.

> ***Author's Note:*** *During 2004 I had the opportunity to ride with a friend, Chief William Doty, Battalion Chief 1-A, I had the opportunity to visit the hazmat station and talk with the officers and firefighters of Engine 60 and Ladder 19, who make up the hazmat team on A Shift, I sensed a great deal of pride and dedication in their quarters and equipment. I rode with Chief Doty on several occasions responding to calls while visiting Philadelphia landmarks.*

Hazmat 1's quarters are laid out all on one floor except for a second-level storage area for hazmat supplies in the engine bay. The engine bay has locker space for bunker gear, a storage room, and decontamination station. Living quarters include separate bath and locker rooms for men and women, bunk room, offices for each company, a conference room, kitchen, and watch desk. The building also houses a police station attached to the west providing police coverage for South Philadelphia.

The Hazmat Station is located in Battalion 1 at 24th and Ritner Streets, the area of highest hazmat exposure in the city and bounded by Market Street in Center City on the North, the Navy Yard on the South, Front Street on the East, and 25th street on the West. Hazmat 1 & 2 respond to incidents throughout Philadelphia. Hazmat 1 is a 2016 KME Kovatch (Figure 5.117). Hazmat 2 was built in 2006 and is a Freightliner Utilimaster (Figure 5.118). In addition to the hazardous materials unit, the station

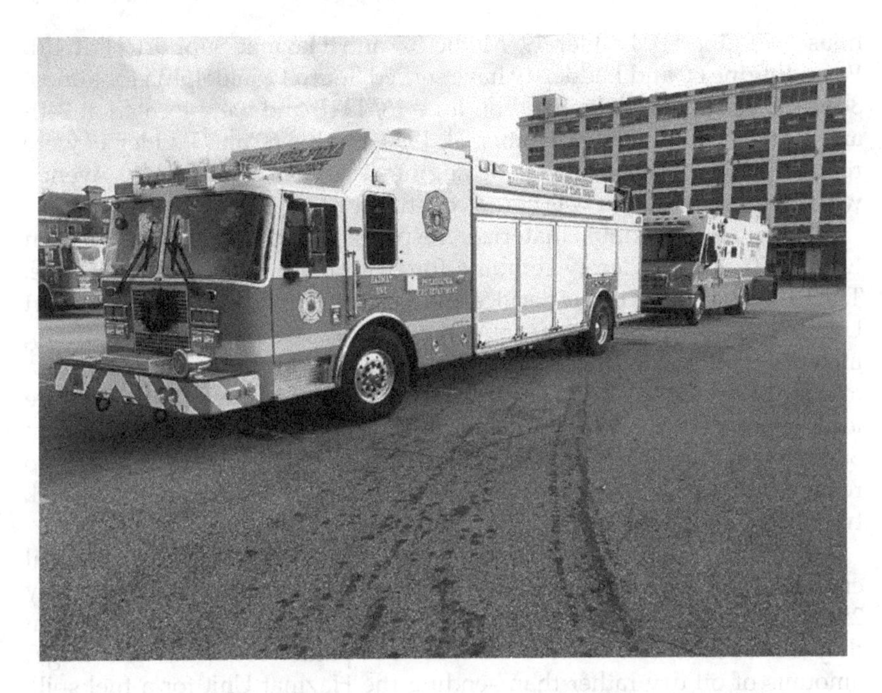

Figure 5.117 Hazmat 1 & 2 respond to incidents throughout Philadelphia.

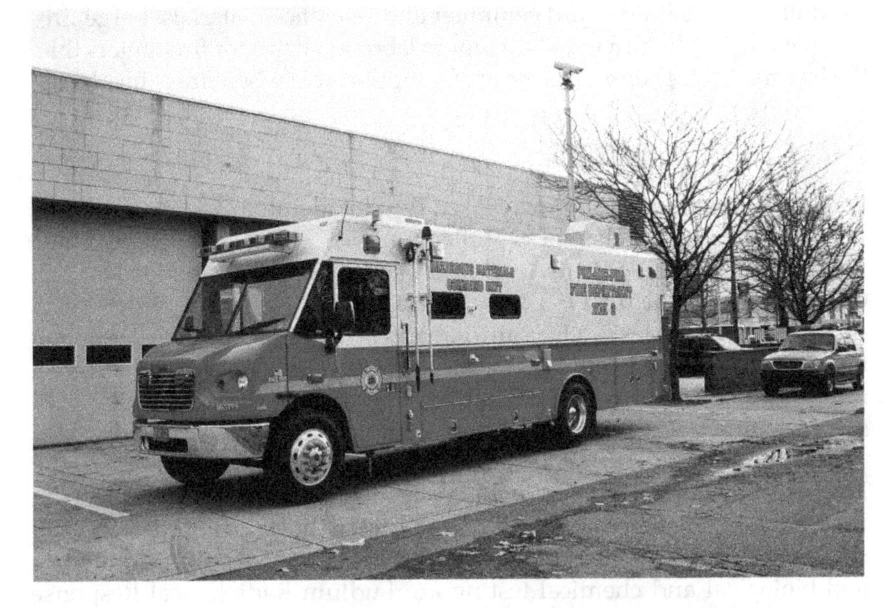

Figure 5.118 Hazmat 2 was built in 2006 and is a Freightliner Utilimaster.

houses Engine 60, Ladder 19, Medic 37, and Hazmat Support Unit 101. Foam Engine 60 and Ladder 19 have served South Philadelphia for almost 30 years. They were organized on June 16, 1921, and were located at 24th and Ritner Streets. On September 24, 1990, Engine 60 and Ladder 19 were relocated to the former house of Engine 47 at 3031 Grays Ferry Avenue while their new station was being constructed.

When a hazardous materials response is requested, crews from Engine 60 and Ladder 19 combine forces to place Hazmat 1 in service. The ladder is placed out of service and Engine 60, Hazmat 1, and Support Unit 101 make up the Hazardous Materials Task Force. Initial manning includes 2 officers, 8 firefighters, and Battalion Chief 1 and his Aid, for a total of 12 personnel. Backup hazardous materials-trained firefighters are located at Stations 1, 10, 24, and 49. These personnel fill in at the Hazmat Station when assigned crew members are off and also are available to respond to an incident scene if needed. Firefighters in Philadelphia work two 10 h days followed by two 14 h nights and 4 days off.

Statistics for hazmat responses within Philadelphia do not include local engine runs for hydrocarbon fuel spills. Each engine carries approximately 25 lb of oil dry material for cleaning up fuel spills. If the spill is larger than they can handle, one of the 7 "Depot" companies responds with larger amounts of oil dry rather than sending the Hazmat Unit for a fuel spill. The "Depot" stations are strategically located throughout the city.

Most incidents are small and usually involve unknown materials. Philadelphia uses the "Haz-Cat" system for identifying unknown materials in addition to their on-board computer and reference materials. Large fires involving hazardous materials are more labor intensive for firefighters than the hazmat unit. Their role is more of a support role when fire is involved.

PPE, Equipment and Training

Personnel Protective Equipment used by Philadelphia Hazmat for Level A are the DuPont Tychem Tk, Trelleborg: APS & TLU, Kappler Tychem Responder, and the DuPont: AcidMater and for Level B protection DuPont: Chemrel and Tychem CPF 3 & Tychem CPF 4. Respiratory protection is provided by M95 Mask, NBC canister, and Scott Air-Pak 50 w/60 min bottles (self-contained breathing apparatus, SCBA) (All Level A suits have a fitting to be connected to the Scott Air-Pak to a supplied air system). In suit communications is accomplished with Epic Com 3.

Monitoring instruments & identification equipment used by Philadelphia's Hazmat Team includes the "Haz-Cat" system for identifying unknown materials in addition to their on-board computer and reference materials. Other portable monitoring devices for air monitoring and biological and chemical testing are Ludlum Radiological Response Kit 2241-3, Philadelphia Rad Kit II: Ludlum Model 3 w/44-9 Probe and an Eberline RO-2 Ion Chamber, Philadelphia RAD Kit I: CDV-700 and CDV-715, ThermoElectron: IdentiFinder Na-I, and MGP: MGP2000 Pager/Dosimeter. Terrorist Agent monitors include Rapid Analyte Monitoring Platform (RAMP): Anthrax, BoTox, Ricin & Small Pox, HazCat Anthrax Kit, M-9 tape, M-8 paper, and M-256 Kits, APD-2000, and Chemical Agent Monitors (CAM).

Other monitoring instruments and testing materials include pH Paper, Cl_2 & NH_3 ToxiMeters (Chlorine & Anhydrous Ammonia) MSA: 5-Star CGI with Cl_2 & NH_3 Sensors, Oxidizer Paper, MSA: 5-Star CGI & MSA Orion, Draeger Colorimetric Tubes, Rae Systems: MiniRAE2000 PID, Photovac MicroFID, ThermoEberline TVA-1000 FID/PID, Rae Systems: SentryRAE, Rae Systems: AreaRAE, Spilfyter: Chemical Classifier Strips and Inficon: Hapsite (Gas Chromatograph/Mass Spectrometer).

Philadelphia's hazardous materials team members receive the Environmental Protection Agency (EPA) Level III Technician Course, Chemistry of Hazardous Materials for Firefighters, Radiological Monitoring, and a five level in-house training program. The five levels consist of Level I – Understanding Hazmat which is 32h, Level II – Personnel Protective Equipment (PPE) 16h, Level III – Decontamination 8h, Level IV – Plugging and Patching 40h, and Level V – Monitoring instruments and Meters 16h. All Philadelphia firefighters and medics are trained to the operations level.

Hazardous Materials Exposures

Most of Philadelphia's hazardous materials facilities are located on the south and east sides and consist of large refineries, chemical plants, and transportation routes. These facilities include Ashland Chemical, Sun Oil Company, Allied Chemical and Rhom & Hass. Port facilities are also present where super tankers are loaded and unloaded as well as smaller intermodal containers from ships. Many of these intermodal contain hazardous materials and are transferred from the ships to trucks and rail flat cars for transportation to other parts of the country. Several major highways pass through Philadelphia including Interstate highways 76 and 95 and U.S. Highways 1 and 13. Conrail and CSX are the major railroads that serve the city and routes pass through the North, South, and Central portions of Philadelphia.

Two major rivers also border Philadelphia, on the East the Delaware, and on the South the Schuylkill. Many of the hazardous materials transported through the city travel the river to or from the many chemical facilities located on the rivers banks. Dangerous chemicals located in and transported through Philadelphia include sulfuric acid, ammonia, formaldehyde, ethylene oxide (which is an ether), propane, benzene, hydrochloric acid, chlorine, and a wide variety of petroleum products are manufactured, stored, and shipped through the city.

Incidents

Gulf Oil Refinery Fire

On August 17, 1975, eight Philadelphia firefighters made the ultimate sacrifice at an 11 alarm hazardous materials incident and fire at the former Gulf Oil Refinery on the city's South side (Figure 5.119). While filling a foam tank on Engine 133 the foam layer on hydrocarbon compounds broke down, the product reignited trapping and engulfing three firefighters in flame. Five more firefighters went to rescue the others and all 8 died in the line of duty. Thirty-three firefighters were also injured and two were burned so badly they would never return to firefighting. This incident was the only time that Philadelphia firefighters lost their lives dealing with hazardous materials.

One Meridian Plaza Fire

While it wasn't technically a hazardous materials incident, three firefighters did lose their lives during a high-profile fire at the One Meridian Plaza high-rise fire in Center Philadelphia. This fire was caused by a chemical reaction of improperly stored linseed oil soaked rags that spontaneously combusted.

Figure 5.119 On August 17, 1975, eight Philadelphia firefighters made the ultimate sacrifice at an 11 alarm hazardous materials incident and fire at the former Gulf Oil Refinery on the city's South side.

***Prolog:** Philadelphia hazmat deploys to every Eagles home game to provide security for potential terrorist attacks. My friend Chief Mike Roeshman invited me to attend a game with the hazmat team in 2009. This was my very first time at an NFL game. We arrived well before anyone else and prepared for the inspections and monitoring. I had a tour of the command post area where security cameras are monitored and security for the game is headquartered. It was great watching the visiting team arrive, the teams warm up and the pre-game and during the game security implemented. Another great adventure in Philly.*

Philadelphia Hazmat at Eagles Games

While fans are busy tailgating and preparing for an NFL football adventure at the "Linc" Home of the Philadelphia Eagles, Philadelphia's Fire Department Hazardous Materials Team is working to keep them safe from Weapons of Mass Destruction (WMD) and other hazardous materials. The City of Philadelphia, with a heavy chemical industry presence on its South and East sides, has always taken a proactive approach to hazardous materials response.

My visit to Lincoln Financial Field, nicknamed "The Linc" took place on a cold December day after Christmas when the Eagles took on the Buffalo Bills. This 68,000 seat stadium is located on Philadelphia's South side just North of Philadelphia International Airport and the former Philadelphia Navy Yards, next to Interstate 95 and adjacent to a main railroad line at 11th and Pattison Avenue. The Linc is also located next to the Wachovia Center Stadium where the 76er's of the NBA and the Flyers of the NHL play and across the street from the Citizens Bank Park where the Philadelphia Phillies National League Baseball Team plays. This was my first time to attend a live NFL game and at the same time having the opportunity to look at behind the scenes operations of a pro football game made it a memorable occasion.

Battalion Chief Michael Roeshman of the Hazardous Materials Administrative Unit of the Philadelphia Fire Department was my guide for the day, as well as former student at the National Fire Academy and good friend. We were hosted at the Linc by Leonard Bonacci the Director of Event Services who is the head of overall security for the Linc. Mr. Bonacci was most gracious in letting us freely tour the facility including some of the secure areas not normally seen by the public and access to the playing field and visiting team areas as the Buffalo Bills and Philadelphia Eagles arrived at the stadium. He was sporting his Super Bowl XXXXII Ring given by the Eagles to staff, which is also the same as those worn by team members. Many thanks to Mr. Bonacci for making my visit to the Eagles game possible.

Chief Roeshman brings 32 member teams and the Mobile Lab and Communications Center (HM-2) to all Eagles home games at the Linc as well as the college Army-Navy game. Survey teams are made up of off-duty hazmat team members who volunteer for the duty. Hazmat 1 and other task force units remain in full service in quarters and respond to normal call loads unless needed at the stadium for an incident. Providing hazmat coverage for the games starts on Thursday of game week at a meeting with team officials to go over security measures for each game. Ninety minutes before kick-off Chief Roeshman meets with the NFL officials and that games head referee to instruct them as to what they should do in the event of an incident during the game. They are instructed to stand at midfield at the 50 yards line and await further instructions such as evacuate up the ramp to dressing rooms, evacuate the stadium, etc. They are instructed to look to the hazmat team for directions.

Teamed up with Philadelphia Police Officers and Stadium Security, Hazmat Team Members monitor vehicles entering the stadium property. One team is assigned to monitor player's vehicles and deliveries to the stadium and the other two teams to check fans vehicles entering the stadium's two VIP parking areas. All vehicles are checked with radiation monitors and Photo Ionization Detectors (PIDs). Trunks are opened and the rear storage areas of SUV's and minivans are checked as well. Security personnel check the undersides of vehicles with mirrors for Improvised Explosive Devices (IUD's) or other contraband. Once all of the vehicles have been checked, hazmat team members are assigned to patrol throughout the stadium monitoring the air for the potential release of hazardous materials or terrorist agents. Following the kick-off the survey teams are assigned to the North End Zone, West Stands, East Stands, and the South End Zone. They continually walk their assigned areas with chemical and radiation detectors throughout the game.

One of the assets utilized by the Philadelphia Hazmat Team at the Eagles games is their new Mobile Laboratory Hazmat 2. This mobile lab is stationed outside the stadium in an area where the Eagles have provided special electrical hook-ups during games. One hazmat team member is assigned to the lab to maintain it operationally throughout the game and additional personnel would be added to the unit during an emergency. The Hazmat Task Force would respond to the stadium for additional resources. Hazmat 2, a specially designed vehicle is equipped with a Gas Chromatograph/Mass Spectrometer, Microscope, on-board Glove Box, fume hood, built-in weather station, telescoping zoom camera, computer work stations, radio and telephone equipment and other portable monitoring, and mitigation equipment.

A Gas Chromatograph/Mass Spectrometer is usually only found in an analytical laboratory and not in the field. With this capability the Philadelphia Hazmat Team can analyze unknown materials in the field as long as they have a test sample of the material already programmed into the device. The Glove Box is a biological and chemical sealed area where samples can be handed in through a port to the outside and isolated from other parts of the vehicle for handling unknown materials without contaminating the unit. They use the microscope to view biological materials to help determine their credibility as a potential hazard (*Firehouse Magazine*).

> ***Author's Note:*** *Chief Roeshman has since retired from the Philadelphia Fire Department, lives in Philly and we remain friends. Thanks for all the memories and friendship Mike.*

Rapid City, South Dakota: Hazmat Response in the Black Hills of South Dakota

Rapid City is the second largest city in the State of South Dakota with a resident population of 77,525 in 2020. In addition to the resident population, Rapid City is also a major tourist destination with tens of thousands of people visiting during the tourist season. The city is known as the "Gateway to the Black Hills" and "Star of the West". Rapid City is nestled on the Eastern slope of the Black Hills Mountain range, home of Mount Rushmore, the Crazy Horse Monument, and Black Hills Gold. Rapid City is also home to Ellsworth Air Force Base which was home to B-17, B-25, B-29, and B-52 bombers and numerous missiles during the cold war. It is presently home to the 28th Bomb Wing and the B-1B Lancer aircraft. Carrying the largest payload of both guided and unguided weapons in the Air Force inventory, the multi-mission B-1 is the backbone of America's long-range bomber force.

Rapid City sits at an elevation of 3,202 feet in the shadow of *Harney Peak,* which at 7,242 feet (2,207 m), and is the highest point east of the *Rocky Mountains.* Rapid City enjoys an unexpected climate, free of the icy blizzards and scorching summers typical of much of the rest of the Dakotas. Summers are warm but dry and autumn is noted for its delightful "Indian summer" weather. Mild, sunny days are common throughout the winter and occasional "Chinook" or warm winds frequently follow a stint of snowy weather. Snowfall is normally light with the greatest monthly average less than 8 inches. Spring is characterized by wide variations in temperature and occasionally some wet snowfall. Low humidity levels, infrequent precipitation, and northwesterly winds prevail in the city.

Fire Department History

Organized fire protection began in Rapid City during March 1886, when 56 volunteers formed the Rapid City Hose Company #1. With hose carts and a hook and ladder, the group was amply prepared to protect the town's 400 residents. In 1915 the City's first fire station was built at 610

Main St. and remained in service until 1975. In 1948 the Rapid City Fire Department changed from a volunteer organization to a full-paid department with 14 firefighters. In 1972 when a flash flood inundated the City on the night of June 9, 238 people lost their lives, including three firefighters. The post-flood period was a time of growth and renewal for the City and the fire department, with a new headquarters station being built in 1975, as well as an increase to five substations. A sixth substation was built in 2002 and opened in January 2003. The opening of this new Station located off Highway 16 added 12 additional firefighters to the department. It also provided a new home for the hazardous materials team.

Today's Modern Department

Today's modern Rapid City Fire Department consists of 158 personnel under the leadership of Chief Rod Seals. They operate out of 8 fire stations with 4 front-line engine companies (2 in reserve) 2 ladder trucks, 1 rescue, 1 hazmat, 5 brush trucks, 2 crash/rescue trucks at the regional airport, and 12 support vehicles including an air cascade system provided by the county. The fire department response area is approximately 51.01 $miles^2$. On March 1, 2003, the Rapid City Fire Department began providing Advanced Life Support (ALS) to Rapid City and Pennington County and currently operates nine ambulances covering an area of 3,200 $miles^2$. During 2018 the Rapid City Fire Department responded to 17,826 calls for service, of those emergency medical service (EMS) accounted for 12.539 or approximately 70% of their call volume.

Hazmat Team History

Rapid City and Pennington County formed their hazardous materials team as a joint venture in 1987 in response to the Emergency Planning and Community Right-To-Know Act (EPCRA) also known as Sara Title III. Their first hazmat unit (still in service today) was a 1989 Central States Fire International (Crimson) (Figure 5.120). Initially they found they had too much equipment for one unit so the team was split-up into separate decon and hazmat teams.

Hazmat Team

The hazmat unit is housed at Station 6 located at 1930 Promise Road off of Highway 16 in South Rapid City. Station 4 houses the decontamination unit and is located at 700 East Fairmont Blvd. also in South Rapid City. Station 6 personnel operate the hazmat unit and Station 4 personnel are trained to the operations level and conduct decontamination operations. All other firefighters on the department are trained to the operations level

Figure 5.120 Rapid City's first hazmat unit (still in service today) was a 1989 Central States Fire International (Crimson).

as well. Captain Dan Goodart is the only original team member still active on the hazmat team. Rapid City's hazmat team responds to an average of 30 hazmat incidents per year not counting transportation fuel spills. Engine companies carry absorbent materials for use on fuel spills of 5 gallons or less. Spills of greater volume result in the dispatch of the hazmat unit.

One of four state-designated hazardous materials response teams, the Rapid City Hazmat Team, covers the Western half of South Dakota West of the Missouri River, which also includes the Pine Ridge and Rosebud Indian Reservations. Mutual aid is available if needed from Ellsworth Air Force Base located on the Northeast side of Rapid City and the 82nd weapons of mass destruction (WMD)-CST South Dakota National Guard Civil Support Team based in Rapid City. They also provide hazmat response services to the State Capital of Pierre. The state capitol is served by an all-volunteer fire department led by a paid fire chief. South Dakota Task Force (USAR) is comprised personnel from Rapid City, Sioux Falls, Aberdeen, and Watertown Fire Departments. Their components are Hazmat, water rescue, canine search, and collapse rescue.

Pierre has an estimated population of 13,876 making it the second-least-populated state capital after *Montpelier, Vermont.* Pierre is located approximately 170 miles Northeast of Rapid City at the geographical center of the state. Back in 2003 Homeland Security grant funds were used to provide a cache of equipment for the Pierre FD. They were trained to assist Rapid City technicians when they respond to incidents in Pierre. Since that time the state has set up a four official hazmat team network for coverage throughout the state. The teams are located in Sioux Falls, Aberdeen, Watertown, and Rapid City. The South Dakota National Guard provides air transport of the hazmat team to Pierre where they have staged equipment to deal with a hazardous materials emergency.

Also housed at Station 6 are Engine 6, Medic 6, and a Rural Engine. The hazmat unit is not dedicated and when a hazmat call is dispatched personnel from Station 6 provide personnel for the hazmat unit. The department has a total of 25 trained hazmat technicians with 6 on duty per shift. Firefighters in Rapid City work a 24h on 48h off shift schedule. Firefighters are sent to the Pueblo, Colorado 80h basic hazmat technician course. They also receive advanced training at Pueblo, and attend the National Fire Academy Chemistry and Operating Site Practices classes. Monthly in-service training is held in house for team members.

PPE, Equipment and Training

Rapid City Hazmat uses MSA 4,500 psi 60min self-contained breathing apparatus (SCBA) for respiratory protection (firefighters utilize MSA 30min SCBA). Level A Chemical protection is provided by DuPont Tychem TK Commander suits and DuPont CPF4 encapsulated suits for Level B. DuPont Tychem Thermopro flash protection suits are available for use in potentially flammable atmospheres. Motorola Clear Command Bone Mikes are used for in-suit communications. Monitoring equipment for hazardous materials include Travel IR Hazmat Chemical Identifier, Orion Multigas Detector, Sirius Multigas Plus PID Detector, and MiniRAE 2000 Photo Ionization Detector. Instruments used for WMD include APD 2000, Prime Alert Biodetection/Threat Verification System, Ludlum 2241-2 Radiation Detectors, Personal Dosimeters, Draeger Civil Defense System, Draeger Chip Meter System, M8 and M9 Paper, and M256A1 Detection Sampler. They also carry a Search Cam Entrylink Wireless Video System, Airshelter inflatable decon tents, and an Airshelter dress out tent.

Technicians receive 24h of annual refresher training. Quarterly training sessions are held with strong training relationships maintained with local CST, Emergency Management and many other response agencies. The State of South Dakota will soon be releasing a Hazmat Technician certification task book which will allow for consistent and standardized training across the state. Work is also progressing on development of Hazmat Specialist task books and a hazardous materials training trailer which may be utilized by many departments/organizations across the state. All personnel outside of hazmat are trained to the Operations Level.

Research Resources

Resource information on hazardous materials carried on the hazmat unit includes both computer based and hard copy. Computer-based resources include CAMEO, WISER, ChemKnowledge System, TOMES System, and Chemical Reactivity Worksheet. Hard copy resources include CHRIS Manuals, Jane's Chemical Biological Handbook, NIOSH Pocket Guide, Hawley Condensed Chemical Dictionary, Emergency Response to Terrorism Job Aid, Medical Management of Chemical, Radiological and Biological Casualties handbooks, and others.

Hazardous Materials Exposures

Hazmat highway transportation exposures in the Rapid City area include Interstate 90, which is a major East/West transportation corridor across the United States, and State Highway 79. Burlington Northern/Santa Fe, CNX, and DM&E railroads transport hazardous materials through Rapid City. Several natural gas and petroleum pipelines also service the area. Fixed facilities located in or near Rapid City which use or store hazardous materials include petroleum tank farms, propane, anhydrous ammonia, chlorine, cyanide, and hydrochloric acid. The most common local hazmat problem involves hydrocarbon fuels.

The mining industry is one of South Dakota's leading industries with an annual production of $254 million. There are 285 active mines in the state, most of which utilize hazardous materials in their operations. These hazardous materials are shipped and stored on site at the mines. Gold, silver, iron ore, and cement are among the leading materials mined in the state. Cyanide is used in gold mining operations and is shipped throughout the state. Major incidents handled by the Rapid City Hazmat Team have included a propane tanker rollover in Hill City, methyl lithium, picric acid at the South Dakota School of Mines, and a cyanide truck rollover.

Incidents

Vast majority of response include hydrocarbon spills, gas leaks, and/or suspected gas leaks. Most recent major incident (Fall 2018) included a tank rollover in Spearfish, SD, which is located approximately 45 min from Rapid City near the Wyoming border. The tanker was carrying approximately 10,000 gallons of diesel and gasoline. Minimal amount of fuel was spilled. RCFD Hazmat team was requested to assist with grounding/bonding, drilling and transferring of fuels to alternate transportation. Incident lasted approximately 12 h.

Tilford, SD September 8, 2018 Propane Tank Explosion

Firefighters from Sturgis Fire Department received a call for a house fire in Tilford, located between Rapid City and Sturgis at 4:00 p.m. for a reported structure fire. Rapid City Fire Department responded on mutual aid. Heavy smoke was visible as firefighters approached the scene and they found a fully involved single family dwelling upon arrival. Initial attempts to battle the fire were hampered by the neighborhoods narrow gravel streets, congestion of nearby homes, sheds and garages, parked vehicles an over growth of summer grasses and underbrush. Witnesses said there were several very large propane tanks on the property.

The property where the fire was located further hampered firefighting efforts by downed power lines and several propane tanks while fighting the fire. Fire spread from the house to several out buildings and impinged upon a 500 gallon propane tank that sat on the south side of the structure. It was reported that the tank had been recently filled. While fighting the fires a BLEVE occurred involving a propane tank that failed due to exposure to the fire. Chief Fischer was attempting to move a Sturgis Volunteer Fire Department vehicle, a Chevrolet Suburban parked in a driveway parked north of the burning home, when the portion of the exploding tank cleared the burning home, a fire engine, and the Suburban and struck Fischer, killing him instantly. Sturgis Assistant Fire Chief David Fischer was a 22 year veteran of the fire department, and also worked for Sturgis Ambulance Service. He was also a member of the South Dakota National Guard's 82nd Civil Support Team. Staff Sgt. Fischer was 43 years old. Chief Fischer was struck by a large portion of the propane tank when it exploded. It was determined that the fire was started by an elderly man who was in bed smoking while on oxygen. He perished in the fire. It is unclear whether he died from the fire in the dwelling or the propane tank explosion. Reportedly the man could not get around very well on his own. Another woman in the house escaped unharmed (*Firehouse Magazine*).

Reno, Nevada Hazmat Team: Protecting the "The Biggest Little City in the World"

Reno is the county seat of Washoe County Nevada with an estimated population of 260,258 in 2020 and is the third most populated city in Nevada behind Las Vegas and Henderson. Reno is located in western Nevada near the California border 26 miles north of the State Capitol of Carson City and 22 miles Northeast of Lake Tahoe, California, in the high desert. It is located next to the City of Sparks with an estimated population of 108,334 in 2020. The cities are often referred to as the Twin Cities. Reno is known as the Biggest Little City in the World. Reno's climate ranges from lows in the 20s in the winter to highs in the low 90s during the summer. Reno experiences lots of sunshine and rain fall averages less than an inch per month year around.

Fire Department History

Fire protection in Reno began with formation of Reno Hook & Ladder Company No. 1 in 1868 and continued with organization and reorganization of various volunteer engine and hose companies under the authority of Washoe County. The City of Reno was incorporated in 1903 by the Nevada Legislature also creating the Reno Fire Department with a career staff. Shortly after, the new Reno City Council authorized purchase of two new steam fire engines which coincided with the retirement of an 1874 steam fire engine that had reportedly worn out. In 1916, the Reno City Council was urged by the fire chief to consider purchase of motorized fire apparatus. This resulted in the acquisition of two new Seagraves fire engines and initiated retirement of the Reno Fire Department's stable of horses and horse-drawn equipment.

In 1917, the City Council approved purchase of motorized apparatus from American LaFrance that included the city's first ladder truck (85 foot) which was recently located by a Reno fire captain and is in the process of being restored; a pumper and a chemical-hose car, which now reposes in the collection of the National Automobile Museum two blocks from Reno Fire Department headquarters in downtown Reno.

Today's Modern Fire Department

David Cochran is the Chief of the Reno Fire Department with 248 uniformed personnel operating from 14 full-time career fire stations and manning 21 engines, (5 are paramedic engines), 8 wild land fire engines, 4 truck companies, 2 medic units, 3 water tenders (tankers) 3 light rescue, 2 heavy rescue, 3 air cascades, 1 hazmat unit, 1 decontamination trailer, 1 mobile command unit, and 3 water rescue units (with boats and kayaks). An additional 17 personnel are assigned to the fire prevention and fire investigation bureaus and there are 20 support staff including mechanics, supply, warehouse, and administrative support personnel. Staffing is four on both engines and trucks. Firefighters work a schedule of 48 hours on and 96 hours off. The work week averages 56 hours. All firefighters are EMT-I trained but only provide medical first response and support until ambulance personnel arrive.

The goal of the Reno Fire Department is to have at least one EMT-I trained person on every engine company. Ambulance transport is provided by a private ambulance company with advanced life support capability. During 2018, Reno fire department (FD) responded to 41,727 calls for service, of which 28,093 were emergency medical service (EMS) and 913 were fires. Fire Protection at Reno International Airport is provided by a separate fire department that works directly for the Reno-Tahoe Airport Authority. The Reno Fire Department and Truckee Meadows

Fire Protection District were consolidated in 2000 and the Reno Fire Department administers the fire district.

The Reno and the Truckee Meadows Fire Protection District covers an area of 650 miles^2 where the most population of Washoe County lives, but also responds anywhere within the 6,000 miles^2 of the county if needed. Washoe County extends north and borders Idaho and its entire western border is common with California. Reno Fire Department and Truckee Meadows Fire Protection District have a combined annual operating budget of $53.6 million. They cover an area of 6,000 miles^2, which is all of Washoe County. Most of the population of the county is located in an area of 650 miles^2.

Reno Fire also responds to neighboring Story County for mutual aid. Four Reno Firefighters have died in the line of duty, but none since 1948. In addition to the 14 city fire stations, Reno also administers 11 volunteer/auxiliary fire departments in the Truckee Meadows Fire Protection District and other unincorporated areas of Washoe County. These departments operate from 12 fire stations and provide 12 fire engines, 16 wild land fire units, 4 water tenders (tankers), and 2 ambulances.

Hazmat Team History

The Reno Fire Department is part of a Regional Hazardous Materials Response Team. This Triad hazmat team is composed of members from Reno, Truckee Meadows Fire Protection district, and the Sparks Fire Department. Hazmat team response in Reno began in 1986 not because of any major incident but rather because of the general concern of hazmat risks present in the community. No fire department wanted to take on the responsibility alone, which ultimately led to the regional team concept. Costs of operating the team are shared by each member entity.

Hazmat Team

Reno's hazmat unit is a 2001 Freightliner Chassis, custom box with an interior laboratory, and is located at Station #3, 580 W. Moana (Figure 5.121). Additional apparatus at Station #3 includes Truck 3, a 100 2004 E-One Tiller, Engine 3, a Pierce Quantum with a 1,500 gpm pump and 750 gallon water tank, Rescue 3, Battalion 2, and Decon 3. All personnel at Station 3 are hazmat technician trained. Reno's hazmat response is not dedicated; personnel from Station #3 man the hazmat unit and respond with other apparatus as a task force for hazmat incidents. There are 12 technicians plus a chief on duty for hazmat response on any given day at Station #3. Station #13 has been designated as an additional hazmat station with Engine 13 and water tender.

All personnel on duty are hazmat technicians. There are 48 total hazmat technician trained personnel on the regional hazmat team.

Figure 5.121 Reno's hazmat unit is a 2001 Freightliner Chassis, custom box with an interior laboratory.

Hazmat response is conducted on a tiered basis depending on the analysis of the incident scene by dispatch or first arriving companies. Incidents that occur on the North side of the city result in the response of Engine 13, Truck 3, and the hazmat rigs. On the South side of the city apparatus from Station 3 responds only. In 2018, they responded to 422 calls. All apparatus carries absorbents and 6-gas meters. Hazmat is called in when a substance is above the Reportable Quantity (RQ), an unknown material, a threat white powder etc., or medical symptoms are noted.

- Level I Incidents involve the response of all on-duty Reno Hazmat Personnel.
- Level II Incidents involve all on-duty Reno Hazmat Personnel plus Sparks Hazmat Personnel.
- Level III Incidents involve all on-duty personnel from Reno and Sparks and all available off-duty personnel. There is no close by mutual aid available to assist the Regional Hazmat Team. The hazmat team responds to an average of 4–6 calls per month, mostly small spills and unusual odors.

PPE, Equipment and Training

Personal protective equipment (PPE) for hazmat team members includes Trelborg, Kappler, and DuPont for Level A & B incidents. The DuPont Tychem Reflector Level A suit is one of the primary suits they use because of its versatility. The Reflector is a single skin multi-layered construction with flash protection on the outside and a visor that almost goes all around the suit for optimum visibility. The exterior of the suit is also reported to be abrasion resistant. It is advertised to be the first Level A suit that is certified to National Fire Protection Association (NFPA) 1991 (2005). Respiratory protection is provided with Scott 4.5 self-contained breathing apparatus (SCBA) utilizing 1 one hour bottles. Positive Pressure Air-Purifying Respirators (PAPR) and Cartridge Respirators are also available for weapons of mass destruction (WMD) type incidents. In-suit communication is provided by Scott Bluetooth though 800 MHz radios.

Reno Hazmat's Operational Procedures place a heavy emphasis on analysis and monitoring of incident scenes to try and conclude incidents more quickly and shrink the required distances of Hot Zones (Figure 5.122). Using this process, they have been able to reduce the average incident time from 6 hours to 45 minutes. Hazmat team members are trained to be analysis "experts". One of the primary tools for on-scene analysis is a laboratory grade portable Inflicon Hapsite Chromatograph Gas Spectrometer used to identify unknown liquids and gases. This is an expensive instrument and very few departments have this resource available to them because of the cost.

Units are known to be located at the Kennedy Space Center in Florida and the New York City Hazmat Team is reported to have one as well. Additional monitoring equipment available includes Radiation monitors, Area Rae, Chlorine and CO detectors 4-gas meters, HazCat kits,

Figure 5.122 Reno Hazmat's Operational Procedures place a heavy emphasis on analysis and monitoring of incident scenes to try and conclude incidents more quickly and shrink the required distances of Hot Zones.

Foxbough-HID/PID for organic vapors, Guardian Reader for Biologicals and the M256 kits for WMD. The Hazcat Kit involves chemical analysis to identify unknown materials or identify chemical families they belong to. All trucks, rescue units and chiefs carry 4-gas meters. Reno has their own maintenance and calibration shop operated by hazmat technicians for testing, calibration and maintenance of their monitoring equipment. The facility is located at Station #3.

Initial training for hazmat team members consists of 2 weeks of the Chemistry of Hazardous Materials, 1 week of chemical analysis, 3 weeks of technical operations, which includes monitoring, tactics and incident command. In-service training for team members consists of 8 hours per month. Most training is conducted in house. All firefighters in Reno are trained to a minimum of hazardous materials operations level.

Reference Resources

Computer-related and hard copy reference information carried on the hazmat unit includes the following: NIOSH pocket Guide, ERG, Hawley's Condensed Chemical Dictionary, the Coast Guard Chris Manual, CAMEO, TOXNET, Farm Chemical Handbook, IFSTA manuals relating to hazmat response, and a MSDS Database.

Hazardous Materials Exposures

Reno is located on a Union Pacific Railroad main route; Interstate 80 and Highway 395 are the major highway transportation routes through the city. As a result of its location, Reno is an avenue for much of the east/west traffic in and out of California in the U.S. Shipments of radioactive materials are also common on along Interstate 80. Highway 395 is a major route leading to U.S. Army and U.S. Navy ammunition depots located 60 miles North and 100 miles South of Reno. Along with being a major warehousing and distribution center, the Reno/Tahoe airport is becoming a major air cargo facility.

Daily, thousands of shipments of hazardous materials move safely through the Truckee Meadows area. Major pipelines and fuel farms carrying chemicals from the California bay area are also located in Reno. The North side of Reno near Station #13 is a heavily industrialized area with the Tesla Giga Factory, RR Donnelly & Company printing operation, JC Penny West Coast Warehouse, and Sierra Chemical Company, which is a wholesaler that takes bulk chemicals and places them in smaller packages. Nevada is a state where a lot of mining takes place. Railcar loads of cyanide and explosives are brought into the state to support the mining industry. Other common hazmat exposures encountered in the Reno area include chlorine, phosgene, pesticides, liquefied petroleum gases, oxidizers, and hydrocarbon fuels (*Firehouse Magazine*).

Sacramento, California Metro Hazmat Team

Sacramento is the capital city of the State of California and county seat of Sacramento County. The city is located at the confluence of the Sacramento River and the American River in Northern California's Sacramento Valley. The city's estimated 2018 population of 521,769 makes it the sixth largest city in California. Sacramento has a geographic area of 100 miles^2 of which is 98 miles^2 of land and 2 miles^2 of water. It is the core of the Sacramento metropolitan area which has an estimated population of 2,242,542 in 2018.

Sacramento County is the central county of the Sacramento metropolitan area. It covers about 994 miles^2 with an estimated population of 1,567,490 in 2020. The county is located in the northern portion of the Central Valley on into Gold Country. Sacramento County extends from the low delta lands between the Sacramento River and San Joaquin River including Suisun Bay, north to about 10 miles beyond the State Capitol and east into the foothills of the Sierra Nevada Mountains. The southernmost portion of Sacramento County has direct access to San Francisco Bay.

Fire District History

Sacramento Metro Fire District was formed in 2000 when the American Fire District and Sacramento County Fire District merged. It is the seventh largest fire department in the State of California. Sixteen fire departments make up the Sacramento Metro Fire District. They protect an area of 417 miles^2 and a population of approximately 750,000 people outside the city limits of Sacramento, California, the State Capital.

Today's Modern Department

Under the leadership of Fire Chief Todd Harms, the Sacramento Metro Fire Department has nearly 716 career of which are 460 paramedics and 19 reserve uniformed personnel operating 36 engine companies, 7 truck companies (3 tiller/tractor and 2 rear mount), 18 medic units, 1 hazardous materials company, and 1 technical rescue company from 41 Stations. Hazmat includes one hazmat truck, one decontamination Unit, three aircraft rescue and firefighting (ARFF), and two foam trailers. Other equipment includes two helicopters, six water tenders, two dozers, and two rescue boats. Sacramento operates dedicated paramedic engine companies in addition to medic units. During 2018, they responded to 96,495 calls for service including 51,971 medical transports.

Hazmat Team

Sacramento Metro Fire District started its own hazardous materials team in July 2003, prior to that they contracted with Sacramento City. The hazardous materials unit is a Type I Hazmat Team and responded to 401 hazardous materials incidents in 2018. Statistics for hazmat responses within do not include local engine runs for hydrocarbon fuel spills. Hazmat responses are categorized into three levels. Fuel spills are Level 1 incidents and the hazmat team does not respond unless a Level 2 or 3 incident occurs. Each engine company carries a limited amount of absorbent material for small hydrocarbon spills.

Regional Type I decon assets have been located within the Sacramento, Yolo and Placer County Regions. Type I decon assets have the ability to decon non-ambulatory patients. Units have lights, water heating, shelter, and modesty garments for 250 persons. The hazardous materials unit is a 2004 Pierce custom made on a rescue chassis (Figure 5.123). It is located at Station 109, 5634 Robertson Ave., Carmichael, CA, along with Engine 109. Hazmat 109 is a dual company and carries all of the equipment of a truck company (minus the aerial device) and operates as Truck 109 on non hazmat alarms. A replacement apparatus is on order with Pierce Manufacturing (expected spring 2020) and will include most of the same features of the current apparatus, including the on-board lab.

Truck/Hazmat 109 has a light tower, television capability, and a lab. One of the truly unique aspects of Hazmat 109 is that it has a fully functioning lab on board for analysis of both chemical and biological materials.

Figure 5.123 The hazardous materials unit is a 2004 Pierce custom made on a rescue chassis.

Lab equipment includes a fume hood, its own air conditioning and heating systems, refrigerator, pass through from the outside for samples, microscope along with other typical laboratory equipment. There are three computers in the command center that have their own wireless network. Internet access is also wireless, and fax and plume modeling are also available. Photographic capability allows for surveillance of the incident scene from the camera outside of the hazmat vehicle or from a portable camera taken into the incident scene and video is transmitted back to the command center for viewing. Cabinet doors in the command center are made of the same material as dry marker boards and can be used for writing on. Engine 110 located at 1616 Mission Avenue is also a designated part of the hazardous materials response. Equipment carried on board includes decontamination, entry PPE, respiratory protection, chlorine kits A, B, and C, patching and plugging, and miscellaneous tools.

There are 55 members of the hazardous materials team. Engine 109 and Hazmat/Truck 109 have seven personnel and a coordinator on duty and Engine 110 has three personnel who serve as the decontamination company.

PPE, Equipment and Training

Level A is DuPont Kappler Tychem Responder, Brigade Commander Flash Protection and DuPont/Kappler Tychem BR, Saint-Gobain One Suit Flash and Level B is Tyvec CPF 3/4. Respiratory Protection is provided by Scott with 60 min bottles for hazmat and Draeger PAPRs and Scott APRs. Engine companies carry bags for weapons of mass destruction (WMD) for personnel along with Mark-I auto-injectors, which are nerve agent antidotes. In-suit communications is accomplished with MSA built in system 800 MHz, Bluetooth headset.

Monitoring instruments & identification equipment used by the Sacramento Metro Hazmat Team include pH Paper, MultiRAE Plus combines a PID (Photo ionization Detector) with the standard four gases of a confined space monitor (O_2, LEL, and two toxic gas sensors) in one compact monitor with sampling pump, Mini RAE's PID PPM, portable RAE's, Infra Red Photospectomitry, Sensor IR, Micro Cap (in lab), TruDefender FT IR Spectrometer, First Defender RM RI spectrometer, Guardian Reader, HazCat 2.0 Pro, and HazMat ID 360 Pro.

Truck companies carry four-gas monitors and special operations officers carry APD 2000, Radiation detection, Canberra UltraRadiac Plus (beta/gamma), Ludlum 2241 (alpha/beta/gamma), FLIR IndentiFinder R400 (gamma/neutron), and Hazcat kits. Radiation instruments for Gamma radiation are carried on all engines and trucks along with dosimeters. Terrorist Agent monitors include APD 2000, detects chemical warfare agents, pepper spray, and mace. Hazcat WMD, M-8, M-9 papers, and the M-256 Military detection kits. Personnel Protective Equipment for Hazardous materials technicians at the Sacramento Metro Fire District are trained as specialists with over 240h of training and get refresher training from the California Specialized Training Institute (CSTI).

Research Resources

Research Resources include CAMEO (Computer-Aided Management of Emergency Operations) and various other hard copy reference books are available in the command section of the hazardous materials units.

Hazardous Materials Exposures

Interstates 5 and 80 and US Highway 50 are the primary transportation routes through the fire district. Barge traffic on the Sacramento River and Union Pacific Railroad has a large railroad switching yard west of the Mississippi River making up the remaining transportation exposures in the Metro Fire District. There are also natural gas pipelines. American River Tank Farms are located along the river. Anhydrous ammonia is used in cold storage and agriculture, and chlorine is used for water and sewage treatment. There is a heavy computer industry presence with its associated chemical exposures. Bulk propane can be found in the area as well as cryogenic and bottles gases. A nuclear reactor is nearby and a plant that manufactures solid rocket fuel.

A regional WMD working group including the fire department, police department, sheriff's office, and FBI deals with issues of terrorism. In nearby Sacramento City California Task Force 7, Urban Search and Rescue Team (USAR) is headquartered and available for local response as needed (*Firehouse Magazine*).

Saint Paul Minnesota Hazmat Team

Saint Paul is the capital and second most populous city in the State of Minnesota. It is the county seat of Ramsey County, the smallest and most densely populated county in Minnesota, with an estimated population of 559,594 in 2020. The city has a geographical area of 56 miles^2 of which 52 miles^2 are land and 4 miles^2 are water. Saint Paul is located on the east bank of the Mississippi River in the area surrounding its point of confluence with the Minnesota River, and adjoins Minneapolis the state's largest city. Known as the "Twin Cities" they form the core of the Minneapolis–Saint Paul Metropolitan area with about 3.6 million residents.

Fire Department History

On November 6, 1854, Pioneer Hook and Ladder Company No. 1 was formed with a volunteer crew of 18 men; so started the proud tradition of organized fire companies in the City of Saint Paul.

> ***Author's Note:*** *The following account is taken directly from a Memorial Display in the lobby of the William and Alfred Godette Memorial Building.*

On September 10, 2010, Saint Paul Fire Department Headquarters Building located at 1,000 W. 7th Street was commissioned and named the William and Alfred Godette Memorial Building. The tradition of firefighting careers being carried on within family circles is as prevalent in today's fire service as it was over a century ago. William and Alfred Godette were brothers who served on the Saint Paul Fire Department in the late 1800s and early 1900s.

William Godette joined the Department in 1885, and became the first black Lieutenant and the first black Captain in Department History (Figure 5.124). William was an exceptional leader and was highly respected throughout the Department. He led all black crews at Station 22 and Station 9 and served the Department for 41 years.

Alfred Godette was appointed a fireman in 1909 and was later promoted to Pipeman. Alfred worked at Fire Station 22 and lived on Albemarle Street with his brother William. Pipeman Godette was killed in the line of duty responding to a fire alarm in December 1921 (Figure 5.125).

William and Alfred Godette were respected and valued members of the Department at a time when the Fire Service and American Society were racially segregated. They were pioneers in the Department at the tie and were active in their neighborhood and the church. Alfred's line of duty death was the ultimate sacrifice made by a family that was committed to the Department, Community, and the profession of Public Service.

The Godette family's connection to the community, their devoted service to the Fire Department and their neighbors, and the sacrifices they made for others exemplify the core values of the Saint Paul Fire Department and should serve to inspire firefighters everywhere.

Today's Modern Department

Today's Saint Paul Fire Department operates out of 16 fire stations under the leadership of Acting Chief Butch Inks. Station 18 is the oldest, built in 1890, and is still in service. Saint Paul has 434 uniformed personnel who provide staffing for 16 engines, 7 trucks (3 towers and 4 straight sticks), 3 squads, and 11 medic units. Companies with a Medic and Engine

Figure 5.124 William Godette joined the Department in 1885 and became the first black Lieutenant and the first black Captain in Department History.

are not dedicated. Whichever apparatus is required for an emergency, the other is taken out of service. Additional apparatus includes 2 hazmat units, 1-38 foot fireboat and one technical rescue. Engines and trucks are staffed with 4 personnel and rescue squads with 5 personnel. Firefighters work a shift schedule of 24 hours on duty and 24 off for 4 cycles then they have 6 days off. This is followed by another 4 cycles and they have 4 days off. Fire responses include 3 engines (1 for Rapid Intervention Team (RIT)) 1 truck and 1 chief. Alarm bells response 2 engines and a truck or squad and 1 chief. Squad 3 is housed at Station 1 and performs technical rescue and works with state police Aviation Rescue Team (ART). Squads carry 36 air bottles and have a pump and small water tank, up to 500 gallons.

Figure 5.125 Pipeman Godette was killed in the line of duty responding to a fire alarm in December 1921.

Hazmat Team

Saint Paul Hazmat operates out of two stations, Station 4 located at 505 Payne Ave. and Station 14 located at 111 Snelling Ave. North. Station 4 houses Engine 4, Medic 4, Squad 1, and Hazmat 1 (Figure 5.126). Station 14 houses Engine 14, Medic 14, Squad 2, and Hazmat 2 (Figure 5.127). Hazmat 1 is a Chemical Assessment Team (CAT) and Hazmat 2 is the Emergency Response Team (ERT). Decon Unit and foam truck are located at Station 10. Foam is also available from a refinery. Engine 15 by the airport carries 100 gallons of foam.

Squads also carry 30 gallons of foam, which can be used with their on-board pump and tank. Squads and ladder trucks carry oil dry for clean-ups at vehicle accidents. Air monitors are located on trucks and squads.

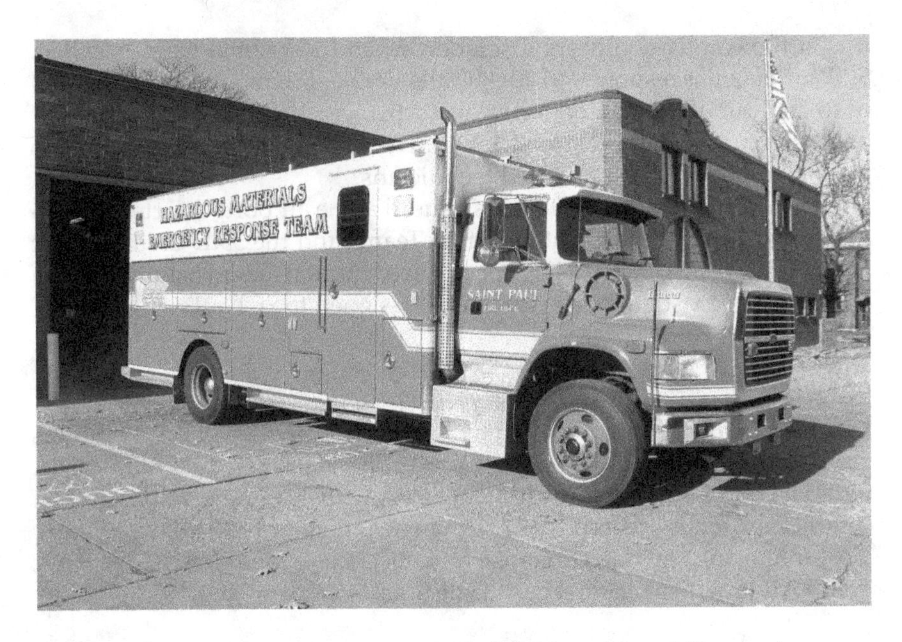

Figure 5.126 Station 4 houses Engine 4, Medic 4, Squad 1, and Hazmat 1.

Figure 5.127 Station 14 houses Engine 14, Medic 14, Squad 2, and Hazmat 2.

It is at the company officers discretion when the hazmat team is called. St. Paul Hazmat responds to 70–80 calls per year. Firefighters are rotated in to and out of the hazmat units every five years. Captains are the only ones who can make a career in hazmat. The Chief of Department appoints team members who are certified technicians.

The 55th Civil Support Team is housed at the air port. Quarterly meetings are held with the 7 State Hazmat Teams and the 55th Civil Support Team. Historically, St. Paul has not required mutual aid on hazmat calls in the city. Two coast guard personnel are assigned to St. Paul.

PPE, Equipment and Training

Level A Chemical Suits provided by Kappler 500 and Level B Splash Suits are also from Kappler. Respiratory protection is provided by MSA SCBA with 1 h bottles. Firefighters use MSA SCBA with 30 minute bottles. In-suit communication is accomplished with bone mikes and 800 MHz radios.

Monitoring and detection instruments include Multi RAE, Area RAE, Toxi RAE Pros, Sensi gold 4 gas, Eagle 5-gas, LCD's.

Reference Resources

CAMEO, ALOHA, MARPLOT, Peak Software, and WISER.

Hazardous Materials Exposures

Interstates 35 and 94 barge traffic on the Mississippi River, Burlington Northern Santa FE (BNSF) and Canadian Pacific (CP) Railroads, rail yards, one major gas line and MSP Airport. Flint Hills Refinery, 3M Company (largest fixed facility). At the University of Minnesota 2-Bio-Safety Level 3 Labs, Hawkins Chemical (2 facilities), UNIVAR Chemical Transfer Warehouse (mixes and blend chemicals). Known chemical hazards include Acids, Nitric and HCL, limited anhydrous ammonia.

Incidents

Pillsbury/General Mills Plant August 11, 2003, 21:58

Anhydrous Ammonia Release

The Chanhassen Fire Department responded to the above address on the date of August 11, 2003, for the report of a large Ammonia leak (Figure 5.128). Enroute, responding command officers requested a weather report from dispatch and took note of the wind direction (northerly) by looking at various flags throughout town. Dispatch was advised of the safe response route for incoming emergency vehicles. Also enroute, dispatch advised

Figure 5.128 The Chanhassen Fire Department responded to the above address on the date of August 11, 2003, for the report of a large Ammonia leak. (Courtesy: Saint Paul, MN Fire Department.)

that they were receiving calls from the Stone Creek neighborhood of an Ammonia smell. Assistant Chief, Greg Hayes, was the first to arrive and set up a command post at the corner of McGlynn Dr. and Audubon Rd.

The Carver County Sheriff's Deputy arrived and agreed to start shutting down all traffic around the plant. The initial distance around the plan was estimated to be 1/2–3/4 of a mile away. Sergeant Williams from the Sheriff's arrived and took command for the law enforcement side. Battalion Chief Dale Gregory was assigned Staging Officer and set up the staging area at the entrance to Paisley Park Studios. Assistant Chief Mark Littfin was assigned to investigate the leak and find the facility representatives. During the event Assistant Chief Littfin acted as the liaison between the command post and the facilities hazmat team. Assistant Chief Littfin coordinated operations with Joe Moran with General Mills.

Fire Chief John Wolff took incident command and Assistant Chief Greg Hayes was assigned fire/hazmat operations. A quick review of strategic and tactical objectives ensued with immediate priorities placed on requesting additional hazmat, fire department, and emergency medical service (EMS) resources from a variety of agencies.

Response Objectives:

- Identify the nature and extent of the leak,
- Determine and deal with life safety issues with employees and downwind businesses/residents, and
- Contain/mitigate the hazard.

The Leak

At approximately 21:50 hours, an employee completing routine rounds during shift change opened the door to Engine Room 1 and visually identified a vapor cloud of Ammonia in the room. The "King Valve" was activated. This valve isolates the six (6) main Ammonia storage tanks and other parts of the delivery system. At this point the maximum amount of Ammonia that could be released was 20,000 lb (3,571.42 gallons) of Anhydrous Ammonia.

At this point, a full plant evacuation was initiated. The time of the evacuation was 2,155.

After making entry the General Mills personnel found Ammonia leaking from insulation around a liquid pipe. After some time, the source was found which had an approximately 1¼ inch long horizontal slice in the line. General Mills personnel isolated the pipe through the use of valves and punched a hole in the bottom of the pipe to drain the liquid Ammonia onto the floor in order to dilute it with water. During the entries, the emergency ventilation system was activated and Ammonia was released outside the building from the roof. Unfortunately, this led to the Ammonia being spread to the rest of the building through the roof top ventilation systems.

According to on-scene General Mills resources the maximum spill possible was 20,000 lb of the 80,000 lb on-scene due to the "king" valve shut off of the various holding tanks. An exact amount of spilled product was unknown throughout the event. But command operated on the basis that a major leak occurred. Post event analysis determined 13,000 lb leaked from the system. During Assistant Chief Littfin's size up of the building he reported a "strong smell of Ammonia" on the southwest side of the facility and that employees were covering their nose and mouths.

The Building

The building is a very large production and food storage plant in the center of the City of Chanhassen. The main entrance of the facility faces west with the engine rooms for Ammonia off the north side of the facility. Engine room #1 which had the leak is approximately 32 feet × 24 feet with a height of 24 feet 6 inches. This room houses pumps for the ammonia system, electrical components, and the buildings sprinkler and fire pump system. There are 2 exits out of the room, one on the north wall directly to the exterior of the building, and the other in the south west corner of the room leading into the production area. During the incident the north door was used as an entry point for personnel and ventilation.

Plant Evacuation

The plant was evacuated by General Mills. The initial evacuation was started at approximately 21:55 hours. Upon arrival of the incident command team, all employees were out of the building. According to General

Mills, all employees were accounted for and were in their designated spots on either the west or east side of the building. Approximately 1 hour into the incident, reports came to the command post that employees were "wondering around" near the "hot zone". General Mills plant management agreed to move all evacuated personnel to the day care center adjacent to the command post.

The location of the day care center parking provided a safer environment for the evacuees as it removed them from any potential vapor cloud, ensured their safety and security, and made it easier to tend to their medical and human needs. General Mills remained in-charge of all of the plant evacuation activities and cooperated fully with public safety personnel. Approximately 16 employees were evaluated by EMS personnel with 9 being transported to local hospitals. No serious injuries were reported and no overnight stays were required. Portable latrines and busses from Southwest Metro were brought in to make employees comfortable. Due to the hazardous nature of the immediate environment, employees were not allowed to return to the plant to get their keys or go in to the parking lot until approximately 1–1:30 a.m. After these areas were cleared by air monitoring teams, employees were escorted in groups of 10–12 to return the locker room area and out to the parking lot. All production employees were clear of the scene by 2–2:15 a.m.

Downwind Evacuation

Command assigned the first in response unit engine 211 to investigate reports from dispatch that residents downwind were complaining of a strong smell of ammonia. Engine 211 was assigned to go to the Stone Creek neighborhood and evaluate the extent of downwind ammonia present and to communicate the evacuation option to downwind residents. Specifically, they were to advise homeowners in the immediate area of the release that they could stay in their homes or evacuate to the Chanhassen Recreation Center. During the initial part of the incident 50 people chose to evacuate to the recreation center. After about 20 min, the engine company was asked to return to staging to retrieve an air monitor and evaluate the neighborhood. The following are some of the key points from the operation:

- The crew reported a strong smell of Ammonia and used Level D with SCBA to start notifying residents of the situation.
- Air monitoring in the area showed levels from 97 to over 200 Parts Per Million (PPM).
- The highest readings were in the area of Bluff View Court and Stone creek Drive.
- After 45 min to 1 h, the company was returned to staging.
- Air monitoring was continued through the Carver County Hazardous Materials Team which found no measurable amounts of Ammonia in the surrounding neighborhoods.

Containment/Mitigation Activities

The leak was in engine room #1 and from the exterior of the facility it was noted that the room was "full" of Ammonia. An insulated pipe failed releasing the ammonia into the plant (Figure 5.129). The conditions in the room during the leak were described as very limited visibility (0–2 foot) with the emergency ventilation system running. In addition to supporting the General Mills entry teams, incident managers following Department of Transportation (DOT) guidelines set up 2 elevated master streams off the north side building on Coulter Boulevard to contain the ammonia plume that was venting out of Engine Room number 1. Ladder trucks from Eden Prairie and Chaska were used to affect this operation. It is estimated that over 56,000 gallons of water was used to knock down the escaping ammonia gas.

The outcome of this activity had an immediate effect on the downwind ammonia exposure levels in the Stone Creek neighborhood. Prior to this activity, ammonia levels were above 10% of the immediately dangerous to life and health (IDLH) in certain parts of the downwind area. After this operation was initiated, downwind ammonia levels were negligible. The General Mills Hazmat Team was in-charge of containing the leak. Members made multiple (5+) entries into the hot zone.

The Carver County Hazardous Materials Team also made entries with the General Mills team with the primary mission to back up the entry

Figure 5.129 An insulated pipe failed releasing the ammonia into the plant. (Courtesy: Saint Paul, MN Fire Department.)

team. Support was also given to the General Mills team through extra personnel, air monitoring, and logistics support. Throughout the event, the Carver County Team made 3 separate entries with General Mills Teams. During the entries to find and control the leak, Assistant Chief Littfin has a 2nd gross/ emergency decontamination station set-up just outside engine room #1.

Through the entire incident mutual aid was received from:

- Excelsior Fire- 1 Engine
- Chaska Fire – 1 Ladder & Command Staff
- Carver County Hazardous Material Group (11 Carver County Fire Departments)
- State of Minnesota Chemical Assessment Team (Hopkins Fire)
- Eden Prairie Fire- 2 ladders, Engine & Command Staff
- Minnetonka Fire- 1 Ladder
- Shakopee Fire- 1 Ladder, 1 Engine & 1 Rescue
- Bloomington Fire - 1 Air Truck & Command Staff
- Carver County Red Cross
- Salvation Army

Facts from the Incident;

- 13,000 lb of Ammonia was released from the corroded pipe.
- Measurable levels of Ammonia (200 ppm) were identified in the neighborhoods to the South West of the facility.
- As estimated, 50 people were evacuated to the Chanhassen Recreation Center.
- 350 employees were evacuated from the facility.
- There were 16 individuals seen be EMS for possible exposures to Anhydrous Ammonia.
- The leak took approximately 2 h and 25 min to control.
- Units were on-scene for 6 h.
- There were 20+ agencies involved in the public safety response.
- 5 Ambulances.
- 12 Paramedics.
- 10 Police vehicles.
- 12 Police Officers.
- 1 CCSO support personnel.
- 4 Chief/staff resources.
- 6 Ladder trucks.
- 4 Engines.
- 2 Rescue.
- 2 Hazmat Units.
- 3 Other resources (state duty officer, Gerber, etc.).
- Total personnel 100 (estimated) (St. Paul Fire Department).

Salt Lake City Hazmat Team

Salt Lake City (often shortened to Salt Lake and abbreviated SLC) is the Capitol of the State of Utah and the most populous municipality. It is the core of the Salt Lake metropolitan area, which has a population estimated to be 1,169,000 in 2020. The region is a corridor of contiguous urban and suburban development stretched along an approximately 120 mile segment of the Wasatch Front comprising a population of 2,625,696 (where 80% of Utah's population is located). Salt Lake is located in NE Utah at the base of the Wasatch Mountains on the East side. The city is bordered on the west by the Great Salt Lake for which the city gets its name. Salt Lake is the world headquarters of The Church of Jesus Christ of Latter-day Saints (LDS Church). The city was founded in 1847 by Brigham Young, and other followers of the church who were seeking to escape religious persecution in the midwestern United States. Salt Lake was host to the 2002 Winter Olympics.

Fire Department History

Salt Lake Fire Department (SLCFD) was established on October 1, 1883. Prior to that time the department had been in operation as a volunteer organization since 1852. By the 1880s the volunteer system was showing some vulnerabilities. A large fire on June 21, 1852 damaged several structures and caused damages of nearly $100,000 was the driving force for the creation of a career department. The department's first fire vehicles were horse drawn. Change over to motorization and combination gas driven pumps occurred in 1911. The entire department fleet became motorized in 1918 and the horses were retired.

Today's Modern Department

SLCFD under the leadership of Chief Karl Lieb (appointed in January 2017) boasts a uniformed staff of 350 personnel. Firefighters operate out of 14 strategically located stations covering 97 miles^2 with a population of 200,000 and an additional 300,000 during the daytime. Housed in those stations are 11 engines, 3 trucks, 1 heavy rescue, 8 medics, 2 quints, 4 airport fire suppression apparatus, 6 type six axillaries, 2 non typed axillaries. Other equipment includes 1 light SUV medical response team, 1 light, and air truck with rehab cache. SLCFD is an ISO Class-1 fire department and responds to over 30,000 emergencies annually.

Hazmat Team History

SLC proposed to have a hazardous materials team in 1982 following Van Water chlorine incidents, and incidents at the VOPAK facility. Training of personnel began training in 1983. Personnel started responding as needed in 1984. Dedicated apparatus placed in service in 1985.

Hazmat Team

There are two hazardous materials stations in Salt Lake. Fire Station 6 is located at 948 West 800 South and houses Hazmat 6, a 2006 E One with a center hallway and research desk and Quint 6 (Figure 5.130). Fire Station 10 is located at 785 Arapeen Drive and houses Hazmat 10, a 2016 Pierce and Engine 10 (Figure 5.131). A state grant provides funding for the building of Station 10 to protect university research labs. Station 6 responds to over 400 hazmat calls annually out of an average of 2,800 total calls. Station 10 responded to over 250 hazmat calls annually out of an average of 1,450 total calls. Total numbers of calls include fuel spills, gas odors, and leaks. Engine companies carry one 40 lb bag of absorbent for motor vehicle accidents and small spills. Anything over 5 gallons triggers the response of a hazmat team. Mutual aid if needed is available from the Unified Fire Authority, West Valley Fire, and South Davis Metro that are the primary sources.

Also headquartered in Salt Lake is the 85th Civil Support Team (CST) that they train with and have a good working relationship with. Eleven technicians are available on each platoon with a minimum of seven required. There are a total of 33 technicians on the hazmat team. Salt

Figure 5.130 Fire Station 6 is located at 948 West 800 South and houses Hazmat 6, a 2006 E One with a center hallway and research desk and Quint 6.

Figure 5.131 Fire Station 10 is located at 785 Arapeen Drive and houses Hazmat 10, a 2016 Pierce and Engine 10.

Lake's hazmat team was dedicated until 2010 when a Jump staffing system was implemented.

PPE, Equipment and Training

Personnel Protective Equipment (PPE) consists of respiratory protection and chemical protective ensembles. Level A Chemical Protective suits are Trellborg Trellchem, Saint Gobain One suit and Lakeland. Level B includes encapsulating and non-encapsulating from several manufacturers. In-suit communications is provided by MSA Clear Com. Respiratory protection is MSA SCBA with 1 one hour bottles. No other respiratory protection is utilized.

Monitoring and detection equipment includes Multi Rae gas monitors in both standard 4 gas and with PID and separate toxic gas versions, Ludlum and Thermo FH 40 G radiation survey meters.

Hazmat team members are Journeyman Firefighters who have completed a 200 hour Technician training course. All recruit firefighters are trained at the Awareness and Operations Levels during recruit school. This is followed by an Advanced Ops section during their apprenticeship.

Reference Resources

Reference materials for getting information about hazardous materials are primarily electronic based by internet, computer, and smart phones.

Hard copy backup includes CHRIS Manual, Hawley's Condensed Chemical Dictionary, NIOSH Pocket Guide, and DOT's Emergency Response Guide Book. Decontamination is performed solely using a spray and wipe system.

Hazardous Materials Exposures

Major transportation routes are I-80 and I-15. Four Class I railroads operate in Utah although today the state's trackage is mostly under the control of Union Pacific (with its ownership of the Southern Pacific Railroad, Western Pacific Railroad, and Denver & Rio Grande Western Railroad) with the rest operated by BNSF Railway. BNSF has a major hub/rail yard in SLC. Salt Lake is a major Delta airlines hub, but hazardous materials are not allowed on commercial air craft. Fixed facilities include refineries, utilizing pipelines and over the road transportation. Fixed facilities include one of the world's largest producers of sulfuric acid and various other industrial users, manufactures, storage and transportation companies. A nuclear reactor and MRI equipment are located at the University of Utah.

Hazmat Incidents

University of Utah in 1975 firefighters were exposed to heavy metal vapors. Since that time, 68 have died from the exposure. Two are still living and suffer from neuropathy, diabetes, or alcoholism.

Other incidents have included:

- 2010 Red Butte pipeline oil release,
- 2010 John Glen organic peroxide incident,
- 2006 LA Chemical hydrochloric acid spill, 2003 Vopac tank fire,
- 1991 Thatcher sulfur dioxide release, and
- 1981 Van Waters Rodgers fire with chlorine tanks.

Salt Lake Valley Hazardous Materials Alliance

This alliance was formed around 2011 because the valley's hazmat response agencies resources were being stretched during lean years after 2007 and meets monthly. It is made up of 12 agencies involved in hazardous materials response. They coordinate resources and training and have developed of common SOPs. One of their accomplishments is the establishment of an annual Hazmat Technician School. It is the single source for training and certifying new hazmat technicians employed by member agencies. Member agencies take turns hosting "Hazmat Wednesday" an interagency drill held the last Wednesday of each month.

Member Agencies include every agency in the Salt Lake Valley, plus Park City Fire and South Davis Metro which border the valley to the East and North plus other government agencies:

- Salt Lake City Fire
- West Valley City Fire
- Unified Fire Authority
- South Salt Lake Fire
- Murray City
- Sandy City Fire
- South Jordan Fire
- West Jordan Fire
- South Davis Metro Fire
- Park City Fire
- 85th CST (UT National Guard)
- Utah State Fire Marshal Office
- Salt Lake County Health Department

(Salt Lake Fire Department)

San Diego Hazmat Team

Prolog: *Following several years of teaching I had built up enough frequent flyer miles to actually travel somewhere for the sole purpose of visiting hazmat teams, in this case San Diego and Yuma, AZ. It just so happens that there are a number of great things to see in Southern California and South Western Arizona along the way. To get to Yuma from San Diego I went via San Juan Capistrano, Palm Springs, Joshua Tree National Park and one of the garden spots and often one of the hottest places in the United States, Blythe, California. Crossed the border into Arizona and caught U.S. 95 South to Yuma. I will stop there, because I described my Yuma adventures in the Yuma Hazmat Section coming up.*

When I left Yuma I headed West on Interstate 8. I had to stop at the Imperial Sand Dunes Recreation area. This was an incredible site to see. It looked like an African desert you would expect camels to ride over the horizon at any moment. I got off the Interstate 8 with my car and followed a sandy road until it looked like it was no longer a road. However my though process was not quick enough to realize I had already left the road and was stuck in the middle of know where. Didn't have a cell phone, but it likely wouldn't have worked anyway. Fortunately two Border Patrol Agents had been watching me because it turns out I was only a couple of hundred feet from the Mexican Border.

They thought I might be looking to pick up some drugs or someone trying to enter the country illegally. After I had explained what had happened and why I was in the area, they were a great couple of guys. They called me a tow truck and stayed with me until the tow arrived. Had a great couple of hours discussing the border and that part of California and learned a lot about what the Border Agents encounter all the time. The tow truck arrived and was soon on my way to visit San Diego Hazmat and arrived on time for my appointment in spite of my adventure getting there. After an enjoyable time with HM 1 my next adventure was to explore San Diego. Saw old San Diego, the downtown area, went on an air craft carrier that is a museum on the water front, went to the beach and had a really good visit. Enjoyed San Diego and hopped on my plane a couple of days later and back to Maryland.

San Diego is located in Southern California in the County of San Diego, along the Pacific Coast. It covers an area of 343 square miles with a population in 2020 of 1,447,100. There are 17 miles of coastline extending 3 miles offshore 4,600 acres around Mission Bay Park.

Fire Department History

Prior to 1889, San Diego was protected by volunteer firefighters and during some periods had no organized fire protection at all. In 1872, Engine Company 1 was organized and fully staffed by volunteers. The apparatus consisted of a horse drawn wagon and 12 buckets that had to be hand filled with water to fight a fire. The early volunteer department continued to grow and in 1887 boasted two steam engines, a hose wagon, 11 horses, and 3,500 feet of hose.

A series of tragic fires in the mid-1880s led to the formation of the city's first paid department on August 5, 1889. A.B. Cairns was the first San Diego Fire Chief and served from 1889 to 1905 (Figure 5.132). Equipment consisted of two steam engines, a hose wagon, two hose carts, a hook-and-ladder, and 4,000 feet of hose. Early firefighters were known as "foremen" and "extra men." Each foreman was paid $12.50 per month and an extra men received $10. Engine drivers and hose-carriage drivers were paid $75 per month and engineers were paid $100. Engineers maintained the steamers and rode the rear step of the steamer in order to light the boiler on alarms. Firefighters worked 24 hours a day for 28 days, then had 1 day off. Family members often lived in the stations with the firefighters. In 1917, San Diego became the first fire department in California to have all mechanized equipment. Two years later, the department launched the world's first gasoline-driven fireboat, built entirely in the fire department shop by firefighters.

Figure 5.132 A.B. Cairns was the first San Diego Fire Chief and served from 1889 to 1905. (Courtesy: San Diego Fire Department.)

Today's Modern Department

Today, the department is led by its 18th, Chief Colin Stowell. Fire companies are housed in 52 stations and there also are 9 permanent lifeguard stations (31 during peak season). San Diego Fire has approximately 892 uniformed personnel, 9 permanent life guards, 31 seasonal, 98 permanent life guard personnel, and 246 civilian employees for a total of 1,276. The department runs the following apparatus: 48 engine companies (with an additional 13 in reserve), the fire engine is a pumper which usually carries 500 gallons of water, hose, pump, and 48 feet of ground ladders. The primary task of a fire engine crew is: search and rescue, locate, confine and extinguish fire and, when warranted, respond to 9-1-1 medical incidents.

13 truck companies (with 5 more in reserve): The San Diego Fire-Rescue Department has four types of aerial truck apparatus: Service Aerial, Tractor-Trailer, Ladder Tower, and Elevating Platform or Snorkel. Service Aerial and Tractor-Trailer consist of a straight three- or four-section ladder in lengths of 75–100 feet. A Ladder Tower consists of a heavy duty four-section telescoping ladder and a passenger-carrying platform. An elevating platform or snorkel is an apparatus that consists of a hydraulically operated articulating boom with a passenger platform or basket.

20 paramedic units: The paramedic unit is a mini-emergency room on wheels equipped with among other things, a gurney, bandages, medication, defibrillator, and oxygen. The ambulance is equipped for advanced life support. Personnel are trained to handle any type of life-threatening emergency in the field. The objective is to stabilize and transport victims to the closest appropriate facility. The Medic Rescue Rig serves as both an ambulance and a mini-rescue rig. It carries vehicle rescue equipment such as the "Jaws of Life."

Miscellaneous equipment and apparatus includes 2 hazmat units, 2 explosive device units, 1 foam tender, 1 Environmental Response Team unit, a foam unit, 6 airport crash/fire rescue vehicles, 1 helicopter; 3 light-and-air units, 11 brush units, 2 water tenders (tankers), a mobile communication unit; 28 lifeguard vehicles, and 5 all-terrain vehicles. It is also home to California Urban Search and Rescue Task Force 8, with 2 units.

San Diego Fire and Rescue responded to 161,575 incidents in 2019. Fires were 129,915 EMS and rescue were calls make up about 80.4% of total alarms. San Diego's EMS system partners San Diego Fire and Rural/Metro, a privately owned ambulance company. Fire calls account for approximately 2.8% and other types of alarms account for the other 0.1%. Every engine and truck company in the city has a paramedic on board. San Diego Fire provides fire protection and hazmat response to San Diego International Airport by contract.

Hazmat Team History

The San Diego Fire-Rescue Department began organized hazmat response in the city in 1982 with an engine company assigned to run hazmat calls. In 1984, it obtained its first hazmat rig, a 1985 International Harvester Model 1954 utility-body Hazardous Material Emergency Response vehicle equipped with an interior lab. The box and cab were custom built by Super Vac. The unit was designed by the "father" of hazmat response in San Diego, Captain Craig Black, who retired some years ago as a battalion chief.

Figure 5.133 The Hazmat Program is equipped with two primary hazardous materials response units; HM 1 and HM 2.

Figure 5.134 Each apparatus is equipped with a mobile mini-laboratory.

Hazmat Team

The Hazmat Program is equipped with two primary hazardous materials response units; HM 1 and HM 2 (Figure 5.133). Each of these "Big Red" hazmat units is a specialized emergency response vehicle equipped to handle hazardous material incidents (chemical spills, fuel spills, compressed gas releases, etc.) and is staffed with specially trained personnel. Each apparatus is equipped with a mobile mini-laboratory (Figure 5.134), which allows the Hazardous Materials Technicians and Specialists to identify unknown substances and "suspicious" materials on site. Each unit carries a wide range of personal protective ensembles, also known as personal protective equipment (PPE), specifically designed to be worn in the presence of hazardous environments that are immediately dangerous to life and property. This specialized PPE allows the Hazmat Techs to work for extended periods in Immediately Dangerous to Life or Health (IDLH) atmospheres.

In 1994, the present hazardous materials team was formed in a joint effort with San Diego County's Department of Environmental Health. The San Diego County Department of Environmental Health Hazardous Incident Response Team (DEH-HIRT) was founded in 1981 and is composed of 10 California state-certified hazardous material specialists. The team serves all unincorporated San Diego County areas, 18 municipalities, 2 military bases, and 5 Indian reservations. The DEH-HIRT responds to over 400 hazmat-related requests for service each year. DEH-HIRT also responds jointly with the San Diego Fire-Rescue Department Hazardous Incident Response Team to investigate and mitigate chemically related emergencies or complaints. Together, they provide hazmat response

Figure 5.135 Emergency response vehicle is Environmental Response Team Unit.

throughout San Diego County. Their response vehicle is Environmental Response Team Unit (Figure 5.135).

Station 45 became the home to the San Diego Fire Department Hazardous Materials Team in 2015 and is located at 9366 Friars Road. It operates two independent hazmat units designated as Hazmat 1 and Hazmat 2 along with an Environmental Response Team unit. HM 1 is a 2004 KME placed in service in February 2005. Both vehicles have analytical labs located in the rear and the KME also has an air cascade system. Station 44 is not a dedicated hazmat station; Engine 45 and Truck 45 also reside there. When a hazmat call is received, personnel from Engine 45 and Truck 45 operate the hazmat units (each hazmat unit responds with four members). In addition, the county Department of Environmental Health responds with two to four personnel during the day (they are on call at night). Twenty additional technician-level personnel are assigned to other companies throughout the city.

The San Diego Fire-Rescue Department provides a bomb squad that works with the arson team throughout the city and county. There are no law enforcement bomb squads in the area. Mutual aid is available from the federal fire department located at North Island Navy Base and Camp Pendleton Marine Base. All hazmat technicians in San Diego must complete the 160-h state training course prior to being assigned to the hazmat station.

Most of the equipment carried on the hazmat vehicles is typical for hazmat and weapons of mass destruction (WMD) response situations. Each has a portable decontamination tent and use Sandia National Laboratories decontamination foam for WMD incidents.

Decon Foam

"Researchers at the Department of Energy's Sandia National Laboratories have created a type of foam that begins neutralizing both chemical and biological agents in minutes. Because it is not harmful to people, it can be dispensed on the incident scene immediately, even before casualties are evacuated. The foam, comprised of a cocktail of ordinary substances found in common household products, neutralizes chemical agents in much the same way a detergent lifts away an oily spot from a stained shirt. Its surfactants (like those in hair conditioner) and mild oxidizing substances (like those found in toothpaste) begin to chemically digest the chemical agent, seeking out the phosphate or sulfide bonds holding the molecules together and chopping the molecules into nontoxic pieces. How the foam kills spores (bacteria in a rugged, dormant state) still is not well understood. The researchers suspect the surfactants poke holes in the spore's protein armor, allowing the oxidizing agents to attack the genetic material inside." – Sandia National Laboratories, 1999.

Thermal imaging cameras are used for hazmat responses, as are video cameras that allow video to be transmitted from the "hot zone" back to the command post or other location for review. Cameras for still photographs are also carried on the units. Inside the crew cab is a command center with laptop computers, fax machine, broadband Internet access, two telephones, a satellite phone, radios, computer resources such as TOMES, CAMEO, and ORIS, and hard copy reference materials that include Emergency Response Guidebook and The Merck Index. Each unit also has a weather station for monitoring conditions during an incident.

PPE, Equipment and Training

Personal protective equipment (PPE) used by San Diego is primarily Tyborg for Level A and Tychem 1000 for Level B. Self-contained breathing apparatus (SCBA) is manufactured by Interspiro with 1 h bottles. In-suit communication is provided by Interspiro. For WMD response situations, the crews use positive-pressure air-purifying respirators (PAPRs) and MSA Millennium canister masks. Monitoring and detection equipment carried on the hazmat units includes: BTA biological test kits, APD2000 chemical agent monitors, multichannel analyzer for radiation, Joint Chemical Agent Detector (JCAD) for 12 WMD materials (military technology), MultiRAE and MiniRAE detectors, Draeger CDS civil defense set, Ludlum radiological detector, Hazard Categorization (Haz-Cat) kit, M-8 and M-9 chemical agent detection papers, M-256A chemical agent detector kit, pH Paper, pH meter, and BioCapture air sampler and detection systems. All companies also carry auto injectors containing 2-PAM chloride and atropine as antidotes for nerve agent exposure (a stockpile for civilian casualties is kept at Balboa Naval Hospital).

Currently, San Diego is working to form a Metropolitan Medical Strike Team (MMST) that will include fire, EMS, hazmat, and law enforcement personnel. Assistance for WMD incidents is available from the 9th National Guard Civil Support Team, based in Los Alamitos, which is 86 miles away. San Diego's hazmat team is a member of the Joint Hazardous Assessment Team (J-HAT), whose other members include the FBI, Civil Support Team, and police SWAT team. J-HAT provides a security assessment of major events taking place in San Diego. J-HAT is under the command of the police department, except when a hazmat incident occurs during an event, then the hazmat team is in command. The J-HAT mission is to assess an event for potential hazards and provide on-scene response if a WMD incident, criminal act, or hazmat emergency were to occur during an event.

Hazardous Materials Exposures

Transportation exposures for potential hazmat incidents in the San Diego area include Interstates 5, 8, 15, and 805 and State Route 163. Hazardous materials are also transported out of Mexico (the Mexican border is 16 miles south of San Diego). A U.S. Border Patrol transportation holding area is located just north of the Mexican border in San Diego County, where there is a potential for leaks and spills to occur. The Burlington Northern & Santa Fe Railroad, which has rail yards in San Diego, transports hazardous materials through the city and county. San Diego is a major marine shipping point and hazardous materials go in and out of the port area on boats, trucks, and trains. Pipelines and tank farms are another source of hazardous materials exposure in the city and county.

San Diego is home to San Diego State University and the University of California, San Diego. These facilities have many research laboratories that pose potential hazmat response dangers. Ordnance and hazardous materials associated with military facilities in the city and county also pose risks. Commercial installations in the response area store large amounts of chemicals, including propane, chlorine, anhydrous ammonia (associated with many cold-storage facilities related to the vegetable industry), pesticides (agricultural industry), plating shops, and clandestine drug labs. Because of San Diego's location, nearly every type of hazardous material could be shipped to or through the area. Hazmat team members must be ready to deal with a diverse group of chemicals (*Firehouse Magazine*).

Incidents

Standard Oil Company Fire

Shortly before noon on Sunday October 5, 1913, a large black mushroom-shaped cloud of smoke reached skyward from the Standard Oil Company tank yard on the waterfront (Figure 5.136). A spark from a passing

Figure 5.136 Shortly before noon on Sunday October 5, 1913, a large black mushroom-shaped cloud of smoke reached skyward from the Standard Oil Company tank yard on the waterfront. (Courtesy: San Diego Public Library.)

locomotive was blamed for starting a fire in a 250,000 gallon tank of distillate oil. As the oil burned it threw sparks skyward which rained down on several other tanks nearby. A tank holding 1,500,000 gallons of black oil ignited erupting into a towering ball of flame.

Firefighters were left to do nothing more than spray water on the other tanks to keep them cool. Firefighters were relieved at their hoses and allowed to get something to eat, quickly returning to take their positions again. Spectators gathered as the firefighters tried to keep the black oil from exploding, but the heat eventually caused a third tank, holding 250,000 gallons of gasoline, to explode with such force that it was heard in La Mesa. Steam driven, the burning oil soared into the air and rained down on adjoining lumber yards.

Firemen dragged what hose lines they could spare from the Standard Oil yard and re-grouped in the lumber yards. The Fire Department sent out a General Alarm. Spectators on the end of the Standard Oil dock had to be rescued by boats as the pier caught fire. Firefighters battled the flames from Sunday through Tuesday, when crews were finally released (*Firehouse Magazine*).

Acetylene Factory Explosion

Five persons are in hospitals as the result of an explosion of acetylene gas that wrecked the plant of the Pacific Acetylene company (Figure 5.137). At Thirtieth and Main Streets and a collision between a fire engine and an automobile as the firemen were answering the alarm. The explosion, resulting from unknown causes, demolished the gas plant, doing damage estimated at $36,000. So great was the force of the blast that a nine inch brick wall was thrown out, pieces of the roof were hurled 100 feet into the air, and windows in nearby houses were shattered.

Figure 5.137 Five persons are in hospitals as the result of an explosion of acetylene gas that wrecked the plant of the Pacific Acetylene company. (Courtesy: San Diego Public Library.)

Responding to the alarm of fire that followed the blast, squad company No. 1 collided with a roadster at Twelfth and Market Streets. H. B. Haley, fire engineer, received concussion of the brain and possible internal injuries, while H. T. Card, driver of the car sustained abrasions and lacerations, and Rosalie Melaney, who was riding with him, is suffering from concussion of the brain and possible broken back. Several others of the nine firemen riding on the truck received minor injuries.

Oakland Tribune California 1927-08-08

Saskatoon, Saskatchewan, Canada

Prolog: *My trip to Saskatoon was to attend the Western Canadian Hazardous Materials Conference as the Keynote Speaker. This was my first overnight stay in Canada. I had made several day trips crossing over from Niagara Falls and Buffalo, New York, and Colebrook, New Hampshire, but this would be a whole new experience. I wasn't sure what Western Canada would be like although I expected it might be colder than Maryland! I was pleasantly surprised that the weather was unseasonably warm part of the time I was there and the farm lands surrounding Saskatoon looked just like home in Nebraska.*

While visiting Canada I thought this would be a good opportunity to visit a Canadian Hazardous Materials team as well. I contacted the Saskatoon Department of Fire and Protective Services and made arrangements to visit their hazardous materials team.

Dangerous Goods Response North of the Border

Saskatchewan is known as a prairie province located in the West Central region of Canada covering an area of 246,134 miles^2 (588,276 km^2) with an estimated population of 1,168,123 in 2020. Most of the population of Saskatchewan resides in the Southern half of the province. The province's name comes from the Saskatchewan River, whose name is derived from its Cree language designation: *kisiskāciwani-sīpiy,* meaning "swift flowing river". The Saskatchewan River flows through the center of the City of Saskatoon. Saskatoon is the largest city in Saskatchewan with an estimated population of 324,721 in 2020. Other major cities include Regina, the provincial capital, Prince Albert, and Moose Jaw.

Fire Department History

Organized firefighting in Saskatoon began around 1882 and consisted of bucket brigades and large barrels of water taken from the river on wagons. Around the turn of the century hand carts with mechanical hand pumps replaced the bucket brigades. In 1903 a steamer was purchased for Saskatoon's volunteers, along with a few hundred feet of hose and a brass nozzle. During the mid-1990s Fire Chief D.M. Leyden had 15 volunteers under his direction as the city population exploded. Along with the increase in population came an increase in fire calls. The first permanent fire station was constructed in December 1908 at the corner of 4th Avenue and 23rd Street.

Chief Thomas Health from Eastern Canada was hired as the department's first paid fire chief in 1909. Horses continued to pull fire apparatus into the late 1920s. It is unclear when Saskatoon made the transition from volunteer fire department to career, however, on May 29, 1918, the International Association of Firefighters granted a Charter to Local 80, the Saskatoon Firefighters Union. During 1946 a three platoon system was implemented for firefighters.

Today's Modern Department

Today's modern Saskatoon Department of Fire and Protective Services under the command of Chief Morgan Hacki responds from eight fire stations (a ninth is planned). They respond with 280 personnel and firefighters work 2 – 10 h days followed by 2–14 h nights and then have 2 days off. This schedule is repeated one time and then firefighters have 6 days off. Firefighters average a 42.5 h work week. There are 12 frontline with a minimum crew of 4 and 4 reserve engine companies, 2 truck companies with a minimum crew of 4 and 1 in reserve. Miscellaneous apparatus includes two tenders, two rural trucks, a dive rescue team and two rescue boats, trench rescue team and hazardous materials team. The city is divided into four districts with a district chief over each. During 2017, they responded to 1,491 fire calls and over 5,700 medical and rescue calls. Saskatoon Fire does not provide transportation for medical calls but each front-line apparatus is staffed with two Primary Care Paramedics (PCP). Each department engine company is equipped with an automatic external defibrillator (AED). Ambulance service is provided by the Saskatoon Regional Health Agency which transports patients to Saskatoon's three medical centers.

Dangerous Goods Team History

> ***Author's Note:*** *In Canada, what we call hazardous materials in the United States are called Dangerous Goods.*

Prior to September 2001, Saskatoon had a limited Dangerous Goods (Hazardous Materials) response capability. Their team was formed around 1992 following a risk assessment of the city which indicated large amounts of manufacturing and storage of Dangerous Goods in the area. The first Dangerous Goods response vehicle was a combination rescue/ Dangerous Goods response unit. In the beginning, there were 24–36 team members and they were trained using the Surviving the Hazardous Materials Incident curriculum from Emergency Response.

Following September 2001 the Canadian government initiated a CBRN JEPP (Joint Emergency Preparedness Program) with funding available for first responders. With that funding Saskatoon purchased their Dangerous Goods trailer, the decontamination trailer and upgraded protective clothing and monitors and decontamination ability. Since terrorist response is very similar to Dangerous Goods response, the addition of the CBRN equipment has greatly enhanced Saskatoon's Dangerous Goods response capability.

Dangerous Goods Team

Saskatoon Station 7 is where Dangerous Goods response vehicles and equipment are located. Station 9 is the decontamination station that operate a decontamination trailer. Dangerous Goods units are not dedicated so personnel from other companies man the Dangerous Goods apparatus and equipment when a call comes in. There are a total of 43 Dangerous Goods (Hazardous Materials) Technicians assigned to the Saskatoon Department of Fire and Protective Services. There are a minimum of eight per battalion. There are a total of 9–12 Dangerous Goods Technicians on duty on any given shift. All other firefighters in the city are trained to the Operations Level. Dispatchers are trained to the Awareness Level.

Station 4, located at 2106 Faithfull Avenue houses Engine 4, which is a 1998 E-One with a 5,000 lpm (1,250 gpm) pump. Engine 4 is first out on all Dangerous Goods incidents and carries monitoring equipment, foam educator, and an information library (Figure 5.138).

Once on-scene Engine 4's crew would determine if additional assistance is required. If the call comes in as a chemical spill, all Dangerous Goods apparatus would be dispatched initially. There are two Dangerous Goods Technicians on duty at Station 4. Station 7, located at 3550 Wanuskew Road, houses HM 40 and the Dangerous Goods trailer and pull vehicle (Figure 5.139). The trailer is a 28 foot Cargo Wells trailer with a command post in the front and a dress area in back along with storage for fire gear, chemical protective clothing, and monitors. The pull vehicle is Hazmat 7, which is a 2018 International MV 607 quad cab 1 ton truck with a service body for hazmat equipment. Hazmat/Rescue vehicle, which is a 1991 International, and it also responds to Motor Vehicle Accidents (MVAs) on the North side of the city. There are four Dangerous Goods Technicians on duty at Station 7 per shift. In addition to the Dangerous Goods apparatus, Station 7 also houses Engine 7, which is a 2000 E-One Cyclone II 6,000 lpm (1,250 gpm) pumper.

Station 9, located at 870 Attridge Drive, is the second Dangerous Goods station and houses the decontamination trailer, which is a 24 foot American pace car hauler trailer (Figure 5.140). It has been converted by

Figure 5.138 Engine 4 is first out on all Dangerous Goods incidents and carries monitoring equipment, foam educator, and an information library.

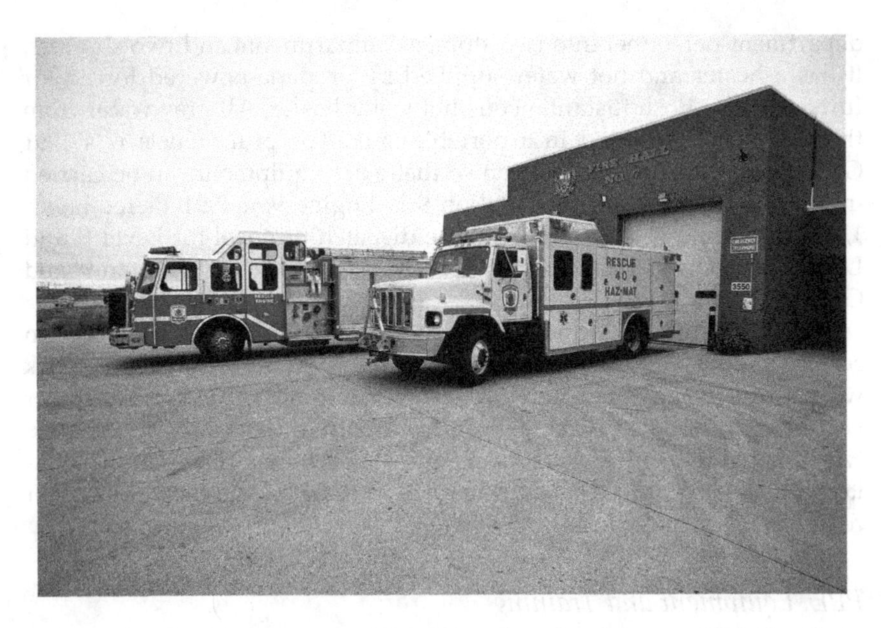

Figure 5.139 Station 7, located at 3550 Wanuskew Road, houses HM 40 and the Dangerous Goods trailer.

Figure 5.140 Station 9, located at 870 Attridge Drive, is the second Dangerous Goods station and houses the decontamination trailer, which is a 24 foot American pace car hauler trailer.

department personnel into two storage compartments and two showers. It has a heater and hot water supplied by propane-powered forced air furnace and a Bosh instantaneous hot water heater. All gray water from the showers is collected in a portable tank. The pull vehicle is a 1991 GMC Crew Cab 4 × 4 with a cap so that extra equipment can be carried in the back. Also housed at Station 9 is Engine 9, a 1991 Pierce Lance 1,250 gpm pumper (Figure 5.141). The dispatch protocol for Level II and Level III responses includes the district engine, Hazmat 7, Decon 9, and Command unit.

Their hazmat truck and trailer are being upgraded and should be in service in 2019. Upgrading the International MV607 SBA quad cab truck with a service body is to store all hazmat equipment. The hazmat trailer will have an updated office area in front with a dress area in the back, which will come with lockers for chemical suits, boots, gloves, SCBA monitors, LCD monitor for computer information, bathroom, and storage for fire gear, and personal clothing.

PPE, Equipment and Training

Level A chemical protective clothing consists of DuPont Tychem TK made by Lakeland Industries, Kappler and Level B DuPont Tychem BR also made by Lakeland Industries. Tychem suits are used because they

Figure 5.141 Also housed at Station 9 is Engine 9, a 1991 Pierce Lance 1,250 gpm pumper.

are compatible with most of their chemical exposures and they have increased visibility with wide vision panels. Saskatoon's team also carries Ansell chemical gloves and tingly and Bata boots, coveralls and Kappler cool vests. In-suit communication is accomplished with MSA push-to-talk bone phones that clip onto back of helmet.

Respiratory protection is provided by MSA Advantage 1,000 full face air-purifying respirators (APR) and MSA G1–60 min SCBA. Computer-based resources include PEAC software Cameo, wireless internet, University Chemistry Department, and Canutec (the Canadian equivalent to the United States CHEMTREC).

Monitoring equipment carried by the Dangerous Goods team includes Industrial Scientific iTx four gas monitors, RKI Eagle hydrocarbon monitors, pH/Oxidizer paper, Spilfyter test strip, Dragger Hazmat Simultest sets I, II, and III, Civil Defense Set I & V, the Clan lab tube set, Smith detection infrared spectroscopy Hazmat ID with the repeat IR and the extract IR kits to complement the hazmat ID and ACE ID Roman spectrometry monitor and radiation monitors supplied by Environmental Instruments Canada in Saskatoon.

All front-line apparatus in the city carry carbon monoxide monitors along with peat sorb for chemical and fuel spills, plug and dike for plugging and patching fuel tanks. Each front-line apparatus also carries a response bag with forms and information books they might need to mitigate of set up a Dangerous Goods incident in the first 5–10 min. CO responses are handled by engine companies. Fuel spills smaller than 200 L (45 gallons) are handled by engine companies as well. Saskatoon utilizes five levels of dangerous goods response:

- Level I response dispatches the district engine. These responses include needle pick-ups, fuel spills under 200 L (45 gallons), CO responses, pepper spray calls outside, or odd smells with no signs and symptoms (burning eyes, coughing, and redness of skin).
- Level IA response dispatches district engine along with aerials for natural gas leaks, and fuel spills larger than 45 gallons (possible LEL atmosphere present). Aerials carry an MSA four gas monitor along with an Eagle hydrocarbon monitor for PPM detection.
- Level II response dispatches the district engine, Hazmat 7, Decon 9, and Command 9. Examples would include unknown chemical spills, ammonia, chlorine releases on highway or at commercial facilities, fires at auto body or similar businesses, or white powder calls at target locations.
- Level IIA responses involve the dispatch district engine, and Hazmat 7 for reported pepper spray calls inside a structure. (Biocleanz with a fogger is used to take away the smell or pepper spray odor from the building, so people can return to building.)

- Level III response dispatches district engine, Hazmat 7, Decon 9, Command 9, and Battalion Chief. Examples of Level III incidents which are large hazmat incidents which include train derailments with mass leakage, large chemical releases from chemical plants where evacuation is required, plane crash and chemical plume crossing jurisdictional boundaries. The emergency management director is notified of any incidents Level II or above.

Mutual aid is provided by the chemical manufacturer's response team and other private contractors. Chemists at the University of Saskatchewan are also available to provide technical information about chemicals and how to handle neutralization and decontamination. The Saskatoon team also works closely with the Royal Canadian Mounted Police (RCMP) and the City police provincial clandestine drug lab team on incidents.

Dangerous Goods Exposures

Saskatoon is located in the heart of the agricultural belt and has several chemical and fertilizer distribution facilities in the area. Also located in the city is the University of Saskatchewan with its related chemical exposures. City water treatment plants utilize chlorine. There are two major chemical facilities on the city's north side that have sodium chlorate, liquid chlorine, sodium hypochlorite, acids, acrylonitrile, sodium hydroxide, and other chemicals on site. Most commonly encountered chemicals at incidents in the past have included anhydrous ammonia, chlorine, natural gas, and hydrocarbon fuel spills. Two major railroads service Saskatoon, the Canadian Pacific on the Northeast side, and the Canadian National on the Southwest side. Major Dangerous Goods transportation routes bring truck traffic to the North end of the city. Eight major transportation routes feed through the city. So far there have been no major dangerous goods incidents in the City of Saskatoon. Personnel in the Saskatoon Department of Fire and Protective Services from the top down have bought into the Dangerous Goods Program and it's that dedication and support that has made their program so successful.

Seattle Hazardous Materials Team

Seattle is the largest city in the Northwest with a population of 783,137 (2020), 1.5 million (daytime population) and 3.7 million in the metropolitan area. The population density is 6,039 people per square mile. Located in the State of Washington between the Puget Sound and Lake Washington, Seattle is located 96 miles South of the border with Canada in King County. Seattle covers an area of approximately 83.6 miles^2 with 193 miles of waterfront lying between the Olympic Mountains on the

West and the Cascade Mountains on the East. Seattle has a mild climate with over 200 cloudy days per year on average. On clear days, spectacular views of the active volcano Mt. Rainier can be seen to the Southeast of the city. Seattle is located in one of the most scenic areas of the United States. Seattle and surrounding areas are home to Boeing Aircraft, many technology companies including Microsoft, and a history of ship building and lumber industries.

Fire Department History

During its early days, Seattle had been served off and on by several volunteer fire companies starting in the 1870s. In April 1884 through passage of an ordinance the City of Seattle took charge of the volunteer fire service in the city. Initial firefighting equipment in the volunteer service included hose and a hose reel, and a used hand-operated pump engine bought from Sacramento, California. By 1884 the volunteers had acquired two steam-powered fire engines and a hook and ladder.

Seattle's paid Fire Department was officially formed in October 1889 just a few months after the Great Seattle fire of June 6, 1889, which destroyed Seattle's volunteer fire stations and 125 acres (25 city blocks) of homes and businesses worth an estimated $8 million. Most of the structures burned were constructed with wood. Following the fire no new wooden structures were allowed in the burned out area. No statistics were kept on any injuries or deaths. It is estimated that over 1 million rats were killed in the fire. The first Seattle Fire Chief appointed was Gardner Kellogg who commanded a staff of 32 firefighters. Seattle Fire Department celebrated the 130th Anniversary of the founding of their department in 2019.

Today's Modern Department

Seattle's Fire Department has evolved into a modern emergency services department under the current leadership of Chief Harold D. Scoggins. A force of 1,000 uniformed personnel operate out of 33 stations with 32 engine companies, 11 truck companies, 1 Rescue, 5 Aid Units (basic life support), 7 Medic Units (advanced life support), 2 air cascade vehicles, 4 fire boats, and 2 hose tenders. All firefighters are Emergency Medical Technician (EMT) Basic Certified and 81 are Certified Paramedics. Seattle Fire Department responded to 91,716 emergency calls in 2019. Of those responses, 20% were fires and 80% were EMS calls. Operations firefighters work a 24h shift schedule based on a 7day rotation. They work one 24h shift and have two 24 shifts off and then they work a 24h shift and have four 24h shifts off. The cycle then repeats itself.

Pioneers Pre-Hospital EMS

Seattle Fire Department was one of the pioneers in the field of pre-hospital EMS in the fire service. During 1970s, Seattle began their Medic One Program when its first group of firefighters were trained as Paramedics in cooperation with Harborview Medical Center and the University of Washington. Providing service to Seattle for over 25 years, their Medic One Program has evolved into one of the most respected in the world.

Hazmat Team History

Seattle Fire Department formed its hazardous materials team in 1980. Motivating factors included the general movement of the fire service in the United States to organized hazmat response and the recognition of the need for specialized training and equipment to safely respond to hazmat emergencies.

Hazmat Team

It is a non-dedicated team that is staffed by on-duty crews at Station 10 located at 400 S. Washington Street. Station 10 houses Hazmat 10, Engine 10, Ladder 1 and Aid 5, Aid 10, Air 10, Squad 10 (utility/special event), Haz80 (back up Hazmat rig), and Haz1 (Figure 5.142). Engine 10,

Figure 5.142 Station 10 houses Hazmat 10, Engine 10, Ladder 1 and Aid 5, Aid 10, Air 10, Squad 10 (utility/special event), Haz80 (back up Hazmat rig), and Haz1. (Courtesy: Seattle Fire Department.)

Ladder 1, Aid 5, and the Hazmat Unit respond together as the Hazmat Team. Typically there are 11 hazmat team members on duty per shift at Station 10. Station 27 located at 1000 S. Myrtle Street houses and provides personnel for the Decontamination Unit. There are 24 Technician Level personnel assigned to Station 27. One officer and three firefighters staff the Decontamination Unit. All members assigned to Station 27 are Decontamination Technicians and have almost as much training as those assigned to Station 10. In addition to the Decontamination Unit, Engine 27 and two tractor trailer apparatus, one for Urban Search and Rescue Team, and one for the Metropolitan Medical Strike Team are housed at Station 27. The hazardous materials unit is a 2017 Pierce Arrow XT – 40 feet with Command Cab and technical space mid-ship (Figure 5.143). They basically replicated their previous rig, with minor changes only. This unit is equipped with an inverter charger to provide un-interruptible power for the command cab, monitors, and computers.

Docking stations can run 3–4 h without the engine on the unit running. Two 12,000 V each quiet generators are also installed on the top of the unit. When the inverter system needs power the generators automatically turn on. A private company has installed and maintains weather stations around the city. The hazmat team has monitors to access the real-time

Figure 5.143 The hazardous materials unit is a 2017 Pierce Arrow XT – 40 feet with Command Cab and technical space mid-ship. (Courtesy: Seattle Fire Department.)

information from the units command cab. They also have two portable weather stations of their own that can be deployed at an incident scene.

Seattle's hazmat team has responded to 41 hazmat calls in 2017. This does not include fuel spills, gas odors, and leaks handled by other companies with operations level personnel. Engine and truck companies do not carry absorbent materials; these are provided from the City's Department of Engineering. Seattle Department of Engineering is separate from the fire department and is responsible for street maintenance and traffic routing. The hazmat team may be called to fuel spills depending on the type of material spilled and the amount. This is determined by the responding engine or truck company. Seattle does not use incident levels. Each incident results in the team response and special assistance is called for specific responses such as white powder and drug lab incidents.

PPE, Equipment and Training

Personnel protective equipment (PPE) used by Seattle Hazmat for hazardous materials exposures includes Trellborg Level A, Kappler CPF3, and Kappler CPF4 Level B. Respiratory protection is provided utilizing Scott open circuit self-contained breathing apparatus (SCBA), powered air-purifying respirators (PAPR), and APR depending on the circumstances of the incident. For in-suit communications, Seattle uses Motorola XTS 5000 450MHz w/bone mic. Firefighters and medics throughout the department are trained to the Operations Level with annual refresher training provided by the department. Approximately 150 other firefighters in the city, special operations personnel, are trained to the technician level in addition to the 56 total assigned to the Hazmat Response Team. Technicians are required to take an 80h technician class and 128h on the job training to obtain certification.

Mutual aid for hazardous materials incidents is provided as needed from Boeing Aircraft Company, the National Guard WMD Civil Support Team, and Region 3 Hazmat, Tukwila.

Hazardous Materials Exposures

Transportation hazardous materials exposures include the Port of Seattle, which provides temporary storage and shipment of materials from all hazard classes, and Interstates 5 and 90. Rail traffic passes through the heart of the city including through a tunnel under the downtown corridor. Olympic Pipeline Company has underground pipelines that pass through the city as well. Olympic pipeline is a 400-mile interstate pipeline system that includes 12, 14, 16, and 20 inch pipelines. The pipeline runs along a 299-mile corridor from Blaine, Washington to Portland, Oregon. The system transports gasoline, diesel, and jet fuel. This fuel originates at four Puget Sound refineries and is delivered to Seattle's Harbor Island,

Seattle-Tacoma International Airport. Fixed facility hazardous materials exposures include Harbor Island a major petroleum distribution facility, multiple manufacturing and storage facilities, bio technical industries, and numerous hospitals.

Incidents

Major hazardous materials incidents in Seattle have typically involved chlorine, ammonia, gasoline, ship fires, high-rise building evacuations, explosives, and biological materials. Recently, the Seattle Hazmat Team was called to the Port of Seattle when U.S. Customs dogs got what appeared to be a positive hit for explosives on two containers that had just arrived from Pakistan. Seattle Hazmat has a full chemical, biological, radiological, nuclear and explosive (CBRNE) detection and mitigation capability utilizing their mass victim extraction MMRS equipment from Station 27. They also have equipment and materials pre-positioned in critical areas for major emergencies (*Firehouse Magazine*).

Sedgwick County Kansas Fire District 1 Hazmat Task Force

Sedgwick County, Kansas, is rich in Native American history. Originally a camping ground of the Osage and Wichita Indian tribes, Sedgwick County is considered to have been discovered and explored by Francisco Coronado and Kit Carson. The area was later settled by M. DuTissenet, a Frenchman operating under the direction of the governor of Louisiana, in 1719. DuTissenet brought with him soldiers, traders, and hunters who soon populated and transformed the region from prairie to a land of opportunity. Sedgwick County was officially established nearly 150 years later on February 26, 1867, and bears the name of Civil War hero Major General John Sedgwick of the Union Army who was killed during the battle of Spotsylvania Courthouse in Virginia.

Fire Department History

Sedgwick County Fire Department (SCFD) officially known as Sedgwick Fire District Number 1, became fully paid in January 1955 with 4 stations. Before that time, fire protection in the county was provided by volunteer departments. They were initially formed to protect wooden bridges in the county. Fire protection in the county prior to 1955 was very limited. Three Sedgwick County firefighters have lost their lives in the line of duty, Gerald Lee Lloyd 1973, responding to a fire when his tender was involved in an accident; Todd David Colton 1990, from heat exhaustion while fighting a brush fire; and Lieutenant Byron Wayne Johnson 2007, was electrocuted while fighting a brush fire.

Today's Modern Department

Today Sedgwick County Fire Department Fire District 1, operates out of 9 fire stations under the command of Interim Chief Douglas Williams. They respond to approximately 691 miles^2 within the county. Several smaller cities in the county provide primary fire protection to their jurisdictions and are backed up by Fire District 1, when requested. Fire District 1, responds to over 9,000 calls for service each year. Of those almost 50% are emergency medical service (EMS) related. There are 137 uniformed personnel on the department, with 44 on duty each shift. Firefighters work 24 hours on and have 48 hours off between shifts.

Fire District 1, operates five engine companies, 3 quints, 8 tenders (Figure 5.144), 6 squads (basic life support, BLS medical calls first responder), 4 brush trucks, 1 heavy rescue, 1 air/light unit, 1 hazmat unit and 2 boats. The department also operates hazmat and Technical Rescue Teams (TRT). Rescue includes confined space, high-angle, swift-water, trench cave-in, building collapse and SCUBA. They respond jointly with their counterparts from the Wichita Fire Department. EMS Advanced Life Support (ALS) is provided by the county, but is not part of the fire department. The fire department does house some of the ALS units in their stations.

Figure 5.144 Sedgwick County Fire District 1, operates 5 engine companies, 3 quints, 8 tenders.

Hazmat Team

Station 34 is the hazardous materials station and is located at 334 N. Main Street in Haysville, Kansas (Figure 5.145). It was built in 2014 and the Hazardous Materials Unit (HMU) (Hazmat 34) has a first response area of 91 miles^2 of urban and rural areas. Fire District 1's HMU can respond as a standalone unit or as part of a Hazardous Materials Response Team (HMRT) (Task Force). The task force is composed of County Unit Hazmat 34, Wichita Fire Department Units and McConnell Air Force Base (MAFB). McConnell Air Force base responds on request to all Level III hazmat incidents or larger.

Technical support is provided by Wichita Environmental Health Department and Sedgwick County Emergency Management. If additional resources are needed mutual aid is available from the Salina, Kansas Regional Hazardous Materials Team. Hazardous materials response in Sedgwick County became a concept in 1980 and was formally established in 1984 with all team members assigned to one station. The mission of the team is to handle the emergency side of chemical releases and Weapons of Mass Destruction (WMD) responses. Hazmat responses by Sedgwick County and the Wichita Fire Department are based on four levels:

Level I: one that can be handled by first arriving units and requires only Hazmat operations-level-trained personnel.
Level II: one that is beyond the capabilities of operations-level-trained personnel, in addition to the first arriving operations level unit, it requires one unit from the HMRT.

Figure 5.145 Station 34 is the hazardous materials station and is located at 334 N. Main Street in Haysville, Kansas.

Level III: one that is beyond the capabilities of a single HMRT unit requiring a full HMRT response.
Level IV: one that is a response to white powder (anthrax threats or scares). This is the same as a Level I unless it is deemed to be a credible threat at which time it will be upgraded to a Level III incident.

Fire District 1's HMRT is not dedicated, personnel at Station 34 man the hazmat unit, but also respond with other apparatus to fire, medical, rescue, and service calls. County wide there are 20 trained hazmat technicians on the department. A minimum of five hazardous materials technicians are on duty at all times and all other firefighters in the county are trained to the operations level. The balance of technician-level personnel needed for a response are assigned to Wichita Station 10 and respond with Wichita units. They have 50 department members trained to the technician level.

Fire District 1 also operates under an agreement with the Kansas State Fire Marshal's Office and will deploy to hazardous materials incidents anywhere in the state of Kansas when requested. They will send a minimum of 5 technician-level personnel and one person from Sedgwick County Emergency Management as a liaison with the local jurisdiction. During 2014 there were 79 hazardous materials responses by the Sedgwick County Fire Department. Of those, 24 required the response of the HMRT task force. The total also includes fuel spills, gas odors, and leaks.

Engine and Quint companies carry absorbent materials. County Health Department Hazmat 99 carries absorbent materials as well. Spills of over 25 gallons require the response of the HMU. Foam Types A and B carried by county units include Phos-check Type A and AR-AFFF Type B. Applicators are the compressed air foam system (CAFS) application system along with a Foam-Pro system using Type A and B. All SCFD front-line pumping apparatus are equipped with foam systems. Some carry CAFS systems as well. Each unit has onboard foam tanks of Class A and Class B foam. Extra foam supplies are stored at the North and South Headquarters. McConnell Air Force Base Fire Department has additional foam available through mutual aid. Also through mutual aid the Wichita Airport and McConnell Air Force Base have airport crash trucks that are available to respond as needed.

Hazmat 34 is built on a 2012 Spartan Gladiator Chassis and the body was built by SVI in Fort Collins, CO (Figure 5.146). It is powered by a 380 HP Cummins diesel engine coupled with an Allison transmission. The unit is 36 feet 4 inch long and 12 feet wide and includes a slide out work area that expands the size of the command center. Also on the unit is a 10 kW PTO generator to supply electrical needs and the light tower. Hazmat 34 was paid for by Homeland Security Funds and is required to provide WMD

Figure 5.146 Hazmat 34 is built on a 2012 Spartan Gladiator Chassis and the body was built by SVI in Fort Collins, CO.

response to 19 additional counties: Barber, Barton, Butler, Comanche, Cowley, Edwards, Harper, Harvey, Kingman, Kiowa, Marion, McPherson, Pawnee, Pratt, Reno, Rice, Stafford, and Sumner. Decontamination is conducted with typical equipment and supplies including wet/dry capability and a decontamination tent.

PPE, Equipment and Training

Personnel protective equipment (PPE) is composed of Level A Trelleborg-Tellchem VPS, Level A Kappler 41550 and Level B Kappler CPF2 chemical protective suits. In-suit communication is provided by Motorola Voice deucer push-to-talk ear pieces used in conjunction with Motorola XTS5000R radios.

Monitoring instruments carried on Hazmat 34 include, Ahura First Defender (various liquid and solid products; Hazmat ID; PPB Rae Plus (volatile organic compounds); Multi Rae (O_2, H_2S, VOC, LEL, CO); Area Rae (O_2, H_2S, LEL, CO); Identi-finder (for radiation detection); Rad Dual Probe meter (Alpha, Beta, Gamma rays); Eagle Multi gas; Weather Station; M-8/9 Paper (chemical agent detector); M-256 Kit (nerve/blister agents; and Bio Check 20/20 Kit (Powder screenings/proteins). They also carry a field screening kit that does not test for specific materials but rather identifies physical and chemical properties of a material.

All personnel assigned to the hazardous materials team receive the International Fire Service Accreditation Congress (IFSAC) Hazmat Technician Level certification. Additional training includes the National Fire Academy Chemistry for Emergency Response, Operating Site Practices and Hazardous Materials Incident Management. Courses including specific areas of WMD response, radiation detection, and integrated command are also provided to personnel.

Research Resources

Research resources carried on Hazmat 34 include, *Emergency Response Guide Book (ERG); Chris manuals; NIOSH Book;* Sax's Dangerous Properties of Industrial Materials; *Emergency Handling of Hazardous Materials; Crop Protection Handbook; Industrial Chemical Safety Manual; The Merck Index; Lewis, Hazardous Chemicals Desk Reference; Tier II Reports for County; Managing Hazardous Materials Incident; Draeger Tube Handbook; Eagle Manual and Correction Factor Book, Relative Response Factors.*

Hazardous Materials Exposures

Transportation routes within Sedgwick County include U.S. 55, Interstate 35, and the Kansas Turnpike. Burlington Northern Santa Fe and Union Pacific have rail lines within the county. Fixed facility exposures are aircraft plants, a chlorine facility with 1,000,000 lb of chlorine on site and 400 plus Tier II facilities. Additional chemicals located at various facilities in the county are propane, anhydrous ammonia, and pesticides (*Firehouse Magazine*).

Somerset/Pulaski County, Kentucky's All-Volunteer Special Response/Ky Haz-Mat 12

Prolog: *Kentucky is like a second home for me. I have taught more National Fire Academy courses in Kentucky than any other state. In the late 1990s, I traveled throughout Kentucky on behalf of Kentucky Emergency Management doing train the trainer classes on a new Department of Transportation (DOT) Emergency Response Guidebook (ERG) course I had developed. Years later the Kentucky Fire Commission adopted the DOT ERG course along with a number of other courses I developed on various Hazmat subjects. I also taught a National Fire Academy Chemistry for Emergency Response Class in the State Capitol of Frankfort (Figure 5.147).*

One of the few state capitals in the United States that does not have a commercial airport. A few weeks after arriving home from the train the

Figure 5.147 Somerset/Pulaski County apparatus at State Capitol of Frankfort. (Courtesy: Somerset/Pulaski County Hazmat.)

> *trainer classes I got an envelope in the mail addressed to the "Honorable Colonel" Robert Burke. My wife happened to get to the mail first and she brought it to me and let the air out of my balloon by making the comment "did you get the recipe." No, I did not, but what was in the envelope, which came from the Commonwealth of Kentucky, Governor Paul E. Patton, was a certificate appointing me as a Kentucky Colonel, the highest award presented by the Commonwealth. I was proud and humbled. So that is why I consider Kentucky my second home.*

Somerset is situated in south-central Kentucky on the northeastern edge of Lake Cumberland, 76 miles south of Lexington, 128 miles southeast of Louisville, and 129 northwest of Knoxville, TN, and is the county seat of Pulaski County. Somerset measures 8 miles^2 and has a population of 11,196. Pulaski County is the second largest county in Kentucky and covers a land area of 653 miles^2 and has a population of 65,155 in 2020.

Hazmat Team History

The Somerset/Pulaski County Special Response Team (SRT) was formed in 2000. Pulaski County had a small hazmat team prior to 2000, but it was disbanded in 1998 due to lack of funding. Original members of the team went to county government to seek funding for the new SRT. Jeff Marcum was the first chief and was replaced by Doug Baker in 2001, who has remained the chief since that time. In addition to Baker, the command staff is comprised an assistant chief who serves as entry sector commander, a major who serves as operations sector commander, two captains who handle logistics and medical and six lieutenants. All officers are cross-trained in all areas of command and operations.

Somerset's SRT is a standalone weapons of mass destruction (WMD)/ hazardous materials team that has no other duties. All personnel, including officers, are volunteers and receive no monetary compensation for their services. It is the only such standalone team in Kentucky. In addition to being standalone, the SRT is relatively autonomous in terms of outside oversight, although it receives input from the Pulaski County Department of Public Safety.

Initially, the team was funded by Pulaski County as a member of the Pulaski County Fire Commission. In 2002, Kentucky Emergency Management initiated a regional response system with 16 regional teams (only 3 of these original teams remain in service; the others are the Blue Grass Emergency Response Team and a northern Kentucky team based in of Alexandria, both based out of fire departments in those areas). The original state-sponsored teams were funded by the federal Department of Homeland Security (DHS) for equipment.

Somerset/Pulaski SRT was chosen to be the host team for the 10 County Rapid Response Team and designated as Hazmat 12. Hazmat 12 is governed by a board of directors with one representative from each of the ten counties. The representatives then elect a board chairman, currently Tiger Robinson from the Pulaski County Department of Public Safety. They cover an area of 3,800 miles^2 with a population of 211,000. Team members are made up of firefighters, law enforcement, former military personnel, paramedics and two members of the Kentucky National Guard's 41st Civil Support Team based in Louisville, Kentucky. They have been in service for 18 years without an injury to any personnel.

Hazmat Team

The SRT is housed in its own 10,000 square foot facility in Somerset (Figure 5.148). Apparatus-related space takes up about 7,000 square feet of the building and there is 3,000 square feet of training space. The SRT has 52 members in Pulaski County, with 45 trained to a minimum of the technician level.

Hazmat 12 has a total of 160 personnel based in the 10-county response area. All team members are put through an extensive background check, including fingerprinting. In the event, mutual aid is required, approximately 12 technicians are available from the Somerset Fire Department, which has 24 paid personnel. Outside Somerset, mutual aid is available from Adair and Boyle counties. Mutual aid personnel are assigned to decontamination.

Apparatus housed at the WMD/hazmat station include: Hazmat 2 is a 1992 Spartan/Supervac rescue (a reserve apparatus) that carries decontamination equipment and personal protective equipment (PPE); this

Figure 5.148 The SRT is housed in its own 10,000 square foot facility in Somerset. (Courtesy: Somerset/Pulaski County Hazmat.)

unit is available to be used by Somerset Fire Department technicians for response to an incident in within Pulaski County in the event the Pulaski County SRT is out of the area on a response as Hazmat 12. Hazmat 3 is a 2002–2012 E-ONE Cyclone II rescue that carries all primary hazmat and WMD response equipment, including instrumentation and PPE. Hazmat 7 is a 2005 E-ONE walk-in rescue for decontamination and testing equipment. Hazmat 4 is a 1997 International/Hackney rescue that carries decontamination equipment. Hazmat 9 is a 2003 Ford E350 used for rehab. Hazmat 1 is a 2009 Chevy Suburban command and communications unit equipped for satellite communications.

Miscellaneous equipment and apparatus: Kawasaki Mule for entry-team transport downrange, 16 foot barrel trailer for spills, 28 foot rehab and mule transport trailer, 28 foot mass-decontamination shower trailer, 18 foot equipment trailer, 24 foot command and rehab trailer, 4,000 watt light pod for night operations. All of the above units are SRT units that also respond as Hazmat 12 in the regional response system. Additionally, two units that are part of Hazmat 12 are stationed in Adair and Boyle counties. They are 2003 Ford F-250s with fully equipped Level A trailers.

PPE, Equipment and Training

PPE carried by the Somerset SRT includes MSA Night stalker CBRN self-contained breathing apparatus (SCBA). Other breathing protection is both MSA and Scott cartridge respirators. Level A protective clothing is

DuPont Tychem with flashover protection. Level B is also DuPont Tychem suits. Tychem coveralls with attached hoods and boots are available as well. Tingley hazmat boots and Petzl Vertex hardhats round out the protective suit ensemble. In-suit communications system is accomplished with two-way radios.

Monitoring instruments carried by the SRT include: First Defender mass spectrometer; Smith Detection Hazmat, and Gas ID; Draeger manual and auto gas detection kits; HAZMATCAD chemical agent detector; Ludlum survey meter; Sensit gold CGI; HAZCAT chemical identification kit; Golden Engineering portable x-ray machine; and Raytek Rayngaer ST pro thermal laser.

The minimum level of training for support personnel is the Operations Level. Personnel have taken additional specialty courses in radiation and response to chemical, ordnance/explosive, biological, and radiological/nuclear (COBRA) incidents at the Center for Domestic Preparedness in Aniston, AL.

Research Resources

Chemical research is primarily conducted using the Internet.

Hazardous Materials Exposures

Hazmat exposures in the SRT response area include the Norfolk Southern Railroad, U.S. Highway 27 and U.S. Highway 80. Fixed-facility hazards are present in chlorine at water treatment plants, propane, pesticides, and a plant where large quantities of ammonium nitrate fertilizer are stored. Hazmat 12 is also a backup for the counties involved in the Chemical Stockpile Emergency Preparedness Program for the Blue Grass Arsenal, where military chemical agents are stored. They would cover south and west evacuation routes in the event of a release of chemical agent at the Blue Grass Arsenal.

The Somerset Fire Department responds to fuel spills and odor complaints within the city. Units carry small amounts of absorbent materials and handle spills up to 25 gallons of gasoline and 50 gallons of diesel fuel. The SRT is called in for larger spills. SRT responds to an average of 15–20 large-scale hazmat and WMD incidents per year.

Incidents

Several major incidents have occurred over the past few years. Sodium hydroxide was spilled on U.S. Highway 27 in McCreary County on April 4, 2013, when a tractor-trailer rolled over into the ditch line. Initially, only eight to 10 gallons of sodium hydroxide were spilled. Hazmat crews established a decontamination corridor and ordered ten truckloads of dirt to build a dike in the event the trailer of sodium hydroxide were to rupture during off-loading. This would keep the sodium hydroxide out of nearby streams downhill from the incident. These preparations would prove to be invaluable as the incident progressed.

Around midnight, the damaged tanker ruptured, emptying between 2,800 and 3,000 gallons of sodium hydroxide. Two clean-up workers were covered with the caustic chemical and had to be decontaminated and rushed to a local hospital. They were wearing protective clothing, but sustained facial burns and vision problems. After almost 24 h on scene, crews completed mitigation of the spill, which included removing contaminated soil and reopening the highway (*Firehouse Magazine*).

Yonkers, New York Hazmat Team

Yonkers is the fourth largest city in New York State covering a geographical area of 20.3 miles^2 with a population of over 199,021 in 2020 and located on the Hudson River on the Northern border of Bronx, New York. Yonkers is the only land connection to New York City. Yonkers was the setting for the movies "Hello Dolly" and "Lost in Yonkers". Home at one time to actor James Cagney of Yankee Doodle Dandy and other movie and Broadway show fame, Otis Elevator Company inventors of the first elevators, which made construction of high rise buildings possible, the first golf course in the United States, St. Andrews Golf Club, and the location of the invention of the first completely synthetic plastic Bakelite. The first FM radio broadcast originated from Yonkers as well. Much of the early industry is gone and Yonkers has become somewhat a suburban bedroom community of New York City. Currently, Yonkers is home to the Yonkers Raceway and *Kawasaki Heavy Industries*, manufacturers of subway cars.

Fire Department History

Firefighting in Yonkers began in 1852 when the first fire company, Protection Engine Co. 1 was formed. Next year in 1853 the Hope Hook & Ladder Company 1 was formed in addition to the Lady Washington Engine Co. 2. During 1855 all fire companies in Yonkers were privately owned and not under the jurisdiction of village authorities. The village appropriated funds to purchase the fire apparatus from the private citizens who owned the fire companies. From 1868 until the late 1890s, 14 additional fire companies were formed including the City of Yonkers fire department in 1896.

By 1888, the volunteer fire companies in Yonkers had grown to 386 firemen. In 1895 fire commissioners were authorized by the Yonkers Common Council after the city charter was revised. Six career firemen were appointed on August 6, 1896, and staffed the Palisade Avenue firehouse. Three career firemen were appointed to the Hope Hook & Ladder Co. 1 and three were appointed to the Lady Washington Engine Co. 2, both quartered in the same firehouse.

On August 27, 1896, James J. Mulchey was elected the first career Chief of the Department. By 1916 the entire fire department of motorized and had over 126 firemen. During 2019 the Yonkers Fire Department will celebrate 123 years as a career fire department.

Today's Modern Department

Yonkers Fire Department response is divided into two battalions with a chief over each and a third safety battalion chief. The safety battalion chief operates citywide 24/7 and responds to all structure fires, auto accidents on the interstate and parkways, technical rescue in addition to hazmat response. Yonkers Fire Department does not provide transport for emergency medical service (EMS) but they do have a first responder program to render aid and Automatic External Defibrillator (AED) until paramedic ambulances arrive. EMS transport is provided by a private company and ambulances are staffed with paramedics. Yonkers Fire Department has 314 uniformed personnel led by Chief of Personnel Thomas F. Fitzpatrick Jr. They operate from 12 fire stations in 2 battalions and respond to over 15,000 alarms each year. Each company normally operates with four personnel, officer, pump operator, and two firefighters. A House watch is assigned for each tour, and the person is responsible for answering the phone greeting visitors and other duties as needed. Yonkers Fire Department has grown from one engine and truck company with 6 firefighters to 11 engine companies, 6 truck companies, heavy rescue, air cascade, foam unit, collapse unit, hazardous materials team, mobile command unit, special operations bus, field communications unit, technical support trailer, EMS support trailer and fire boat. During 2008 three new engines, 307, 310, 312, three new trucks, 70, 73, 74, and Rescue 1 were added to the Yonkers fleet. All three engines are 2007 American La Frances with 1,000 gpm pumps and 500 gallon water tanks. Rescue 1 is a 2007 American La France rescue pumper with a 1,000 gpm pump, 500 gallon water tank, 10 kW generator and Lukas Tool. The trucks are Smeal 100 foot Rear Mount Aerial Ladders. When Squad 11 is third or fourth due, they may be assigned search and rescue duties by the battalion chief. When the Rescue is not available, they are assigned to boxes citywide as a replacement. They may also be special called citywide for any incident if required by the incident commander for technical rescue, fires, or hazmat.

Hazardous Materials Team History

During the early 1980s, personnel in the Yonkers Fire Department started thinking about response to hazardous materials. During 1990 the department recognized the need to have a special unit to respond to the rising numbers of hazardous materials incident and to comply with OSHA regulations for hazmat response. Six officers were trained to the technician level. All other firefighters in the city were trained to the operations level. The team was started and commissioned as Squad 1 during June 1992 and continued for 7–8 years as a two person squad. One officer and one firefighter manned the unit 24/7 when called upon for serious hazmat incidents while remaining in their positions on engine and ladder companies.

Hazmat Team

There are several unique aspects of the Yonkers Fire Department Hazardous Materials Team that I find very interesting. First of all, every one of the uniformed firefighters on the Yonkers Fire Department are trained as Hazardous Materials Technicians except for the last class of recruits. Another unique feature of the Yonkers hazmat response is that all truck companies along with hazmat Engines 307 and 310 carry "hazmat skids" (plastic tote containers) that contain two level A Responder Suits, two pair of Tingley Boots, two hard hats, two additional portable radios with Scott envoys attached, four extra 60 minute self-contained breathing apparatus (SCBA) bottles, CL2 single sensor gas detectors and Scott 4 gas Scout detectors. Spare bottles and their SCBA are equipped to accommodate a Supplied Airline System, weapons of mass destruction (WMD), and firefighting operations.

Currently the hazmat team operates with a tiered "Task Force" concept. Gas odor response involves the Rescue 1 (Figure 5.149) or Squad 11 plus the first-due engine and ladder with a battalion chief. Carbon monoxide response is a single ladder with gas detection equipment. Minor fuel spills/hazmat investigation involves a single first-due engine response. Large fuel spills, CO incident, or unusual hazmat situations are single engine or single ladder responses and they can request rescue for additional hazmat support.

Full hazmat responses for minimum manpower involve the first-due engine and truck (8 personnel), two hazmat engines (8 personnel), Squad 11 (4 personnel), Rescue 1 (4 personnel), Safety Battalion (1) and Battalion Chiefs (2). Engines 307 and 310 are designated as hazmat engines. Yonkers also has a tractor trailer drawn "beverage" truck that was donated to the fire department that is used for transporting the major hazmat equipment to the incident scene. One battalion chief and the safety battalion chief respond on hazmat incidents as well.

When foam is needed, Foam Unit 13 responds from Station 13 with the crews of Engine 313 and Ladder 73 providing manpower. Foam 13 is a 1994 Sutphen 1,000 gpm pumper and carries 120–5 gallon pails of foam.

Figure 5.149 Yonkers Rescue 1.

Additional foam is stockpiled at Stations 3 and 13. Each engine company in the city carries 4–5 gallon pails of foam and an educator. Engines also carry Narcan Kits for opiod overdose reversal.

Yonkers Hazmat responds to an average of between 400 500 runs per year including fuel spills but not counting gas leaks and odors. Four personnel are assigned to Special Operations M-F daytime for support and they calibrate meters, do equipment operational checks, suit testing, respond to incidents with specialized equipment when requested, repair equipment, maintain supplies, develop standard operating procedures (SOPs), keep reserve fleet ready and specify and order new equipment. During the months of July and August, two of the operations personnel are detailed to the fire boat. Squad 1 has an Intec Video System with a color camera mounted atop a telescoping mast, with zoom, pan and tilt capabilities is part of the rescue. It also has two drop down monitors, one pre-wired to a Bullard TIC receiver for the "T" camera, and the other for a multi-channel hazmat entry link camera.

PPE, Equipment and Training

Yonkers hazmat team members wear Kappler Responder, Responder Plus, and Responder Reflector Level A suits. SCBA is Scott with 60 minute bottles. For fire response ladders, rescue and Squad 11 carry 45 minute bottles and engine companies carry 30 minute bottles. Hazmat also utilizes Scott Positive Air Pressure Respirators (PAPRs) for WMD events. In-suit communications equipment is the SCOTT Envoy. For WMD response, all fire department units, medic units, and police carry Mark I antidote kits for Nerve Agent exposure. Additional units carry 50 kit boxes including Rescue 1, Squad 11, Hazmat Engines 307 and 310, Battalion Chiefs, Safety Battalion, and the support unit carries 300. Yonkers Hazmat uses Coastal Weather Systems and Weatherpak MTX.

Monitoring instruments for hazardous materials and WMD detection are the Rae – MultiRae, Area Rae Rapid Deployment Kit, Scott Scout 3 and 4 gas detectors, Single Sensor MSA Pulsar, Portal Monitor, Canberra Rad 60, APD 2000, Ludlum Response Kits, Defender XL, HazMat ID Command System, Tetracore Bio strips with Alexeter Reader and M-8 paper & M-9 tape.

Hazmat training requirements for all Yonkers firefighters includes New York State Awareness, Operations, and Technician. In-service training includes training on their own equipment, monitoring instruments, and chlorine kits.

Hazardous Materials Exposures

Hazmat transportation exposures in Yonkers include Interstate 87 and State Route 9A. There are no major pipelines that go through the city other than natural gas. Barge traffic uses the Hudson River although Yonkers does not have a port facility. Conrail rail tracks are located through the city along the Hudson River but they do not have a rail yard. Major industry utilizing hazardous materials is sugar refining where they experience frequent acid spills. There are also compressed gas companies and a hospital in Yonkers. Hillview Reservoir is located in Yonkers, which provides 99% of New York City's water supply. Within the site are two water treatment buildings, one with 30 one ton chlorine containers and the other with 50. The one ton containers are also transported through the city to reach the reservoir. Residential neighborhoods are located just a few hundred yards from the reservoir site making them an exposure in the event of a chlorine release from the site (*Firehouse Magazine*).

Yuma, Arizona Hazmat Team

Prolog: *My visit to Yuma, Arizona, occurred during a trip to San Diego, California, to do a story on their hazmat team. I had never been to San Diego before, so it seemed a logical place to use my frequent flyer award ticket to escape the cool October temperatures in Maryland. Looking at a map of California while planning my trip I noticed Yuma, Arizona, was not that far away. Because of its location it seemed like an interesting place to visit. So I contacted Yuma Fire Department about doing a story on their hazmat team while I was there. After arriving in San Diego I traveled north on Interstate 5 towards Palm Springs.*

Traveling through the brush covered mountains it was not hard to imagine how tough brush fires must be in the area. I stayed the night in beautiful Palm Springs and the next morning headed for Yuma. From Palm Springs I traveled east through Blythe, California to U.S. Highway 95 that would take me South to Yuma. I knew that most of my trip was through desert. Earlier trips to Las Vegas and Phoenix had prepared me for some of the desert sites I would encounter so it wasn't a big shock seeing the vast Southern California and Arizona deserts. Temperatures were in the 70's in San Diego when I arrived and had reached the low 90's through the desert.

As I neared, Yuma I began to notice a change in the scenery. Vast desert wasteland had suddenly given way to rich farmland. At first, I thought I might have taken a wrong turn and ended up in Iowa or Nebraska! Was this really Yuma, Arizona, or just a mirage? What happened to the desert? As it turns out it was Yuma, truly is an oasis in the desert. I wondered how they managed to haul in all of that topsoil to grow crops. Yuma sits atop an old river bed, which accounts for the rich farmland.

Many larger cities are recognizable from their skylines, which also usually denote the downtown area of a community. However, driving into Yuma, which is very flat, I could not see the downtown area. After exploring I discovered that downtown Yuma is a historical area of one-, two-, and three-story buildings. The rest of the city is laid out like a vast suburban area of homes and businesses. Yuma's streets were organized in a logical pattern, but I was unfamiliar with where I was going. I continued west until I encountered a fire house. After all, they should be able to tell me where I was and how to get where I needed to be. Turns out I was pretty close, but was just heading in the wrong direction away from Fire Station #2, which was my ultimate destination. Yuma is a city of over 83,000 population which doubles in the winter time as "snow birds" travel down to escape the cold in the Northern United States. Yuma is home to the U.S. Army Yuma Proving ground and the U.S. Marine Corps Air Station Yuma, which is an alternate Shuttle Landing Site with its 5 mile long runway. Yuma is located in extreme Southwest Arizona on the Colorado River near the border with California and just a few miles from the border with Mexico.

The City of Yuma is a beautiful city nestled right in the southwest corner of Arizona, in the Sonoran Desert, Yuma Desert sub-region, bordering California and Mexico. Home to 100,198, with a metropolitan population of 216,872 residents in 2020. The population nearly doubles during the winter season when the "snow birds" arrive from the northern United States.

Fire Department History

Yuma Fire Department was organized in the late 1880s as a volunteer fire department. The original companies used hose carts to fight fires. One of the early hose companies was SPRR Hose Company No. 1. During 1899 the City Council ordered ladders and decided a special volunteer company was needed to handle the ladders. During January 1890, Hook and Ladder and Chemical Company No. 1 of the Yuma Volunteer Fire Department was formed in 1900. During 1918 the Yuma City Council purchased a ford truck with a hose bed added and a chemical extinguisher attached. This was the first motorized apparatus on the department (they still have it today). Volunteers continued to operate the Yuma Fire Department until the 1940s when the transition began to a fully paid fire department.

Today's Modern Fire Department

The modern day Yuma Fire Department operates out of six fire stations and provides fire, emergency medical service (EMS) (advanced life support, ALS), rescue, and hazardous materials response to a coverage area in Yuma of 30 miles^2. Yuma firefighters led my Chief Steve Irr operate 7 engines, 1 truck, and 5 medic units. Engine companies also provide ALS service. Other equipment and apparatus includes mass casualty trailer, rehab van, 2 UTV, Zodiac boat, 2 jet skies, 3000 gallon water tender. The Yuma Fire Department is a full time paid department with 114 uniformed personnel and 10 civilian employees led by Chief Stephen Irr. Firefighters in Yuma work a 56 hour work week with a day on and a day off for three shifts and then four straight days off. They work a total of 9 shifts in 27 days.

Many areas of the county are covered by Rural Metro Fire Department who contracts with the county to provide fire protection. Yuma fire responds to approximately 13,000 emergency calls a year of which about 84% are emergency medical related. Yuma fire operates seven engine companies (two are squirts), one truck, one heavy rescue, three medic units, a Special Operations (hazmat unit), and dive and river rescue teams.

Hazmat Team History

Yuma formed its hazardous materials team in 1994. Yuma fire personnel realized that there were numerous hazardous materials involved in their agricultural community and they needed to be prepared to handle accidents involving them. Agricultural crops include oranges, limes, lemons, and grapefruits, along with dates and vegetables, including most of the lettuce grown in the United States during the winter. Their first response vehicle was an old bread truck.

Hazmat Team

The current hazmat team is located at Station 2 at 3284 South Avenue A. Hazmat team members responded to 50 calls in 2018, which include hydrocarbon fuel spills. Their run statistics have remained within plus or minus 15 calls of the 2018 totals since 2007 with an average yearly total of 52. Engine companies carry emulsifiers, vapor suppressant, and absorb-all type products on their apparatus. Gasoline service stations are allowed to handle spills of 5 gallons or less. Anything larger requires the response of an engine company. If they cannot handle the volume, the hazmat team is called. Hazmat Techs maybe assigned to any company in the city. Two Techs are assigned to Station 2 to maintain equipment, calibrate monitors, and ensure readiness of equipment and vehicle. All other fire department personnel are trained to the operations level and provide support for the hazmat team. The first training class of hazmat technicians in Yuma included eight firefighters.

Presently Yuma's hazardous materials response vehicle is a 2001 Ford 750 that pulls a Wells Fargo Trailer custom built by Yuma firefighters over the period of about 1 year (Figure 5.150). The craftsmanship on the vehicle is just extraordinary. It they hadn't told me they made it themselves I would have thought it came from an apparatus manufacturer. Yuma Fire Department received a "Star" Award from the City of Yuma

Figure 5.150 Presently Yuma's hazardous materials response vehicle is a 2001 Ford 750 that pulls a Wells Fargo Trailer custom built by Yuma firefighters over the period of about 1 year.

for innovation and savings to the city of $300,000 by building the hazmat unit themselves.

Yuma's hazmat response unit is outfitted with an observation deck (Figure 5.151) on top for viewing incidents and typical hazardous materials equipment along with two generators, an air compressor for filling self-contained breathing apparatus (SCBA) bottles, an inside shower and bathroom, floodlights, computers, fax machine, reference books, and is completely air conditioned. Forty-one personnel are trained as members of the Yuma hazardous materials team. Thirteen team members are on duty each shift. When additional assistance is required mutual aid is available from hazmat teams at the Yuma Proving Ground and the Marine Air Station. Automatic aid is provided by the Marine Air Station to an ammonia facility in the city near the air station. Structural response is also provided automatically to city areas near the air station. Additional mutual aid is available from the Imperial Valley of Southern California. Yuma hazmat has been called upon to provide equipment and expertise to deal with hazmat incidents in Mexican border towns as well.

PPE, Equipment and Training

Hazmat personnel in Yuma wear 1 hour Scott SCBA with heads up display for respiratory protection in oxygen deficient atmospheres, toxic gases, or atmospheres of unknown materials. Each department member

Figure 5.151 Yuma's hazmat response unit is outfitted with an observation deck.

is also issued a cartridge respirator for specific types of respiratory hazards. Protective clothing consists of Level A Kappler and Level B Tychem. In-suit communication is provided by face piece mounted devices.

The following monitoring instruments are utilized by personnel wearing proper protective equipment Chlorine Meter, Anhydrous Ammonia Meter, PID, Four Gas Meter (CO, LEL, O_2, and H_2S), Multi RAE, APD 2000, Ludlums Radiation, M-8 paper, M-9 paper, M256A Kits, pH, Oxidizer Paper, and a HazCat Kit.

Training for Yuma Hazmat Team members includes the 200 hour Arizona Technician Course, monthly in-house training, annual refresher, and annual drills. Monthly hands-on training and bi-annual drills.

Reference Resources

Reference materials include Aris (for all companies), CAMEO, MERCK Index, *Railroad Explosives Book, and Coast Guard Chris Manual.* Each engine company also carries the *Emergency Response Guidebook (ERG), NIOSH Pocket Guide to Chemical Hazards, NFPA Fire Protection Guide for Hazardous Materials, and Jane's Chemical & Biological materials book.* MCD's with internet access on every vehicle, Hazmat laptop with printer in HM vehicle.

Hazardous Materials Exposures

Hazardous materials exposures in Yuma revolve around transportation and chemicals associated with the agricultural industry. Anhydrous ammonia is very common with the vegetable and fruit processing and storage. Other exposures include chlorine, sulfuric acid, hydrogen fluoride, toluene, pesticides, liquefied petroleum gas, and ethylene oxide. Interstate 8 is a major east-west transportation route connecting Yuma with California and other locations in Arizona including Phoenix and Tucson. U.S. Highway 95 is a major route north out of Yuma eventually hooking up with Interstate 10 (another major east-west corridor). Hazardous materials are also transported out of Mexico into the Yuma area from the South and West. Union Pacific Railroad serves Yuma and has a rail yard located there.

Incidents

During 1997 a 96% nitric acid release from a 40–50,000 gallon tanker in the rail yard kept hazmat personnel busy for 8 days (Figure 5.152). The incident started when the nitric acid, which is a strong oxidizer at that concentration, dripped onto rail ties soaked with cresol a hydrocarbon based preservative, starting a fire. During off-loading operations the off-loading tanker rolled over because of the soft sand in the rail yard.

Figure 5.152 October 2005 a sulfuric acid tanker collided with a car on Pacific Avenue. The acid tank was not damaged although diesel fuel leaked from the tractors saddle tanks. (Courtesy: Yuma AZ Fire Department.)

Mutual aid was requested and received from the Imperial Valley of California and the Marine Air Station. During February 2005 a tanker carrying JP-5 jet fuel overturned near the off ramp of Interstate 8 at Avenue E. Only about 5 gallons of the 5,000 in the tank was released. Industrial hazmat teams from Phoenix off-loaded the tanker before it was uprighted. Foam units from the Marine Air Station responded and stood by with Yuma Hazmat & Fire personnel in case the fuel found an ignition source. Another close call occurred in October 2005 when a sulfuric acid tanker collided with a car on Pacific Avenue. The acid tank was not damaged although diesel fuel leaked from the tractors saddle tanks.

On August 4, 2009, Yuma Fire Department responded to a report of a commercial building full of smoke. Upon approach firefighters found billowing black smoke and flames coming from the loading dock area of Desert Depot, an agricultural supply warehouse. Four sprinkler heads had activated on the inside bay doors on the loading dock preventing the fire from entering the building. Evacuations homes and businesses occurred in the immediate area around the facility.

On June 7, 2011, firefighters responding to a medical call found a semi-truck leaking an unknown substance in a parking lot. A 270 gallon tote leaking a corrosive liquid. Hazmat responded. The product was contained and a special team responded from California to remove the contained product.

On December 2, 2011, Yuma Fire Department responded to a report of an explosion at the Yuma Express Cooling Company and employees being exposed to ammonia. First arriving units found multiple victims with varying degrees of ammonia exposure. The most severe exposure was immediate transported to a local hospital. Eleven additional employees were transported after evaluation and two were eventually flown out to medical facilities in Phoenix. Due to exposures to contaminated patients, three Yuma Fire Department (YFD) and four Rural Metro personnel were also decontaminated and evaluated for exposure to ammonia.

On October 15, 2015, reports of a strong smell of ammonia began to be received with a report of ammonia possibly leaking from a rail car. Investigating firefighters found no leak in the ammonia car and traced the ammonia leak to True Leaf Farms. YFD Hazardous Materials Technicians, working with True Leaf employees isolated and shut down the source of the leak on the roof of a building at the facility. A preliminary estimate by True Leaf Farms personnel was the leak involved in excess of 100 lb of ammonia.

While hazmat technicians were working to locate and shut down the leak, the operations of several nearby businesses were interrupted. YFD personnel were sent to Love's Travel Stop on Gila Ridge Road due to persons feeling ill from the strong ammonia smell. Five people were checked by paramedics and as a precaution, three were transported to Yuma regional Medical Center for further evaluation.

YFD was also called to the International Paper Company plant to check out two employees. One of those employees was transported to the hospital for additional precautionary evaluation.

On April 29, 2016, Yuma Fire Department responded to Yuma High School due to a report of spilled chemical in the Research Building. YFD personnel found that a bottle containing a chemical had accidently fallen and broken in a storeroom. No one was injured but four Yuma High School employees and one student were evaluated for possible exposure. To avoid spreading fumes from the chemical, as a precaution, the air-conditioning system for the building was shut down and the building was evacuated. The area of exposure was limited to the storage room.

Rehab Unit

Yuma's hazardous materials response unit doubles as a rehab unit for fire department personnel during hot weather fire and other types of incidents. While Yuma may be an Oasis in the desert, they have not been spared the extreme summer heat. Temperatures from mid-July to the end of August range from 120°F in the day time to 106°F at night time. Humidity is usually low in the Yuma area, around 10% except for July and August when it can reach 40%–50%. During the extreme times of heat fan, misters are used to cool personnel, pop-up tents are used to supply shade, and lots of ice and bottled water are carried in the Battalion Chief's Vehicles.

Bibliography

Volume 5

Beatrice Fire Department, Beatrice NE, Visited the Beatrice Department and visited with Chief Brian Daake to collect information for this volume and continued with phone interviews and email exchanges.

Burke, Robert, Beatrice Nebraska Hazmat Team, Visited the Beatrice Department to collect information for this volume and continued with phone interviews and email exchanges with Chief Brian Daake.

Burke, Robert, Charles County Maryland Special Operations, *Firehouse Magazine*, July 1, 2010.

Burke, Robert, Cheyenne, Wyoming, Hazmat Team, *Firehouse.com website*, February 2007.

Burke, Robert, Columbus Nebraska Hazmat Team, Visited the Columbus Department to collect information for this volume and continued with phone interviews and email exchanges with Chief Dan Miller.

Burke, Robert, Corpus Christi, TX, Hazmat Response, *Firehouse Magazine*, May 1, 2016.

Burke, Robert, Dayton, OH, Fire Department Hazmat Team, visited the department to collect information for this volume and continued with phone interviews and email exchanges with Hazmat Chief Denny Bristow.

Burke, Robert, Denver Fire Department Hazmat Team, *Firehouse Magazine*, September 2007.

Burke, Robert, Durham, NC Police Department, Hazmat Response in the 'Research Triangle', *Firehouse Magazine*, April 14, 2013.

Burke, Robert, Fire and Police Combine For Hazmat Response - Part 1, *Firehouse Magazine*, May 1, 2013.

Burke, Robert, Hastings Nebraska Hazmat Team, Visited the Hastings Department to collect information for this volume and continued with phone interviews and email exchanges with Captain Darin Clarke.

Burke, Robert, Hazmat Response in Baltimore City, *Firehouse Magazine*, July 1, 2008.

Burke, Robert, Hazmat Response in Baltimore County, *Firehouse Magazine*, July 1, 2010.

Burke, Robert, Hazmat Response in Honolulu, *Firehouse Magazine*, July 1, 2005.

Burke, Robert, Hazmat Response in Kansas City, KS, *Firehouse Magazine*, January 24, 2011.

Burke, Robert, Hazmat Response in Kansas City, MO, *Firehouse Magazine*, March 8, 2008.

Burke, Robert, Hazmat Response in Philadelphia, *Firehouse Magazine*, December 31, 1998.

Burke, Robert, Hazmat Response in the Heartland, *Firehouse Magazine*, November 30, 2008.

Burke, Robert, Hazmat Response on the "Big Island" of Hawaii, *Firehouse Magazine*, July 22, 2009.

Burke, Robert, Hazmat Studies: Fire and Police Combine For Hazmat Response: Part 2, *Firehouse Magazine*, December 1, 2013.

Burke, Robert, HazMat Team Spotlight: Anchorage Fire Department, *Firehouse Magazine*, May 30, 2006.

Burke, Robert, HazMat Team Spotlight: Anne Arundel County, Maryland, *Firehouse Magazine*, December 8, 2005.

Burke, Robert, HazMat Team Spotlight: Anniston, *Firehouse Magazine*, October 4, 2004.

Burke, Robert, HazMat Team Spotlight: Lincoln Nebraska, *Firehouse.Com Website*, December 7, 2004, Visited the Lincoln Department to collect information for this volume and continued with phone interviews and email exchanges with Captain Francisco Martinez.

Burke, Robert, HazMat Team Spotlight: Louisville, *Firehouse Magazine*, August 17, 2004.

Burke, Robert, Hazmat Team Spotlight: Omaha, NE, Fire-Rescue Department, *Firehouse.Com Website*, February 1, 2008. Visited the Omaha Department to collect information for this volume and continued with phone interviews and email exchanges with Assistant Chief Joseph Salcedo.

Burke, Robert, HazMat Team Spotlight: Philadelphia, *Firehouse.Com Website*, May 11, 2004.

Burke, Robert, Houston's Hazmat Team Marks 25 Years of Service, *Firehouse Magazine*, February 28, 2005.

Burke, Robert, Imperial County, CA Hazardous Emergency Assistance Team (IV-HEAT), Found out about this team from the Yuma Fire Department. Contacted them and gathered information for this team through their web-site, emails and phone conversations with Captain Oscar Robles.

Burke, Robert, Inside Milwaukee's Hazmat Team, *Firehouse Magazine*, May 1, 2019.

Burke, Robert, Inside the Edmond, OK, Hazmat & Training Facilities, *Firehouse Magazine*, July 1, 2015.

Burke, Robert, Inside the Greater Cincinnati Hazmat Unit, *Firehouse Magazine*, May 1, 2017.

Burke, Robert, Inside the Houston Hazmat Team, *Firehouse Magazine*, February 1, 2019.

Burke, Robert, Jacksonville Fire Department: First Hazmat Team in the United States, *Firehouse Magazine*, November 30, 2006.

Burke, Robert, Kansas City's Darkest Day, *Firehouse Magazine*, November 1, 2018.

Burke, Robert, Kingman, AZ Hazmat Response, Visited Kingman to gather information and continued updates through emails and telephone calls with Assistant Chief Chris Angermuller.

Burke, Robert, Meet "Chicago's Twins", *Firehouse Magazine*, September 1, 2005.

Burke, Robert, Memphis' Evolution of Hazmat to All-Hazards Rescue, *Firehouse Magazine*, April 1, 2019.

Burke, Robert, Naval Air Station Corpus Christi, TX, *Firehouse Magazine*, February 1, 2017.

Burke, Robert, New Orleans Fire Department Hazmat Response Following Katrina, *Firehouse Magazine*, March 27, 2006.

Burke, Robert, Norfolk, Nebraska Hazmat Team, Visited the Norfolk Department to collect information for this volume and continued with phone interviews and email exchanges with Chief Scott Cordes.

Burke, Robert, Norfolk, VA, Hazmat Team, Obtained information about the Norfolk Department for this volume with phone interviews and email exchanges with Jim Bailie.'

Burke, Robert, Oklahoma City, Before & After the Bombing, *Firehouse Magazine*, May 1, 2018.

Burke, Robert, Philadelphia's Hazmat Team: On Duty at Eagles Football Games, *Firehouse Magazine*, December 31, 2008.

Burke, Robert, Protecting the Pentagon, *Firehouse Magazine*, May 2, 2011.

Burke, Robert, Protecting the World's Biggest Railroad Yard, *Firehouse Magazine*, October 1, 2013. Visited the North Platte Department to collect information for this volume and continued with phone interviews and email exchanges with Chief Dennis Thompson.

Burke, Robert, Red Willow Webster County Hazmat Team, McCook, NE, visited the Red Willow, Webster County Department to collect information for this volume and continued with phone interviews and email exchanges with Firefighter Billie Cole.

Burke, Robert, Regional Hazmat Team, Ashland, KY, *Firehouse Magazine*, April 1, 2016.

Burke, Robert, Remembering the Gulf Oil Refinery Fire, *Firehouse Magazine*, December 3, 2010.

Burke, Robert, Rescuing the Rescuer: Allegheny Counties Specialized Intervention Team, *Firehouse Magazine*, November 30, 2006.

Burke, Robert, Scottsbluff, NE Hazmat Team, Obtained information about the Scottsbluff Department for this volume with phone interviews and email exchanges with Chief Thomas Schingle.

Burke, Robert, The First Volunteer Hazmat Team in the U.S., *Firehouse Magazine*, September 1, 2018.

Burke, Robert, The Kingman Rail Car BLEVE, *Firehouse Magazine*, July 21, 1998.

Burke, Robert, The Southwest Boulevard Fire: Kansas City Remembers a Tragedy, *Firehouse Magazine*, November 30, 2009.

Burke, Robert, Training & Equipping a Local Hazmat Team, *Firehouse Magazine*, November 30, 2014.

Columbus Fire Department, Columbus, NE, visited the Columbus Department and visited with Chief Dan Miller to collect information for this volume and continued with phone interviews and email exchanges.

Dayton Fire Department, Visited the Dayton Department to collect information for this volume and continued with phone interviews and email exchanges.

Fire Department Hazmat Team, Aurora, IL, Visited the Aurora Department to collect information for this volume and continued with phone interviews and email exchanges.

Fire Department, Norfolk, NE, Visited the Norfolk Department and visited with Chief Scott Cordes to collect information for this volume and continued with phone interviews and email exchanges.

Fire Department, Norfolk, VA, Talked with Firefighter Jim Bailie on the phone and exchanged emails to gather information for the story in this volume. History of the Norfolk Fire Department, received from Jim Bailie Norfolk firefighter.

Gathered information primarily through phone conversations and emails with Deputy Chief, James Ley Retired, and the Milwaukee Fire Historical Society.

Hastings Fire Department, Hastings, NE, Visited the Hastings Department and visited with Captain Darin Clark to collect information for this volume and continued with phone interviews and email exchanges.

Imperial County Fire Department, Talked with Captain Oscar Robles on the phone and exchanged emails to gather information for the story in this volume.

KOTA TV, Rapid City, SD, Firefighter Killed in Blaze, September 8, 2018 Propane Tank Explosion, https://www.kotatv.com/content/news/Firefighter-killed-in-blaze-and-explosion-near-Tilford-492742891.html.

Oakland Tribune California, August 8, 1927.

Red Willow Western Rural Fire Protection District, McCook, NE, Visited the RWWRFPD and visited with Chief Bill Elliott's daughter Billie and her husband Jeff Cole to collect information for this volume and continued with phone interviews and email exchanges.

Saint Paul Fire Department, Visited the Saint Paul Department and visited with various firefighters during a 6 day National Fire Academy Course I was teaching to collect information for this volume.

Saint Paul Fire Department, Visited the Saint Paul Department and visited with various firefighters during a 6 day National Fire Academy Course I was teaching to collect information for this volume.

Salt Lake City Fire Department, visited the Norfolk Department and visited with PIO Audra Sorensen and numerous firefighters and hazmat team members to collect information for this volume and continued with phone interviews and email exchanges.

Scottsbluff Fire Department, Scottsbluff, NE, Did a phone interview with Chief Thomas Schingle to collect information for this volume and continued with phone interviews and email exchanges.

U.S. Geological Survey (USGS), Great M9.2 Alaska Earthquake and Tsunami of March 27, 1964, https://earthquake.usgs.gov/earthquakes/events/alaska1964/.

Index